OHIO STATE UNIVERSITY BIOSCIENCES COLLOQUIA

Analysis of Ecological Systems

THEMES, DATES, AND ORGANIZERS

Genetics and Biogenesis of
Mitochondria and Chloroplasts
5–7 September 1974
C. W. Birky, Jr., P. S. Perlman, and T. J. Byers

Regulatory Biology
4–6 September 1975
J. C. Copeland and G. A. Marzluf

Analysis of Ecological Systems
29 April–1 May 1976
D. J. Horn, G. R. Stairs, and R. D. Mitchell

Plant Cell and Tissue Culture
Principles and Applications
6–9 September 1977
W. R. Sharp, P. O. Larsen, E. F. Paddock,
and V. Raghaven

EDITED BY DAVID J. HORN,

GORDON R. STAIRS, AND

RODGER D. MITCHELL

Analysis of Ecological Systems

OHIO STATE UNIVERSITY PRESS: COLUMBUS

Library of Congress Cataloguing in Publication Data

Biosciences Colloquium, 3d, Ohio State University, 1977.
Analysis of ecological systems.

(Ohio State University biosciences colloquia)
Includes index.
1. Ecology—Congresses. I. Horn, David J. II. Stairs, Gordon R. III. Mitchell, Rodger David, 1926– IV. Title. V. Series: Ohio. State University, Columbus. Ohio State University biosciences colloquia.
QH540.B57 1977 574.5 78-13419
ISBN 0-8142-0276-4

Contents

Introduction

I subscribe to a computer search system that retrieves the titles of books and articles dealing with a variety of subjects. One of them is ecology. At biweekly intervals the computer dutifully sifts through the mountain of titles that has accumulated during the intervening fortnight and compiles my own personal information packet. The results of those searches have been most instructive. I have learned that families, graveyards, dog-bites, the New Testament, nuclear attacks, cosmetics, computers, breast-feeding, and singing leaves, as well as more traditional subjects, have ecologies. I am somewhat dismayed by this, because I have a difficult enough time keeping pace with changes within my own areas of interest without adding the avalanche of new areas that the computer regularly spews forth.

Certainly, the term *ecology* enjoys a currency and breadth of applicability that it lacked in the past. One reason is the national compulsion to append the term to any of our current endeavors. In certain instances this is like adding chrome to a new car: some people will think that it improves its performance. In other cases the appellation *ecology* indicates a reordering of thought and approach. As a consequence, old information is realigned into new formations; previously unrelated observations are unified, and the horizons of a field are expanded. Changes of the latter sort seem to be occurring in the fields of animal behavior, anthropology, and those fields concerned with land use, such as landscape architecture and agriculture. These diverse and often-unrelated applications of ecology to many other areas are paralleled by a vast diversity inside the field itself.

Ecology deals with the properties and interrelationships of organisms and their environments. It is unquestionably the most complex, challenging, and subtle of the sciences, and these attributes are reflected in the activities of its practitioners. Some ecologists place great emphasis on the ecosystem as *the* fundamental unit in ecology. The patterns, processes, and interactions among the populations that are included in such systems are presumed to have functional relationships that preserve and maintain the ecosystem. Indeed, the existence of the system is deemed crucial to the existence of its component parts. This approach attributes homeostatic properties to the ecosystem and is an application of the concept of the individual as an integrated, interdependent unit to the level of the ecosystem. Others view the structure of ecosystems as an outgrowth of interactions among species and the resources that they exploit. Here, competitive interactions are the catalysts of ecosystem structure. Notions of competition coefficients, carrying capacities, niche breadths, and community matrices are fundamental to this approach to ecological problems.

Many ecologists confine their work to local populations of particular species, and for them the role that their populations might play in a community or ecosystem is of secondary importance. It is not uncommon for such population ecologists to interpret the behavior and activities of the organisms as adaptations that ensure the survival of the population or species. (This view has been subjected to much criticism, but it is the most commonly accepted view among non-ecologists).

By contrast, a number of ecologists interpret ecological and behavioral adaptations as mechanisms whereby individuals increase the representation of their genes in subsequent generations. Here, individuals are primary and their actions are driven by self-interest. Although such an approach represents an attempt to integrate evolution and ecology, relatively recent developments may still provide to this approach a unified view of population structure, interactions between species, or community composition.

Thus, ecologists differ in the level at which they approach an ecological problem; but they also differ in fundamental ways with regard to the way in which they view the nature and the structure of ecological systems.

Diversity is also apparent in the ways that ecologists measure the attributes of the systems and organisms that they study. Biomass, calories, or numerical abundance are frequently used to index what, how much, or how many times there are, and reasonable arguments have been put forward for each measure. Interactions and processes have been characterized by rates of exchange, efficiency, changes in numbers of progeny, and alterations in biomass and numbers. Notions of optimality, maximization, cost/benefit, strategy, and tactic have served to focus the discussion of many of these studies.

The importance of different species or populations within an ecosystem has been assessed in a variety of ways. Sheer numbers are often used, but in certain instances the role that a particular species might play is of more importance. For example, while the energy flow from producers to consumers is very low, consumers may act as important regulators of the relationships among producers. It is the functional role of the organism that is being emphasized here, not its abundance.

This overview indicates the kind of diversity that permeates and enlivens the field. Diversity reflects both the youth and vigor of the science. The major developments in ecology have occurred within the last two or three decades. Different approaches and points of view may well represent initial attempts to integrate information, since in many ways it is a science in search of a synthesis. Opportunities to achieve exchange between ecologists of diverse approaches are all too rare. The papers that compose this volume are the outgrowth of a symposium that sought to provide that kind of interaction.

Jerry Downhower

Department of Zoology
Ohio State University

Third Annual Biosciences Colloquium
College of Biological Sciences
Ohio State University
29 April–1 May 1977

ANALYSIS OF ECOLOGICAL SYSTEMS

Organizers

David J. Horn, Department of Entomology, Ohio State University
Gordon R. Stairs, Department of Entomology, Ohio State University
Rodger D. Mitchell, Department of Zoology, Ohio State University

Speakers

Richard C. Lewontin, Museum of Comparative Zoology, Harvard University
Rodger Mitchell, Department of Zoology, Ohio State University
D. G. Embree, Maritimes Forest Research Centre, Fredericton, New Brunswick
Eric R. Pianka, Department of Zoology, University of Texas, Austin
Lawrence E. Gilbert, Department of Zoology, University of Texas, Austin
Gordon H. Orians, Department of Zoology, University of Washington
Philip L. Miller, Department of Biology, San Diego State University
Richard G. Wiegert, Department of Zoology, University of Georgia
David J. Horn, Department of Entomology, Ohio State University

OHIO STATE UNIVERSITY BIOSCIENCES COLLOQUIA

Analysis of Ecological Systems

R. C. LEWONTIN

Fitness, Survival, and Optimality

1

Despite the pious hopes and intellectual convictions of evolutionary geneticists and ecologists, evolutionary genetics and ecology remain essentially separate disciplines, traveling separate paths while politely nodding to each other as they pass. The functional separation of population genetics and ecology is immediately obvious in books on "population biology," as, for example, the superb introductory text by Wilson and Bossert (1971), in which the sections on population genetics, ecology and biogeography are totally independent entities, each standing on its own feet, each quite self-contained in its analysis. The marriage of population genetics and ecology was declared to be the social event of the season some years ago, but it has so far been without issue and, in fact, has yet to be consummated.

Population genetics and ecology each have a separate but parallel phenomenology and separate but remarkably similar theoretical structures. Both are concerned with the same general problem, the way in which the numbers, kinds, and distributions of organisms have come into being and will change. The reason that population genetics and ecology have successfully carried out their separate programs despite what would appear to be their ultimate concern with the same set of natural phenomena lies in the actual structure of the dynamics of community evolution.

THE GENERAL SHAPE OF EVOLUTIONARY ECOLOGY

Community evolution can be represented in a multidimensional space such as figure 1, in which some dimensions are the *numbers* of each species and

R. C. Lewontin, Museum of Comparative Zoology, Harvard University, Cambridge, Mass. 02138

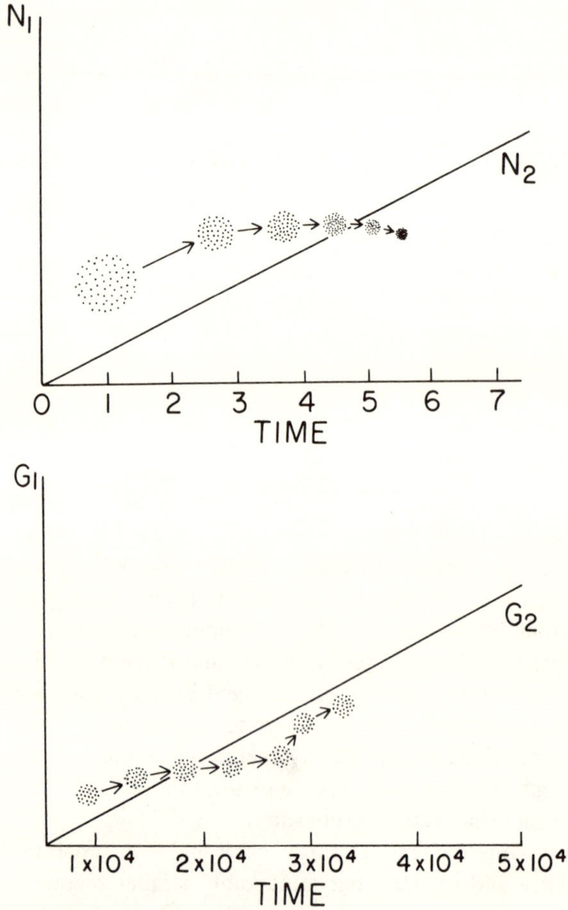

Fig. 1. *Top.* The space of evolution of the ensemble of species abundances. The cluster evolves in the direction of the arrows. *Bottom.* The space of phenotypic and genetic evolution of the community of species.

others are the *phenotypic* or *genotypic* descriptions of each species. A community at any moment in time is represented by a point in this space, completely specified by all of the abundances and descriptions of the component species. Over short periods of time and in replicated habitats, the community description varies because of random environmental fluctuations, chance variations in abundances of species, and chance genetic changes from generation to generation. Then a more realistic representation of the community structure is a cloud of points representing the distribution of community compositions over short periods of time in similar habitats. Population ecology and population genetics are concerned with the location and configuration of the ensemble of points in the space and how the ensemble evolves through time.

The evolution of the ensemble does not occur with equal speed along the dimensions of abundance and of phenotype so that, to a first approximation, the evolution of the community ensemble is decomposable into changes that occur in "ecological time" and those that happen only in "evolutionary time." The evolution of a community ensemble in ecological time is represented in figure 1 (top), a space where dimensions are species abundances only. From general dynamical laws like the logistic equation and its multiple species generalization, as, for example, the Lotka-Volterra equations, more or less rapid changes in abundances are produced from an initial colonization reaching in a short time a stationary distribution, at least in theory, in scores of generations or less. Evolution in the genetical space (fig. 1, bottom) occurs much more slowly, although there may be an occasional rapid episode, and the assumption is made that the slow phenotypic evolution of the community goes on forever without reaching a steady state.

Put in formula terms, the vector of abundance $\mathbf{N}$ and the vector of genetic compositions $\mathbf{G}$ are assumed to evolve by laws of the form

$$\frac{d\mathbf{N}}{dt} = f(\mathbf{N}, \Pi) \tag{1}$$

and

$$\frac{d\mathbf{G}}{dt} = g(\mathbf{G}, \mathbf{W}) \tag{2}$$

where Π is the set of parameters of growth rates and interactions of the species that are determined by their morphologies and physiologies, and $\mathbf{W}$ is the set of fitnesses of the various genotypes determined by the abundances and interactions of the various species. Although $\mathbf{W}$ and Π are treated in each system as parameters, they are in fact functions of the variable of the complementary system. That is,

$$\Pi = h(\mathbf{G}) \tag{3}$$

and

$$\mathbf{W} = \mathrm{H}(\mathbf{N}) \tag{4}$$

so that expressions (1) and (2) are not really independent systems but are linked through their pseudo parameters by (3) and (4). The reason that (1) and (2) have been regarded as describing essentially independent domains is that they operate initially on different time scales. Since genetic evolution is very slow as compared with the initial changes described in expression (1), the vector **W** can be regarded as essentially a constant during the course of change of **N.** In a complementary fashion, the rapid change in **N** leads very quickly to a stationary distribution of abundances so that **W** is regarded as more or less constant, or at least shows no secular changes, over long periods of genetic evolution. Whether this latter assumption is valid, however, depends critically on the forms of expressions (3) and (4). If **W** is a strong function of **N** and if Π is a strong function of **G**, then for most of the evolution of the community, after an initial rapid adjustment of numbers, the further slow evolution of species abundances and genetic compositions will be tightly coupled. The "parameter" sets Π and **W** will themselves be changing at rates that are comparable to the changes in **N** and **G**, while **N** and **G** will no longer be changing on different time scales, but both will be evolving slowly in contact with each other. It is this slow community evolution, when both abundances and genotypes are at a quasi-steady state, that is the domain of integrated evolutionary ecology.

OPTIMALITY PRINCIPLES

In order to study evolutionary ecology, it would be necessary to know the forms of expression (3), which relates species growth rates and species interactions to genetic composition, and expression (4), which relates genetypic fitnesses to species and genotype abundances. Such information is difficult enough to get under controlled laboratory conditions and virtually impossible to acquire in nature. Not the least of the problems is that of *recognizing* different genotypes in nature. There is phenotypic variation of most characters in most species, but we do not, in general, know what the reproductive characteristics of these phenotypes actually are, and even where we can measure the correlation between phenotype and reproduction in one community composition, we do not know how it changes with composition. Finally, we do not know what kind of genotypic variation underlies the phenotypic. At minimum, we must have a measure of heritability in the narrow sense (h^2), and for long-term community evolution we would need to know whether there is overdominance for the character and the rate of generation of new variation by mutation and migration.

Since none of the genetic information exists for natural populations, evolutionary ecologists have had to resort to a form of reasoning that is a short-cut attempt to predict community evolution. This is the use of *optimality* arguments, as distinct from *dynamical* arguments. A dynamical theory of the evolution of a community uses the machinery of expressions (1)–(4) to predict the evolutionary trajectory of the community ensemble on a purely mechanical basis. The differential equations and the initial conditions are sufficient to predict the outcome. By contrast, optimality arguments do not contain a detailed machinery, but are constructed to predict the outcome or direction of community evolution based on a general maximization principle. Optimality arguments begin by postulating a "problem" to be solved by organisms. The "problem" may be heat regulation, finding the most food with the least expenditure of energy, avoiding predators, and so on. Sometimes the problem is asserted to be that of leaving the most genes in future generations; but that is never the actual problem set in the optimality argument itself, since it would require the very genetic information that optimality arguments are designed to dispense with. Rather the "problem" of leaving more genes in future generations is always converted immediately into one of the more specific "problems" of finding mates, heat regulation, food search, and such. Having set the "problem" that the organism is to solve, the evolutionary ecologist then uses an explicit or implicit engineering analysis to prescribe what morphology, physiology, or behavior would "best" "solve" the "problem." It is then asserted that the species or the community will evolve so as to establish that phenotypic manifestation.

Such optimality arguments have become extremely popular in evolutionary biology in the last 15 years, and at present represent the dominant mode of thought. Any issue of the *American Naturalist* or such influential books as E. O. Wilson's *Sociobiology* will provide numerous examples. The reasons for the popularity of optimality arguments are easy to see. First, there is the general appeal that maximization principles have had in natural and social science since the nineteenth century (Fermat's principle of least action, Hamilton's principle, the idea of the "invisible hand" in economics). Second, optimality arguments dispense with the tedious necessity of knowing anything concrete about the genetic basis of variation or about the actual norms of reaction of different genotypes. Third, anyone can play the optimality game, since all that is required is imagination, although optimality arguments may use an explicit and sophisticated engineering analysis. Thus Leigh (1971) has derived the optimal shape of a sponge from an exact hydrodynamical model, on the assumption that the "problem" for a sponge is to process the maximum volume of food-containing water per unit time. On the other hand, Trivers's (1974) theory of parent-offspring conflict is simply a heuristic qualitative argument about how parents and offspring ought to behave toward each other in various circumstances if each is pursuing its own "best" interest.

But optimality arguments are useful only to the extent that they are truly short cuts to the result of the dynamical process and not simply plausible stories that may or may not predict the actual evolutionary events. There are a number of serious problems with such arguments that seem not to have been appreciated by their proponents.

1. How is the "problem" to be solved recognized as opposed to all the other "problems" that are posed for the organism? Why, for the sponge, is the "problem" of water-processing the one for which the sponge's shape has evolved rather than, say, appearing less attractive to predators?

2. Related to 1. Optimality arguments are always *ceteris paribus*. The distribution of food-particle size gathered by a bird may evolve to an energetically most efficient one, *all other things being equal*; but all other things never are, and the most efficient search pattern may expose the nestling to long periods of exposure to predators (see Orians' essay in this volume). Thus the failure of the predicted optimal pattern can always be rationalized by appeal to some other "problem" that must be compromised with.

3. There may be no significant genetic variation in the phenotypic direction postulated. Optimality arguments always assume that if some phenotype would be optimal, the appropriate genotypes will appear.

4. Related to 3. Genetic variation may be of the wrong sort. So, if the "optimal" phenotype happens to be a heterozygote, selection will produce a stable polymorphic species rather than a uniformly optimal one. Optimality theories assume that genetic variation when it appears will be of the appropriate sort to allow fixation of the optimal type. Thus Maynard Smith's Evolutionary Stable Strategies (1972) are phenotypes such that if all members of the population possess them any individual deviant is at a disadvantage. But such phenotypes are "evolutionarily stable" only if they are based on homozygosis (or some special genetic mechanism of fixed heterozygosity).

5. Conversely, some optimal "solutions" are mixtures of phenotypes. But these mixtures are only dynamically feasible under extremely restricted genetic conditions or else they must be based on phenotypic flexibility of homozygous genotypes. Thus most "optimal mixed strategies" are not genetically feasible (Lewontin, 1961).

6. It is not clear what criterion of optimality should be involved. Figure 2 illustrates two contrasting criteria of optimality that have been used in evolutionary theory. The case is that of two species or two individuals in one species. These are "players" in a "game." Each player may play any one of a continuum of "strategies." The strategies of player one are ordered on the ordinate and the strategies of player two on the abscissa. Corresponding to any strategy S_1 played by player 1 and S_2 played by player 2, there is an outcome ("utility" or "payoff") to each player that we suppose can be numerically rated. The payoffs to each player then form surfaces shown in the diagram as topographic maps U_1 and U_2, the payoff functions for players 1 and 2 respectively. We now ask which "strategies" players 1 and 2 should play to op-

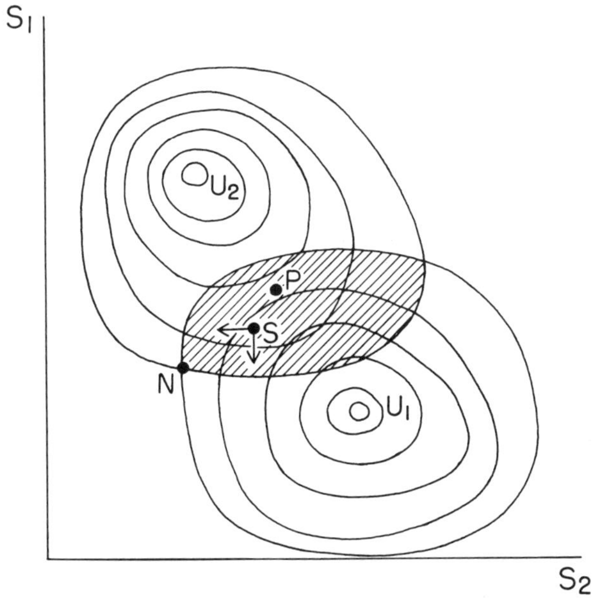

Fig. 2. Nash and Pareto criteria of optimality. S_1 is the strategy of player 1, S_2 the strategy of player 2, U_1 the pay-off surface for player 1, U_2 the pay-off surface for player 2, N the Nash Point, and P the Pareto point.

timize their payoffs. One solution to this game is shown by the point N, the so-called *Nash solution*. The Nash point has the property that if player 1 does not change its strategy from N, it does not pay player 2 to change its strategy from N and *vice versa*. Moreover, this is a stable equilibrium, as the reader can verify.For suppose players 1 and 2 play some different strategy pair, P.S Then, provided player 1 does not change, it will pay player 2 to change in the direction of the horizontal arrow while player 1 should change in the direction of the vertical arrow given that player 2 does not change. The net result is that both will move toward the Nash point. The Evolutionary Stable Strategy (ESS) of Maynard Smith is a Nash optimum by its definition. While the Nash optimum has a dynamic that will drive a pair of species toward it (given that all the genetic requirements discussed above are fulfilled), it has the curious property that *both species can do better simultaneously*. It is easy to see that any strategy pair in the shaded lens-shaped area in figure 2 will result in a greater payoff to *both* players. There is, moreover, a point P, the *Pareto* point, that is in some sense the joint optimum. It is the point such that any further improvement in payoff to player 1 must be at the expense of player 2 and vice versa. It is, thus, the cooperative equilibrium since it is the strategy point that will be achieved if the players are in collusion. It has the property, however, that it will pay each player to change if the other does not, and so puts a premium on cheating. This cooperative equilibrium is the optimality contained in Trivers's theory of reciprocal altruism (1971) and is related to the optimal solutions of Levins's fitness sets (1968).

Here then are two "optimal" solutions with quite different properties. Neither one necessarily exists in any given situation, but both may. Although the ESS (Nash) solution appears to have a dynamic that leads to it, this is only true if there is some way of passing from one pure strategy to another. If populations are genetically heterogeneous, then it is not clear what dynamic changes will follow. In general, the actual dynamic of evolutionary change may correspond not to either Nash or Pareto equilibria but to some quite different point that is not optimal under either criterion.

In many cases, the *n*-person game analogy is itself inappropriate, and each species can be regarded as playing a "game against nature" with variation and unpredictable states of nature being the background against which the species is to choose an "optimal" strategy. In such cases there are a number of solutions none of which are really satisfactory and most of which are not genetically feasible except under unusual circumstances (Lewontin, 1961).

We may sum up the difficulties of optimality arguments by saying that they are useful only when the solutions they propose correspond to the outcome of a dynamical process of genetic evolution, but that each case must be separately verified. If, however, it is necessary to solve the dynamical problem in order to validate the optimality argument, then the optimality technique has added nothing. Only if we can establish by empirical work that there are indeed some properties of species that are generally maximized by natural selection, will we have confidence in maximization or optimization arguments. It is therefore of very considerable interest to try to establish some general principles of evolutionary optimization beginning with the dynamics of natural selection, both in an exact theoretical form and as a general empirical rule.

CONSEQUENCES OF INTER-DEME SELECTION

If we concern ourselves only with the change in frequency of different genotypes within a population as the result of differential Darwinian fitness, we quickly come to the conclusion that no property of the population, as a population, undergoes a process of optimization simply as a consequence of the increase in more-fit genotypes. This seems, at first sight, paradoxical, since the increase in more-fit genotypes ought to result in a more-fit population. Let us consider two cases. In both we begin with a resource-limited population of,

TABLE 1

DEMOGRAPHIC AND GENETIC CHANGES IN HYPOTHETICAL RESOURCE-LIMITED POPULATIONS WITH MUTATIONS TO INCREASED FECUNDITY (a) AND INCREASED RESOURCE EFFICIENCY (b)

Generation	Expected No. of A	W_A	Expected No. of a	W_a	N	λ
			a. Doubled Fecundity			
1	99	1.00	1	2.00	100	1.00
2	98.02	.99	1.98	1.98	100	1.00
3	96.12	.98	3.88	1.96	100	1.00
4	92.53	.96	7.47	1.93	100	1.00
5	86.10	.93	13.90	1.86	100	1.00
6	75.59	.88	24.41	1.76	100	1.00
7	60.76	.80	39.24	1.61	100	1.00
8	43.64	.72	56.36	1.44	100	1.00
9	27.91	.64	72.09	1.28	100	1.00
10	16.22	.58	83.78	1.16	100	1.00
11	8.83	.54	91.17	1.08	100	1.00
12	4.62	.52	95.48	1.04	100	1.00
			b. Doubled Resource Efficiency			
1	99	1.00	1	2.00	100	1.00
2	99	1.00	2	2.00	101	1.01
3	98.02	.99	3.96	1.98	101.98	1.01
4	96.12	.98	7.77	1.96	103.89	1.02
5	92.53	.96	14.96	1.93	107.49	1.04
6	86.10	.93	27.84	1.86	113.95	1.06
7	75.59	.88	48.88	1.76	124.47	1.09
8	60.76	.80	78.56	1.61	139.32	1.12
9	43.64	.72	112.80	1.44	156.44	1.12
10	27.91	.64	144.24	1.28	172.15	1.10
11	16.22	.58	167.60	1.16	183.82	1.07
12	8.83	.54	182.35	1.08	191.18	1.04
13	4.62	.52	190.76	1.04	195.38	1.02

say, 100 individuals of type *A* requiring 1 unit of resource per individual to reach adulthood. In the first case a mutation to *a* occurs that doubles the fecundity of its bearers without altering the efficiency of resource utilization. Table 1a shows the spread of this mutation through the population, together with the absolute fitness of each genotype, W_A and W_a, the population size, N, and the finite rate of increase of the population, λ. Since nothing has changed in the resource utilization, the population size remains constant, λ is unity throughout, and the absolute fitnesses of the genotypes decline from $W_A = 1.00$ and $W_a = 2.00$ to $W_A = .50$ and $W_a = 1.00$. This decline occurs in such a way that the absolute fitness of the population remains exactly at 1.00. Table 1b shows, by contrast, the changes in a population when a mutation is introduced that doubles resource efficiency. Again the fitnesses decline from $W_A = 1.00$ and $W_a = 2.00$ to $W_A = .5$ and $W_a = 1.00$, but the population grows during the gene replacement. At first, the rate of growth increases as evidenced by the increasing λ from 1.00 to 1.12, but when the proportions of A have fallen to .333, the finite rate of increase of the population begins to fall off and finally returns to unity when the population reaches its final size at $N = 200$. Thereafter, there is no difference between this population and the one with doubled fecundity except that it is twice as numerous. Before and after selection in both cases, the population consists entirely of individuals with fitness equal to unity. In both cases all that has happened is that the number of biological units has doubled. In the first case the number of eggs has been doubled, but these units decrease during development to adulthood. In the second case the number of eggs has also been doubled (but not the fecundity per individual), but these units do not decrease up to adulthood. In what sense are these populations more fit after selection than before? That depends upon the particular environmental circumstances. A population with higher fecundity is better buffered against density-independent accidents such as unfavorable temperatures or random egg predation. On the other hand, it is more susceptible to predators who switch their search images to more numerous immature stages, and more likely to suffer if resources for immatures are patchy and variable. The more efficient food processor clearly is at an advantage when resources are temporarily short, but the more dense adult population will be more susceptible to epidemics and to adult predators who are attracted by numbers. It is clear that one cannot begin with the bare demographic and genetic changes that occur by natural selection and make any prediction about the future of the species. Every extant species is fit both in its present state and in any future state in which it is still extant. There is nothing in the process of selection that implies optimization or improvement. Indeed, if the population selected for increased fecundity had been so unfortunate as to be subject to a density-cued seed or egg predator, it might have been *less* numerous and *more* liable to extinction than before the selection process. Yet that fact would not stop the very rapid increase within the population of the more fecund genotype (unless eggs or seeds were produced in well separated clutches). Indeed, the optimality theorist, happening on the scene after selection, would more than likely explain the high fecundity of the population as an adaptation to high predation pressure on the juveniles, whereas the high predation pressure might, in fact, be the *consequence* of the increase in fecundity. The dynamics of natural selection does not include foresight, and there is no *theoretical* principle that assures optimization as a consequence of selection. If, then, we are to have any faith in optimality arguments as a substitute for the actual mechanics of natural selection, we will have to reach for *empirical* generalizations in the actual world of organisms. That is the domain of experimental ecological genetics.

EXPERIMENTAL EVIDENCE FROM DROSOPHILA

The question is whether there is any general correlation between components of fitness and whether genotypes that increase in frequency within populations because of their greater survival and reproduction do, in fact, confer on the population as a whole demographic properties that can be said to be adaptive or optimal. No general answer to such questions is available, but we would like to review some experiments carried out by us at various times that were meant to address these questions. The experiments in question are all laboratory experiments, and they are all on various species of *Drosophila*, so the answers are relevant only to the peculiar biology of that genus and in the special circumstances of laboratory culture. There are no predators, females do not have to search for egg-laying sites, males and females are always within a few centimeters of each other, and a host of other phenomena of the natural history of *Drosophila* are absent or altered in tbe laboratory. *Drosophila* is a holometabolous flying insect, displaying virtually no social behavior throughout most of the genus, with high fecundity and reasonably broad ecological tolerance, at least for the species with which we have worked. All of these facts must be taken into account in generalizing from our results.

Density-Frequency-Fitness Relationships

Lewontin (1955) and Lewontin and Matsuo (1963) carried out experiments on genotypes of *D. melanogaster* and *D. busckii* to determine whether the larval viability of genotypes when interacting only with themselves would predict the outcome of interaction between different genotypes. This is essentially the question of whether the pure population that results from the competitive replacement of one genotype by another is in any sense "better off" than the population before the selection process. In the experiments, counted numbers of fertilized eggs of different genotypes are placed together on rigorously standardized food at different total densities, and the number of adults of each genotype emerging as adults is recorded. Each genotype is cultured alone ("pure" culture) at various densities, and then pairs of genotypes ("mixed" culture) at various mixture proportions are cultured at the same total densities. A typical result is shown in figures 3 and 4. Figure 3 shows the proportion of cultured eggs emerging as adults at various densities for two different genotypes, *Acme-A* and *white* of *D. buskii*, in pure culture and in varied mixed proportions with each other. The most striking feature of the figures is the marked appearance of an optimal survival density at about 32 larvae per vial. This is a common phenomenon in *Drosophila*, probably the result of the conditioning of the food surface by the larvae themselves. Figure

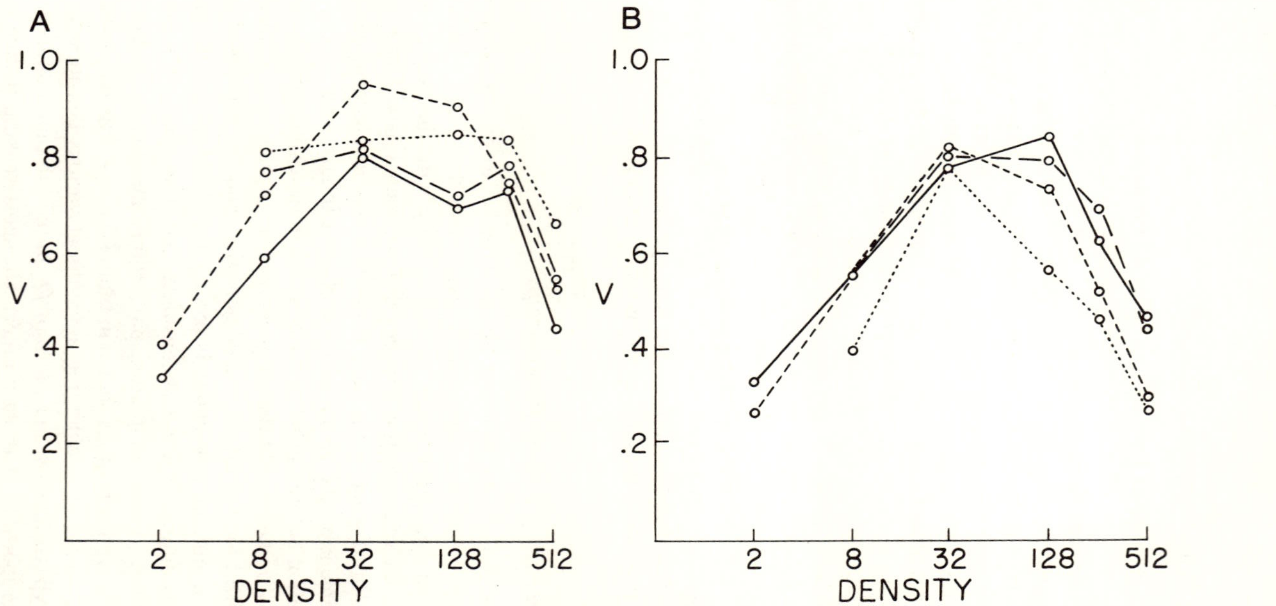

Fig. 3. Absolute viabilities of *Acme-A* and *white*. A. solid line: 100% Acme; long dashes: 75% Acme; short dashes: 50% Acme; dots: 25% Acme. B. solid line: 100% white; long dashes: 75% white; short dashes: 50% white; dots: 25% white. (from Lewontin and Matsuo, 1963).

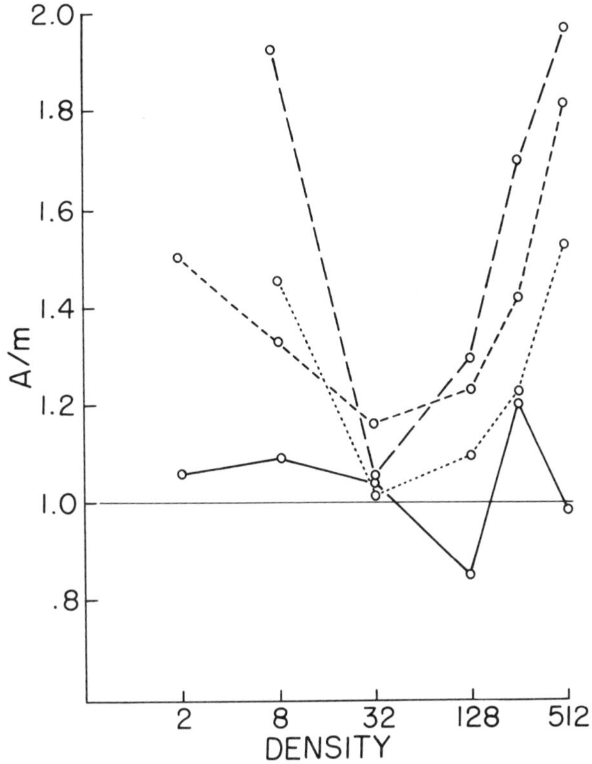

Fig. 4. Relative viabilities of *Acme-A* and *white*. Solid lines: predicted from pure culture; long dashes: observed in 75% Acme culture; short dashes: observed in 50% Acme culture; dots: observed in 25% Acme cultures (from Lewontin and Matsuo, 1963).

4 combines the data in figures 3a and 3b to show the *relative* viabilities of the genotypes in pure and in mixed culture at various densities. The solid curve shows the *predicted* relative survivals based on the separate pure culture viabilities. It appears that there is virtually no difference between genotypes consistent over densities. The other curves show the *actual* relative viabilities from the mixed-culture experiments. Only at the intermediate, optimum, density of 32 larvae per vial does the outcome of competition correspond to the prediction from pure culture. At the less favorable low and high densities, Acme A is clearly superior to white and the superiority becomes *greater* as the proportion of Acme A is *greater*. Yet despite this increasing superiority of Acme A as it gets more common, there is no effect on the survivorship of a pure population because pure Acme A survives no better than pure white. This kind of density-proportion interaction is common in our experiments and

shows that in *Drosophila*, at least, success in larval competition is not an indicator of greater absolute survival.

Correlation of Fitness Components

Birch, Dobzhansky, Elliott and Lewontin (1963) and Dobzhansky, Lewontin and Pavolvsky (1964) studied the age specific mortality and fecundity schedules of various geographic population and chromosomal types of *Drosophila serrata* and *D. pseudoobscura*. From the larval development time, larval survival, age-specific mortality of adults, and age-specific fecundity schedule of adult females, they then calculated the intrinsic rate of increase *r* of each race or genotype population by the standard Lotka equation

$$\sum_x e^{-rx} 1_x m_x = 1$$

Table 2a shows the mean larval viability, mean adult egg production, and mean longevity of adults together with the estimates of *r* for five geographical populations of *D. serrata* at two temperatures. Table 2b gives the same information for laboratory population cages of *D. pseudoobscura* containing

TABLE 2

MEAN VIABILITY (v), AVERAGE FEMALE FECUNDITY (f), AVERAGE LENGTH OF LIFE (l), AND INTRINSIC RATE OF INCREASE (r) FOR VARIOUS NATURAL POPULATIONS OF D. SERRATA AND FREE LABORATORY POPULATION OF D. PSEUDOOBSCURA

a. *D. serrata**

	25° C				20° C			
Population	v	f	l	r	v	f	l	r
Sydney	.62	559	27.1	.233	.66	428	50.1	.124
Brisbane	.64	670	25.5	.249	.70	546	44.6	.156
Cairns	.57	498	25.4	.227	.53	373	42.3	.137
Port Moresby	.61	368	23.2	.210	.64	270	35.2	.109
Rabaul	.64	151	12.1	.135	.61	69	27.9	.051

b. *D. pseudoobscura*†

	25° C				16° C			
Inversion types	v	f	l	r	v	f	l	r
AR + CH	.895	377	33.6	.220	.922	502	72.3	.100
AR + PP	.863	251	34.1	.214	.896	526	53.9	.110
AR (Mather)	.905	261	31.7	.207	.927	416	75.9	.098
AR (Pinion)	.890	265	29.7	.205	.894	548	56.4	.108
CH	.794	249	28.2	.192	.831	370	66.4	.096
PP	.813	118	19.8	.170	.823	337	44.7	.094

* Data from Birch et al., 1963.
† Data from Dobzhansky et al., 1964.

various chromosomal inversion types. These experiments show that adult fecundity and longevity are indeed correlated and that they are the chief contributors to variation in intrinsic rates of increase, while larval viability is unrelated to these measures, although, of course, larval viability is formally one of the determinants of r. It seems likely that in a holometabolous insect the larval and adult stages can be regarded ecologically as two quite different organisms with such different physiological requirements that we may expect larval viability to be uncorrelated with adult fitness components.

Population Fitness Experiments

We have designed and partly carried out a large and comprehensive experiment to try to relate components of fitness of genotypes to fitness properties of populations as a whole. For this purpose, we have used genotypes of *D. melanogaster*. The plan of this experiment can only be sketched out here and a sampling of the results given. Much of the experimental work, which will be reported in detail elsewhere, has been done in collaboration with A. O. Tantawy of the Department of Genetics, Alexandria University, Alexandria, Egypt.

Three genotypes easily detectable in homozygous and heterozygous condition were chosen. They are wild type (+), spineless (SS), a bristle mutant, and aristapedia (SS^a), an antenna mutant. Spineless and aristapedia are pseudoalleles, so their heterozygote SS/SS^a is wild type, but they do not recombine. The net fitness of each genotype is measured in two ways. The genotypes are put together in population cages, and the system is allowed to evolve. From the changes in gene frequency, fitness estimates can be calculated. In these experiments, + eliminates both SS and SS^a when cages are started with all heterozygotes +/SS or $+/SS^a$. After eight generations of selection, the frequency of SS was .142 and .154 in two replicates, and the frequency of SS^a was .021 and .078 in two replicates. In two populations with all three genotypes present, the frequencies after eight generations were $+ = .854$, $SS = .083$, $SS^a = .063$, and $+ = .870$, $Ss = .088$, $SS^a = .042$. Finally, the results of cages containing only SS and SS^a were a stable polymorphic equilibrium at about $SS = .65$ and $SS^a = .35$. All these experiments are in excellent agreement that wild type is much more fit than either SS or SS^a homozygotes, and SS/SS^a heterozygotes are also more fit than the mutant homozygote but not more fit than wild type.

The second measure of net fitness is from the age specific mortality and fecundity schedules of the genotype. These are in qualitative agreement with the population cage estimates except that the fecundity of SS^a/SS^a homozygotes is very low and seems to account for most of the difference between genotypes. There is little viability or longevity variation. This turns out to be an extremely important point in understanding the results.

Having determined the fitnesses of the genotypes tautologically, by evolutionary changes in gene frequency, and physiologically by components of r, we then attempted to measure population fitness of populations of different genotypes from (1) productivity in numbers and biomass, (2) probability of extinction in fluctuating unfavorable environments, and (3) competitive ability with a second species.

Productivity measurements were made in population cages either homozygous for each genotype or stably polymorphic for SS/SS^a. The results are given in table 3. There is no relation between weight per fly and the intra-population fitnesses previously measured, but there is a positive correlation between genotype fitness and total number of flies produced per food cup. The homozygous +/+ population is the most productive, and the SS^a/SS^a population the least. The difference is on the order of 20%, and there seems to be no difference between the polymorphic SS/SS^a population and monomorphic SS/SS. For numbers, then, there is a qualitative agreement with fitness measures.

Extinction rates were measured by creating 100 vial populations of each genotype in each of four environments: constant 25°C (optimal), constant 28°C (high stress), constant 15°C (low stress), and fluctuating 28°C–15°C between successive generations. All populations were propagated by carrying over all live flies to a new culture vial at the end of each generation. The number of newly extinct populations was recorded each generation, and the result is given in figure 5. All populations survived in the constant environments and are not shown. In the fluctuating environments the order of survival

TABLE 3

Average Number of Flies per Food Cup/Cage, Body Weight/Cup, and Average Weight per Fly

Population		Individuals Produced/Cup	Biomass Produced/Food Cup (mg)	Adult Fly Weight (mg)
+/+	A	512.93±57.72	331±4.0	0.644±0.033
	B	560.87±53.23	363±4.2	0.614±0.037
SS/SS	C	449.67±48.45	279±2.7	0.634±0.036
	D	451.93±48.43	298±2.9	0.670±0.036
SS+SS^a	E	432.26±37.75	277±2.7	0.638±0.022
	F	454.06±41.90	295±2.5	0.650±0.022
SS^a/SS^a	G	420.87±47.01	271±2.9	0.654±0.032
	H	411.73±41.17	259±2.8	0.635±0.031

Source: Data of A. O. Tantawy and R. C. Lewontin.

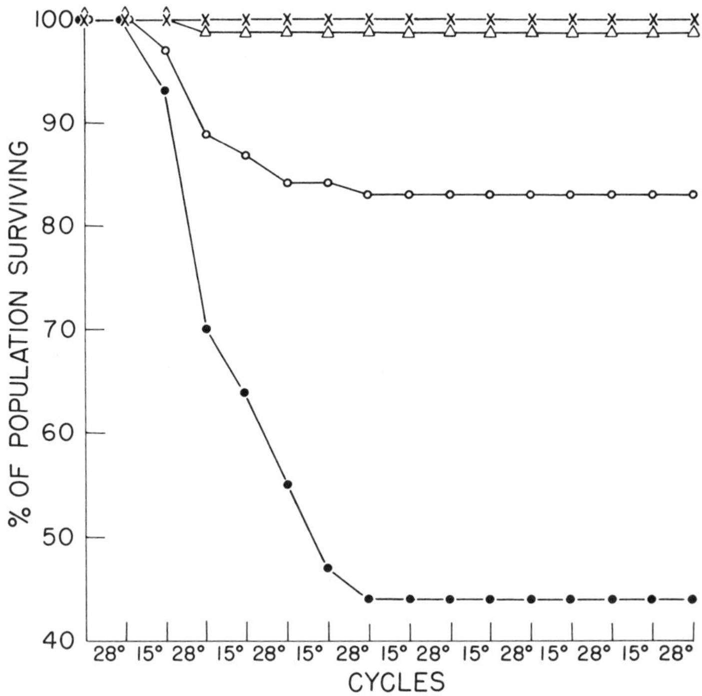

Fig. 5. Proportion of vial populations still surviving at various generations in the cyclic environment. Crosses: +/+; triangles: SS/SS^a; open circles: SS/SS; closed circles: SS^a/SS^a.

is +/+, SS/SS^a, SS/SS and SS^a/SS^a. Again there is excellent agreement of this fitness criterion with the measured intra-population fitnesses.

The species competition experiments were carried out with *D. simulans* as the competitor at three temperatures and under three demographic conditions. In the first, mimicking the invasion of a *simulans* population by *D. melanogaster*, a few fertile melanogaster females were seeded into a demographically stable population of *D. simulans* (few *melanogaster* × many *simulans*). In the second, both species are introduced into a habitat that is unoccupied by seeding a few fertile females of each in a large cage with abundant food (few *melanogaster* × few *simulans*). In the last, two large demographically stable populations are merged, by introducing eggs, larvae, and adults of both species into a new cage (many *melanogaster* × many *simulans*). The results are given in figure 6. Except where indicated, each line is the mean of three replicates. In all cases, +/+, SS/SS, and SS/SS^a populations being the most effective competitor. At the lower temperatures, however, *simulans* eliminates SS^a/SS^a in several cases, especially when there are many *D. simu-*

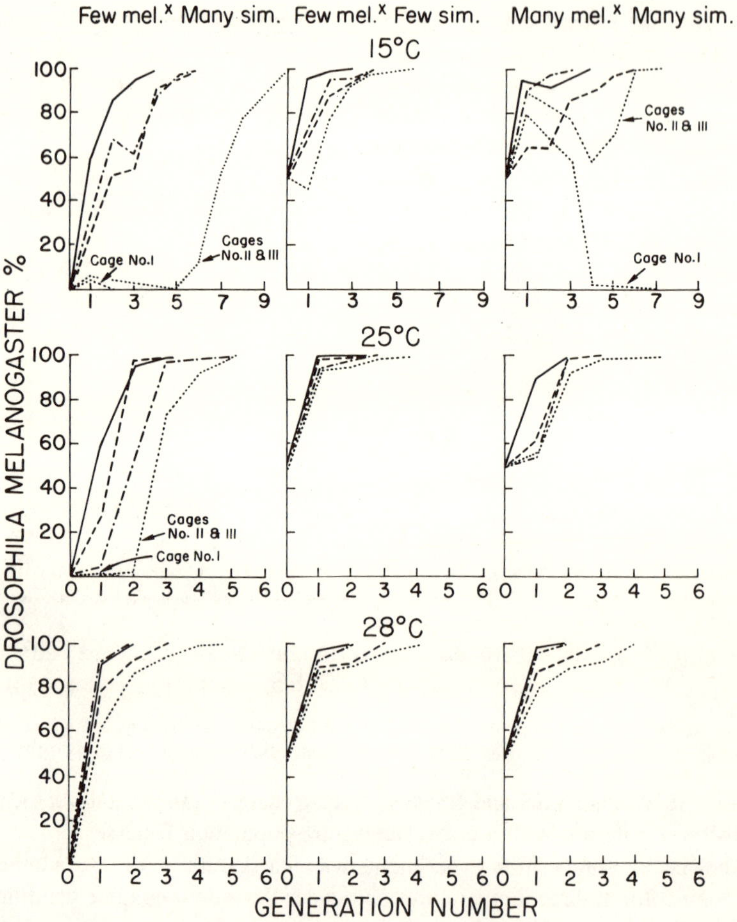

Fig. 6. Percent age of *melanogaster* in competition cages with *D. simulans* at various generations in different temperatures. Solid line: +/+; dashed line: SS/SS; mixed line: SS + SS^a; dotted line: SS^a/SS^a.

lans already present. Once again the results are consonant with the intrapopulation fitness measures.

The general outcome of these experiments is that, no matter how one measures population fitness, there is good agreement with the results of genotype natural selection. This is in sharp contrast with the results of the larval interaction experiments discussed in previous sections. It was pointed out, however, that the chief difference among the genotypes in these experiments lay in *fecundity*, especially the very low fecundity of SS^a/SS^a. Since fecundity is, in general, a noninteractive component of fitness, unlike larval viability, then when fecundity differences are at the basis of the selective differences, we may expect to see them reflected in biomass, extinction rates, and competitive ability. Had the chief difference been in larval viability, with its strong interactions, the result would have been quite different.

I must reiterate, then, that the consequences of natural selection of genotypes, for the life history of populations and species, depends in detail on the particular biology of the species and its form of interaction with others. No empirical generalizations seem possible, and, in the absence of theoretical generalizations, optimality arguments are quite dangerous. At the very least, we must say that the price of optimality arguments is eternal vigilance.

LITERATURE CITED

Birch, L. C., Th. Dobzhansky, P. O. Elliott, and R. C. Lewontin. 1963. Relative fitness of geographic races of *Drosophila serrata*. Evolution 17:72–83.

Dobzhansky, Th., R. C. Lewontin, and O. Pavlovsky. 1964. The capacity for increase in chromosomally polymorphic and monomorphic populations of *Drosophila pseudoobscura*. Heredity 19:597–614.

Leigh, E. 1971. Adaptation and diversity. Freeman, Cooper & Co: San Francisco.

Levins, R. 1968. Evolution in changing environments. Princeton University Press: Princeton, N.J.

Lewontin, R. C. 1955. The effects of population density and composition on viability in *Drosophila melanogaster*. Evolution 9:27–41.

Lewontin, R. C. 1961. Evolution and the theory of games. Journal of Theoretical Biology 1:382–403.

Lewontin, R. C., and Y. Matsuo. 1963. Interaction of genotype determining viability in *Drosophila buschii*. Proc. Nat. Acad. Sci. USA 49:270–278.

Maynard Smith, J. 1972. On Evolution. Edinburgh University Press: Edinburgh.

Trivers, R. L. 1971. The evolution of reciprocal altruism. Quarterly Review of Biology 46:35–57.

Trivers, R. L. 1974. Parent-offspring conflict. American Zooloogist 14:249–264.

Wilson, E. O. 1975. Sociobiology: The new synthesis. Harvard University Press: Cambridge.

Wilson, E. O., and W. H. Bossert. 1971. A primer of population biology. Senauer Associates, Sunderland.

RODGER D. MITCHELL AND MARY B. WILLIAMS

Darwinian Analyses: The New Natural History

2

> How few generalizations there are among systematists; I really suspect there is something absolutely opposed to each other and hostile in the two frames of mind required for systematizing and reasoning on large collections of facts.—Letter from C. Darwin to J. D. Hooker, 25 September 1853

In recent years field biologists have enthusiastically moved natural history from a purely descriptive science into a vigorous analytical discipline. The analysis of data from the large number of first-rate field studies of animal populations has revealed new insights that have given great depth to our understanding of how organisms operate but have not developed the discipline into a science. As enthusiasms waxed, a welter of special models were put forth; in fact, a different model was developed for nearly every species studied. This situation is frustrating for ecologists because it shows that the body of general theory that stimulated the research does not provide a meaningful basis for organizing the insights obtained from individual field studies into useful generalizations.

It is difficult to specify precisely the significance of field biology to ecology, but models from field ecologists seem to be on the verge of adding a whole new dimension to the evidence for Darwinian theory. The initial arguments in support of evolution were based on evidence from geology and biogeography. Stratigraphy established that an orderly replacement of forms must have taken place and that the geographic distribution of species could best be explained as the outcome of a set of geologic events. This body of

Rodger Mitchell, Department of Zoology, Ohio State University, Columbus, Ohio 43210.
Mary B. Williams, Department of Philosophy, Ohio State University, Columbus, Ohio 43210.

evidence in support of evolution stood as the principal evidence for seven decades until a genetic theory for natural selection was published by R. A. Fisher (1929). This theory allowed impressive amounts of genetic evidence to be added to the evidence for evolution.

Now, almost five decades later, field ecologists have developed analyses that provide rigorous quantitative explanations for the way suites of traits act together to determine the fitness of a species. The journals are flooded with papers in which the devices of energy budgets and life tables are used to determine the values of adaptations. Ingenious field studies that specify costs and benefits, define optimal tactics, and measure fitness are often used to develop careful analytic models. These analyses ought to provide the means of adding ecological evidence to the evidence for evolution.

But the more we admire one of these pieces of research, the more likely we are to find the authors giving voice to their sense of frustration in trying to fit the data into general theory. General theory provides inspiration but little, if any, guidance for the field biologist who tries to understand adaptations. It is painful to find first-rate field studies concluding with apologies such as:

> Nonetheless, demonstration of increased K-selection would require that available energy resources be identical or, in experimental conditions, that reduced reproductive frequency is genetic, not phenotypic. Our data are thus inconclusive but . . . consistent with theoretical predictions. (Huey et al., 1974)

Or:

> These findings and other conflicting data show a need for a new theoretical basis for the correlation of body size with seed size in heteromyids. (Smigel and Rosenzweig, 1974)

The weight of abstract mathematical theory can be so great that some biologists dismiss their own careful analyses of extensive field data as irrelevant to theory:

> Since . . . animals on an area change quickly and vary widely . . . the contribution of various factors to determination and regulation of population density must be transitory in time and space. This is not to say that the detailed studies of populations are worthless, but it does point out the difficulty in testing general hypotheses with specific examples. (Slade and Balph 1974)

Cherished folk theories such as the relation of diversity to stability, niche theory, competition theory, and r versus K, to mention the major elements of grand theory in population ecology, have stimulated both fruitful research and a great deal of useful argument, but neither activity has sharpened these theories into falsifiable hypotheses that are useful in the design, interpretation, and synthesis of first-rate detailed analytic studies of natural history.

Are field ecologists simply left with the task of developing a separate analytic model for every species on earth? The art and practice of analyzing field data is an exciting kind of research that often gives real insight into the lives of animals, but the enthusiasm for telling stories about animals wears thin when there is little hope of fitting the research together. The only general themes obvious in this research are the attempts to show that (1) no two species are alike and (2) traits can be shown to be adaptive.

There is an urgent need for a general method of analyzing biological phenomena that will enable us to express in a common framework the important insights obtained in these field studies about the functional relationships between suites of traits and environmental characteristics. Since the term *strategy* appears frequently in the work of these analytic field biologists, we looked at its dictionary definition for a clue to the underlying common framework.

> 1.a) The science and art of employing the psychological, political, and military forces of a nation or group of nations to afford the maximum support of adopted policies in peace or war. b) A variety of or instance of the use of strategy. (*Webster's Third International Dictionary*)

The biological usage differs from the dictionary usage in two significant ways: (1) in most biological studies of strategies, only one component of an adaptive system is considered—feeding, defense, and so on; and (2) whereas the dictionary definition specifies an intermediate goal that plays a part in meeting the ultimate goal of national survival, the biological usage rarely specifies an analogous intermediate goal. There is an obvious trend toward models of biological strategies that treat several components or that specify intermediate goals, and it seems that a general model for biological strategies would provide a useful structure for working with these studies. This paper will present such a model and will indicate how analyses using this structure can solidify intuitive judgments of adaptiveness into rigorous scientific arguments. We will call the models *strategy models* and the analyses *Darwinian analyses*. Many first-rate field studies are partial strategy models; field studies that do a Darwinian analysis of their biological systems are rare, though intuitive judgments of adaptiveness are ubiquitous. It is the Darwinian analyses that will add the results of ecological field studies to the evidence for evolution.

INTRODUCTION OF STRATEGY MODEL TERMINOLOGY

Informally, a strategy model of a species is a model that (1) describes the calorie output of an organism of the species by describing the allocation of the input calories into the various needs of the organism, and (2) describes the calorie input of an organism by describing the resource spectrum from which it draws and the feeding tactics it uses.

The components of the pool of resources drawn on by species S will often be defined as the set of species taken as prey. As an analysis of feeding becomes more sophisticated, it is likely that this taxonomic resource spectrum will be replaced by one that subdivides prey items by the criteria that are relevant to the predator—for example, caterpillars (of one or several species) on the upper leaf surfaces, caterpillars on tree trunks, and so on; the ideal resource spectrum sees the prey as the predator sees them. Thus the way in which prey species are lumped or subdivided into components will be determined by the particular predator-prey system.

The resource spectrum used in a particular study will be specified in the format:

(description of component 1, . . ., description of component n), where the description of component 1 might be, for example, "planktors less than one millimeter."

Then the resources can be described as a resource vector:

$$\mathbf{R}=(R_1, R_2, \ldots, R_n) \tag{1}$$

where R_i is the amount of assimilable calories available in the ith component of the resource spectrum.

The behavior and selectivity of the consumers determines the probability (P_i) of the ith component of the resource spectrum being eaten. These probabilities determine the feeding tactics vector:

$$\mathbf{P}=(P_1, P_2, \ldots, P_n) \tag{2}$$

For a planktor that unselectively filters algal cells randomly dispersed in water, $\mathbf{P}$ is easily specified; the probability of the ith component being eaten is just equal to its relative frequency in the environment. For a predator whose hunting pattern is constantly modified by experience, $\mathbf{P}$ is very difficult to specify; each P_i might then be a function of the availability in the recent past of the ith component compared with the availability of all other components. For some species it may be necessary to specify different feeding-tactics vectors for different phenotypes.

The product R_iP_i gives the amount of calories available from component i. The sum of these products determines the input I of assimilated energy.

$$\Sigma R_iP_i = I \tag{3}$$

The strategy of a species is characterized not only by its resources and its feeding tactics but also by the way in which it allocates the energy it assimilates. For a given study, the components among which the energy is partitioned will be specified by an energy spectrum:

(Description of component 1, . . ., description of component m)

The energy budget of a species can also be expressed as a vector:

$$\mathbf{E} = (E_1, E_2, \ldots, E_m) \tag{4}$$

For some studies E_i will be the quantity of energy used in a particular biological function (e.g., defense, thermoregulation, hunting, courtship, and so on), while for other studies, **E** will show the allocation of energy into just a few comprehensive categories (e.g., maintenance, biomass of parent, and biomass of young). In some situations it may be possible to let E_i be the cost of obtaining a unit of the ith resource, with $E_{n+1}, \ldots E_m$ the costs of reproduction, defense, and so on; it would then be possible to explore the relationship between cost and gain for specific resource components.

The sum of the E_i determines the output of energy O:

$$\sum_{i=1}^{m} E_i = O \tag{5}$$

In the above discussion of the resource vector, the feeding-tactics vector, and the energy vector, we have not mentioned the fact that all of these may change as environmental conditions change. For some studies it may be necessary to express one or more of them as a function of significant environmental conditions. We can now give a formal definition of a strategy model:

> A strategy model of a species S is a model that specifies the resource spectrum and the energy spectrum and gives experimentally determined values for the components of the resource vector **R**, the feeding tactics vector **P**, and the energy vector **E**.

The vectors of the strategy model define the extrinsic resource vector as a component independent of the two intrinsic biological vectors, the feeding-tactics vector and the energy-budget vector. The feeding-tactics vector is the outcome of long-term evolutionary responses to the resource vector and may be modified through short-term learning responses.

The products of the resource vector and feeding-tactics vector define the energy assimilated by an animal, and the way in which the energy is used is given by the energy vector. Hence, the strategy model fits the research of field biologists together with the data on energetics to provide a mathematical model of the natural history of a species. Examples of the study of one component or another are easy to find but, as yet, there is no set of data for a complete strategy model.

EXAMPLES

1. The Use of Behavioral Data to Define **P**

The partitioning of resources among species by spatial specialization is well known in birds; after MacArthur's (1958) study of foraging by warblers, virtually every ecology textbook has included a diagram of where warblers feed in a pine tree as an example of ecological specialization. Williamson (1971) took the analysis of feeding tactics a step beyond the partition by distribution in space when she considered the sequence of movements by the two sexes of red-eyed vireos. Both had identical frequencies for each kind of movement; but a disproportionate number of vertical moves by females resulted in feeding, and these feeding events were followed by a downward movement from which the bird might feed again. Males fed more often from horizontal moves. Table 1 shows how these behavioral differences determine difference in utilization by the sexes of prey resting on upper, vertical, and lower surfaces. Thus if the resource spectrum is (prey on upper surface, prey on vertical surface, prey on lower surface), the resource vector is

$$\mathbf{R} = (R_1, R_2, R_3)$$

TABLE 1

LOCATIONS OF PREY TAKEN BY RED-EYED VIREOS

Position of Prey	Frequency of Feeding Moves	
	Male Vireos	Female Vireos
Upper surface	0.17	0.12
Vertical surface	0.52	0.33
Lower surface	0.31	0.55

NOTE: This is a conversion of direction of attack (Williamson, 1971) to surface on which the prey rested to illustrate how $\mathbf{P}_i$ of males differs from that of females (N for both samples = 200 feeding events).

and the feeding-tactics vectors are:

$$\mathbf{P}_{♀}=(.12,\ .33,\ .55)$$

$$\mathbf{P}_{♂}=(.17,\ .52,\ .31)$$

In this study the feeding-tactics vectors were a finding incidental to the development of models for movement through the environment that specify the areas explored and kinds of contacts made by four different species of birds sharing a habitat; such models are a crucial first step in understanding tactics. The techniques for field study and the analysis of data that are now in hand make it reasonable to expect studies of foraging behavior to go beyond tabulations of differences. Since the average heights at which the two sexes forage are different, their resource vectors would also be different; if the resource vectors were known for these birds, hypotheses about the usefulness of these differences in lessening competition for resources between the sexes could be tested with the aid of the relations:

$$\text{I (female)} = \Sigma R_i P_i = .12R_{1♀} + .33R_{2♀} + .55R_{3♀}$$

$$\text{I (male)} = \Sigma R_i P_i = .17R_{1♂} + .52R_{2♂} + .31R_{3♂}$$

Similarly, hypotheses about the adaptiveness of the different foraging tactics of the four species sharing this habitat could be tested if the resource vectors and the feeding tactics vectors were known.

2. *The Experimental Analysis of **R** and **P***

Tactics that change as a result of learning may be very difficult to measure in the field, where changes in resource availability are likely to complicate the picture. Smith and Sweatman (1974) studied foraging tactics by titmice in an experiment in which **R** was held constant. The available resources consisted of an aviary containing six feeding patches of 256 food cups that were maintained at densities of 0, 4, 8, 12, 16 and 20 mealworms per patch (fig. 1). Within each area the distribution of resources was random. Each bird was given 30 five-minute trials, and the proportion of searching effort devoted to each patch was recorded. The data were pooled into three blocks consisting of the first ten trials, the second ten trials, and the third ten trials. The distribution of search effort among the patches was close to random during the first ten trials, and was significantly nonrandom (at the .01 significance level) during the second ten trials, with a strong concentration of search effort at higher densities; the concentration of search effort at higher densities increased still further during the third ten trials (fig. 2).

For this experiment the resource spectrum was (patch with 0 mealworms, patch with 4 mealworms, . . ., patch with 20 mealworms).

R = (0/256, 4/256, 8/256, 12/256, 16/256, 20/256)

P can be estimated from Figure 2.

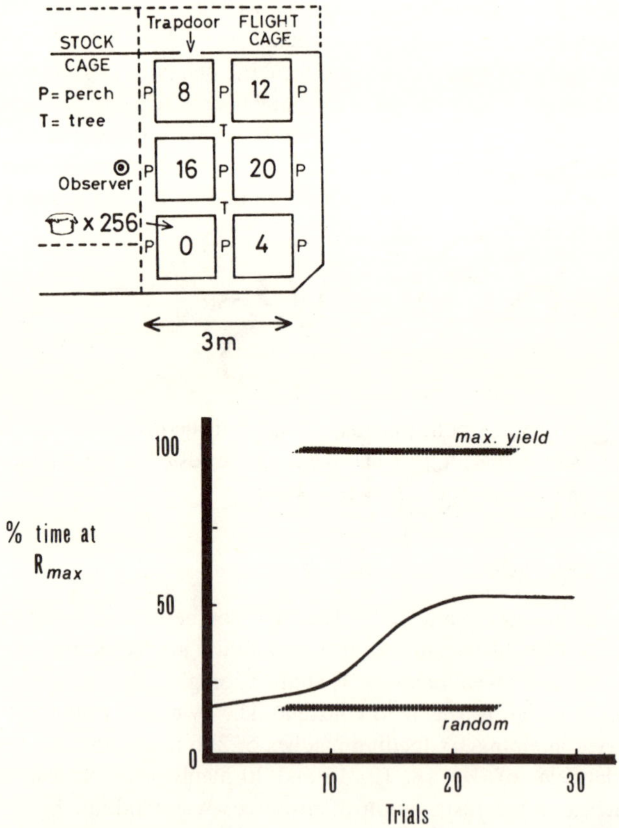

Fig. 1. Diagram of the experimental universe used to study titmice feeding tactics and a graph of the performance of titmice. The numbers in the boxes (which are the six food patches) are the number of food cups out of the 256 in a patch that contained a *Tenebrio* larva. The larvae were placed at random in the cups of a patch. Maximum yield would be expected if titmice learned to spend all the time in the patch with .08 larvae per cup. (Modified from Smith and Sweatman, 1974)

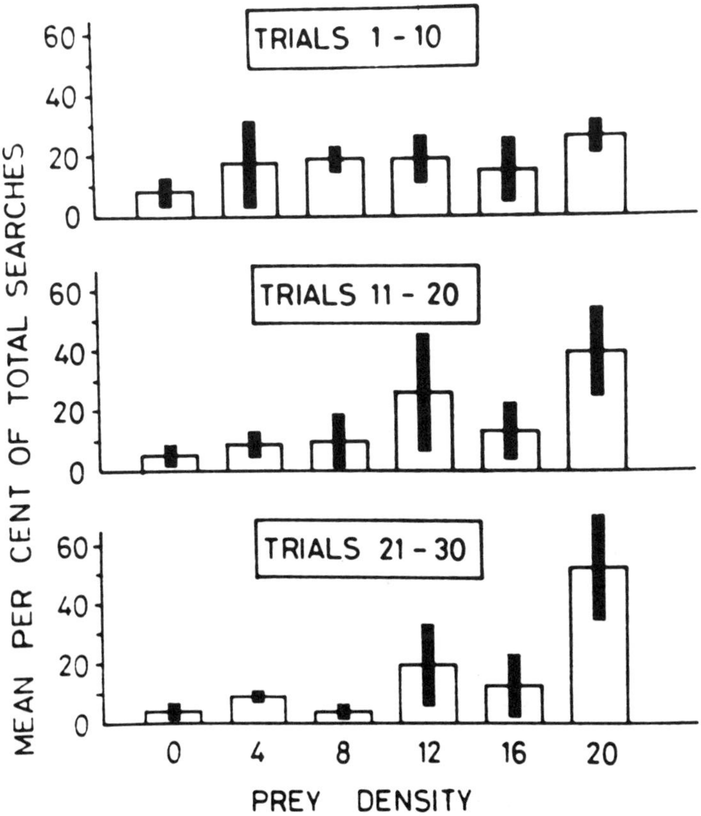

Fig. 2. Pooled responses to prey size by six great tits. The top histogram shows the mean feeding-area position preference during ten 5-minute trials before the start of the experiment. The lower histograms show that the feeding tactics changed over time. (From Smith and Sweatman, 1974)

For first block: $\mathbf{P} = (.09, .16, .18, .16, .15, .22)$

For second block: $\mathbf{P} = (.06, .09, .10, .23, .12, .30)$

For third block: $\mathbf{P} = (.03, .09, .03, .19, .14, .52)$

Clearly **P** changed with learning in a way that increased $I = \Sigma P_i R_i$.

A Darwinian analysis based on these data quickly runs into difficulties caused by the artificiality of the experimental environment, although the artificial environment provides important preliminary information. The experiment shows that titmice can learn to improve their tactics; it is reasonable to believe that this ability was selected for. But notice (figure 1) that they were unable to increase their success rate beyond 50% of the maximum possible success rate that would occur in the patch with 20/256 mealworms; we can explain this by pointing out that in nature resources always vary widely, and hence the exploratory behavior that resulted in poor performance in the experiments is adaptive in nature. Field observations (Smith and Sweatman, 1974) support both the hypothesis that titmice learn to improve their feeding tactics and the explanation that exploratory feeding, which reduced the efficiency in the experimental universe, is, in nature, the basis for comparisons that enable titmice to monitor alternative resources and shift from areas with less than average yields to areas of greater than average resource density. Presumably the innate pattern for the allocation of effort to exploration versus the exploitation of a discovered resource clump has been fixed by selection; further work is needed to test this presumption.

3. *The Analysis of the Limits for* **E**

In a study of vampire bats (*Desmodus*), McNab (1973) presents the data for **E** at three different temperatures and uses it to argue that (p. 142) "vampires are limited in their distribution to a tropical climate because of the combination of food habits that require feeding only at the terminus of a flight and a limit placed on the maximal meal that they can carry in flight." The output spectrum for this study is: (resting metabolism in cave for size x and temperature y, resting and feeding metabolism out of cave for size x and temperature y, urine loss for size x and temperature y, fecal loss for size x and temperature y, flight for size x and temperature y, other activity for size x and temperature y). Clearly **E** is a function of size and temperature; we will denote it by $\mathbf{E}(x,y)$. Measuring in kilocalories per day, he gives the energy vector for a vampire of average size (42 grams) at three temperatures.

$$\mathbf{E}\ (42\text{g},\ 20°\text{C}) = (16.0,\ 1.1,\ 0.6,\ 5.4,\ 1.1,\ 0.6),\ \Sigma E_i = 25.7$$

$$\mathbf{E}\ (42\text{g},\ 10°\text{C}) = (27.9,\ 1.9,\ 1.0,\ 8.6,\ 1.1,\ 0.6),\ \Sigma E_i = 41.1$$

$$\mathbf{E}\ (42\text{g},\ 5°\text{C}) = (33.7,\ 2.4,\ 1.1,\ 10.3,\ 1.1,\ 0.6),\ \Sigma E_i = 49.2$$

The typical daily meal of the vampire at 20°C is 23.8 grams of blood, which provides 25.7 kilocalories. The meal sizes necessary to maintain a 42g vampire at 10°C and 5°C would be 38.1g and 45.6g. *Desmodus* excretes up to 25% of the weight of a meal before flying back to its roost, but even after this reduction the weight increase for maintenance meals at 10°C and 5°C would be 68% and 81%. But a 43% increase in weight prevents flight in a bat of this size. Therefore *Desmodus* cannot survive in an area with minimal temperatures of 10°C or less.

The argument presented above shows why vampires are not found in areas with temperatures lower than 10°C. But why have they not evolved a larger size, enabling them to carry a bigger meal and expand their distribution? They have not, McNab (1973) argues, because an increase in the weight of the bat increases the total rate of metabolism, decreases the maximum % weight increase allowed by wing-loading characteristics, and increases the cost of flight. In effect, then, he argues that the selective landscape determined by this combination of food habits has a maximum at a particular combination of temperature tolerance and size and that the data show that the temperature tolerance and size exhibited by *Desmodus* are at, or very near, this maximum point.

INTRODUCTION OF DARWINIAN ANALYSIS TERMINOLOGY

The field studies used to build a strategy model for the species S deal with the short term responses of populations; the characteristics of the species that determine its resource spectrum, feeding tactics, and energy budget are presumed to be stable. Consequently, these characteristics are not explicitly mentioned in the strategy model. But when the ecologist speaks of the adaptiveness of a particular suite of traits, he is saying that Darwinian selection has fixed this suite of traits because its contribution to fitness is greater than the contribution of the previous suite of traits. Consequently, in a Darwinian analysis we will need to explicitly name the traits under study. This will be done in a trait spectrum:

(Description of trait 1, . . ., Description of trait K)

The trait vector will then be

$$\mathbf{T} = (T_1, T_2, \ldots, T_K) \tag{6}$$

where T_i is one of the possible states of trait i.

The specification of what it is that Darwinian selection is maximizing is a very tricky task. We offer the specification given below somewhat tentatively, but we feel it is at least a useful approximation, which may help in clarifying the situation sufficiently to allow a less tentative specification. Let I_p be the calories stored in the present generation by the previous generation; let I_s be the sum of the inputs over the lifespan of the present generation; let O_s be the sum of the outputs over the lifespan of the present generation, where O_s includes the cost of reproductive tactics but not the energy stored in surviving offspring. (O_s also includes the output caused by death.) Then

$$Q = I_p + I_s - O_s \tag{7}$$

is the energy stored by the present generation in their surviving offspring. It is Q that is maximized by Darwinian selection.

For Q to be maximized by selection on the suite of traits **T**, it must be a function of **T**. And clearly if we say, for example, that powerful leg muscles were selected for because they allowed the predator to catch more prey, we are saying that I_s is a function of the suite of traits that affect speed; similarly, when we say that fur was selected for because it decreased heat loss, we are saying that O_s is a function of the suite of traits that affect heat loss. Therefore (7) should be expressed as:

$$Q(\mathbf{T}) = I_p + I_s(\mathbf{T}) - \mathrm{O_s}\,(\mathbf{T}) \tag{8}$$

Then Darwinian analysis is defined as follows:

A Darwinian analysis of a suite of traits **T** is an analysis that presents evidence that $Q(\mathbf{T})$ has been increased by Darwinian selection on **T**.

EXAMPLES

1. The Use of Gut Content Data

In an investigation of the partitioning of food resources among four sympatric species of the cyprinid genus *Notropis*, Mendelson (1975) collected most of the data necessary to determine **R** and **P** for these species. The resource spectrum he used had two components, benthic organisms and drift organisms. $\mathbf{R}=(R_b, R_d)$ can be estimated from the data from benthos and drift samples given in table 2 together with data giving the average number of assimilable calories per organism for each of these taxa. **P** can be estimated from the gut content (**G**) data of figure 3 and **R** by using the relationships $\mathrm{P}_b=\mathrm{G}_b/\mathrm{R}_b$ and $\mathrm{P}_d=\mathrm{G}_d/\mathrm{R}_d$. (We would actually need Mendelson's raw data to make these estimates.) Since the separation of the gut contents into benthic and drift components is not exact, this estimate of **P** can only be used in investigations for which this inexactness is not crucial. Other data in Mendelson's paper show that **G** and **R** vary seasonally, so table 2 and figure 3 serve primarily to show how data from which **G** and **R** can be calculated can be collected.

TABLE 2

HABITAT CLASSIFICATION OF THE INVERTEBRATE FAUNA BASED ON THE TOTAL NUMBER OF INDIVIDUALS OF EACH TAXON COLLECTED

Taxon	N	% in Drift	% in Benthos
		Drift fauna	
Baetis	490	100.0	0.0
Dixa	6	100.0	0.0
Simulium	9	100.0	0.0
Cricotopus	136	100.0	0.0
Nanocladius	279	100.0	0.0
Psectrocladius	64	100.0	0.0
Cyclopoid copepods	1513	99.6	0.4
Corynoneura	269	97.4	2.6
Thiemaniella	90	94.8	5.2
		Indeterminate fauna	
Gammarus	8242	72.0	28.0
Polypedilum I	96	54.8	45.2
Enchytraeidae	165	54.6	45.4
Chironomid pupae	128	52.6	47.4
Aquatic Acari	8	50.0	50.0
Brillia	50	32.0	68.0
Pentaneura	46	30.6	69.4
		Benthic fauna	
Pycnopsyche	129	15.3	84.7
Microtendipes	45	9.0	91.0
Prodiamesa	37	5.9	94.1
Paratendipes	1280	5.9	94.1
Sialis	180	4.9	95.1
Stictochironomus	74	4.7	95.3
Cryptochironomus	49	4.5	95.5
Hyodrilus	407	4.4	95.6
Micropsectra	12,359	3.7	96.3
Chrysops	39	3.6	96.4
Asellus	534	3.5	96.5
Dicranoia	118	3.4	96.6
Limnodrilus	1504	2.3	97.7
Peloscolex	8	0.0	100.0
Polypedilum II	58	0.0	100.0
Tipulidae "A"	350	0.0	100.0
Hexatoma	39	0.0	100.0

NOTE: Data from the three sampling stations have been combined. N = total number collected (from Mendelson, 1975).

These four species form a series with respect to body shape and mouth position (Mendelson, 1975). *N. atherinoides* and *N. spilopterus* are both laterally compressed with terminal mouths. *N. dorsalis* is dorsoventrally compressed with a ventral mouth; and *N. stramineus* is midway between these extremes in both characteristics. Hubbs (1941) has suggested that lateral compression with terminal mouth is associated with a free-swimming mode of life, and dorsoventral flattening and ventral mouths are associated with a bottom-dwelling and bottom-feeding mode. Clearly it would be useful to be able to

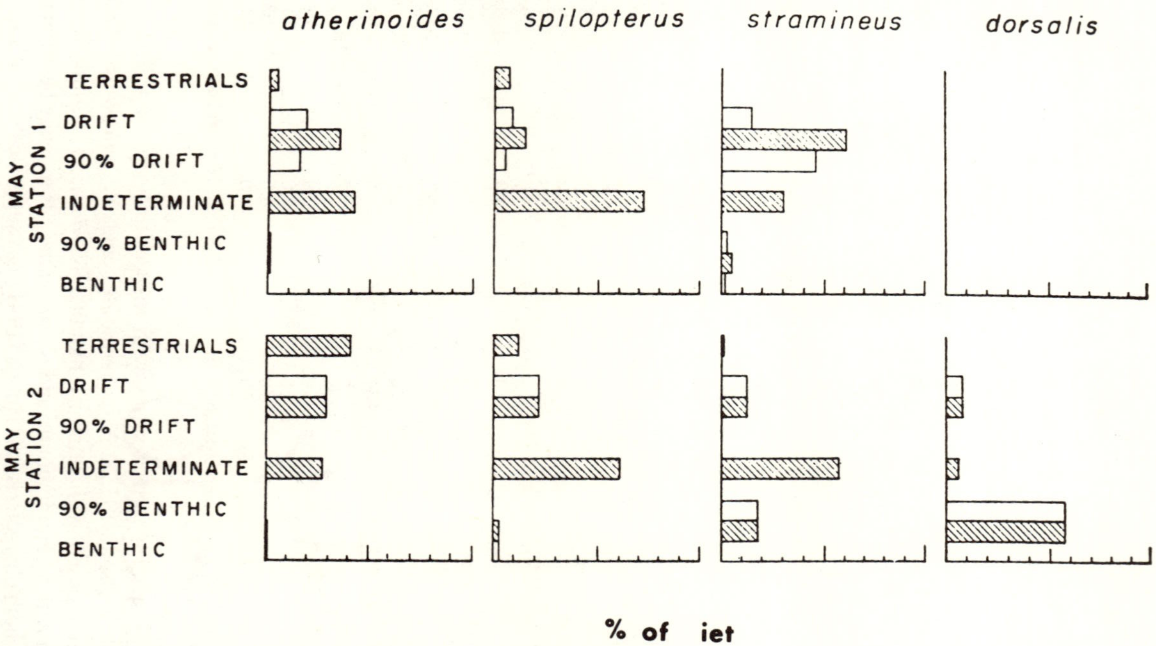

Fig. 3. Percentages of each of six habitat categories in the diets of *Notropis* species collected in May. The percentages given for each species indicate the relative sizes of the R_iP_i for that species. (From Mendelson, 1975)

test the hypothesis that fish of the latter type are more efficient at cropping benthic prey and fish of the former type more efficient at cropping drift prey; evidence for this hypothesis would be evidence for the hypothesis that natural selection has formed the different suites of traits found in these different species. And we can use Mendelson's data to test this hypothesis. For this Darwinian analysis the trait spectrum is: body shape, mouth position, spatial preference. Let H_i denote the body shape of species i and M_i denote the mouth position; if we rank the degree of dorsoventral compression on a scale from 0 to 5, H for *N. antherinoides* and *N. spilopterus* would be 0, for *N. stramineus* it would be 3, and for *N. dorsalis* it would be 5; similarly M for *N. atherinoides* and *N. spilopterus* would be 0, for *N. stramineus* it would be 3, and for *N. dorsalis* it would be 5. Let C_{ib} and C_{id} be the time spent by species i in the benthos *(b)* and in the drift *(d)* respectively; let D_{ib} and D_{id} be efficiency of species i in cropping benthic organisms and drift organisms respectively. And let $\mathbf{P}_i$ be the feeding tactics vector of species i, so that, for example, P_{ib} is the probability of species i capturing benthic organisms: Then

$$P_{ib} = C_{ib}D_{ib} \text{ and } P_{id} = C_{id}D_{id}$$

P_{ib} can be estimated from figure 3, and C_{ib} can be estimated from figure 4. From these we can calculate the efficiency of species i at cropping benthic organisms: $D_{ib} = P_{ib}/C_{ib}$. We can similarly calculate D_{id}. We can then use this to get at least an idea of whether Q is maximized for this guild of fishes by their spatial preferences; that is, is the sum of the Q's for these four species of fish larger than it would be if their spatial preferences were different? If we assume that the energy output does not differ significantly in the two environments, we can estimate the effect on Q of a change in spatial preference by calculating the effect on I_s. C_{ib} is a measure of the spatial preference of species i. So we compare, for example

$$Q(H_1, M_1, C_{1b}) + Q(H_2, M_2, C_{2b})$$

with

$$Q(H_1, M_1, C_{1b} + 10\%) + Q(H_2, M_2, C_{2b} - 10\%)$$

to see if raising by 10% the amount of time species 1 spends in the benthos and simultaneously lowering by 10% the time species 2 spends in the benthos would decrease the sum of the Q's.

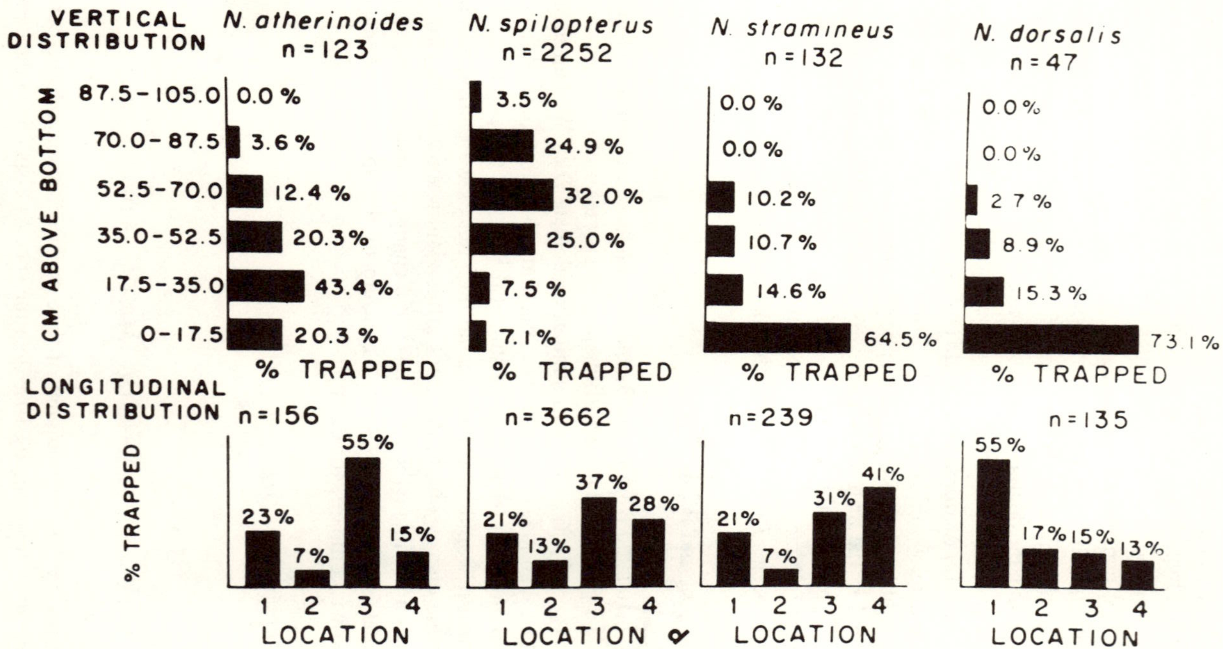

Fig. 4. Vertical distribution of *Notropis* in Roxbury Creek pools. If each of the six habitat categories is predominant at a different depth, then these data provide estimates of the feeding tactics vectors for the four species. (From Mendelson, 1975)

2. The Use of Fecundity Data

The utilization of resources by the bean weevil *Callosbruchus maculatus* is determined solely by oviposition tactics, since the larva matures in the bean on which it was placed and does not feed as an adult. Mitchell (1975) determined these tactics under laboratory conditions that resemble the storage conditions in which the beetle has thrived for a few thousand years. The survivorship for that temperature and humidity regime is a regression on weight when there is one egg per bean and is a constant at higher loads (figure 5). Consequently, the maximal yield would be achieved if the first egg were placed on the largest bean, the next egg on the second largest, and so on. Because the interactions between larvae result in the mean fitness of a second egg on a bean being independent of bean weight, maximal yield in crowded conditions would be obtained by putting second eggs first on the smallest bean, next on the second smallest, and so on. Clearly, the beetle cannot have the perfect knowledge of all the beans in her universe, which would be required to obtain this maximum yield. The oviposition tactics of *C. maculatus* are shown in figure 6.

These observed oviposition tactics can be simulated almost exactly by a model in which the beetle's view of the available resources contains only two beans at a time: the bean she is presently on and the bean she was previously on. Let B_1 and B_0 be the weights of the present and previous beans, and E_1 and E_0 be the egg loads of the present and previous beans; then the beetle sees the present bean as being in one of four categories:

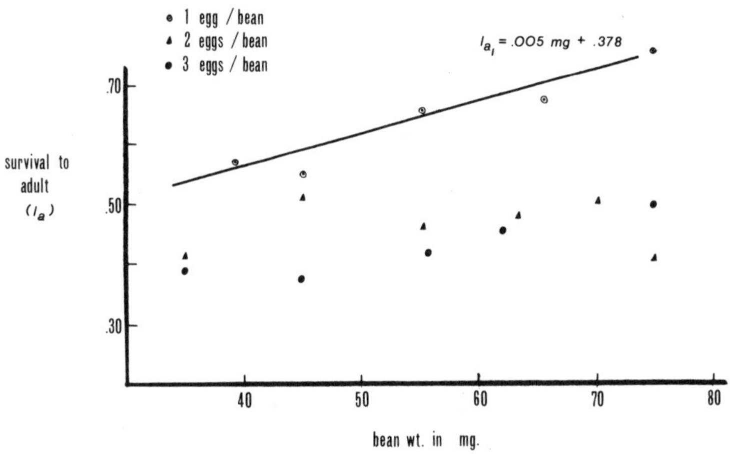

Fig. 5. Survivorship of eggs to the adult stage (1_a) as a function of bean weight and number of eggs per bean. (From Mitchell, 1975)

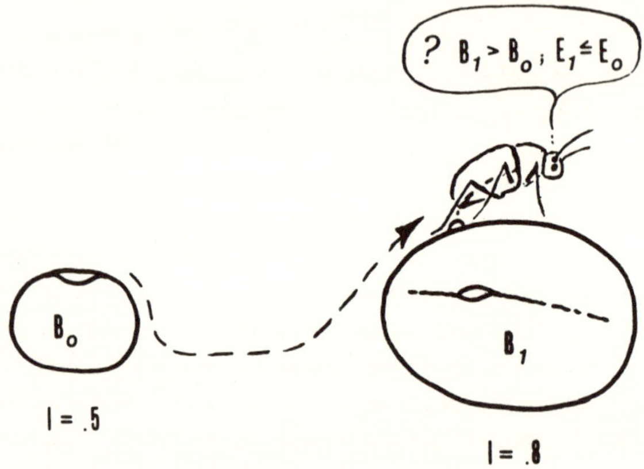

Previous bean smaller → Present bean larger	Present bean with fewer eggs	→ OVIPOSIT
	Number of eggs equal	→ OVIPOSIT
	Present bean with more eggs	→ Reject
Previous bean larger or equal → Present bean smaller	Present bean with fewer eggs	→ OVIPOSIT
	Number of eggs equal	→ Reject
	Present bean with more eggs	→ Reject

Fig. 6. Oviposition tactics of *C. maculatus*. If R is specified, then the above R,P program will simulate the performance of the beetle with respect to the size and egg load of a bean. The expected yield of adults from both observed and simulated oviposition pattern can be predicted from the functions in fig. 5, and they do not differ from each other in either pattern or yield. The simulation is from Mitchell 1975.

> ($B_0 < B_1$ and $E_0 \geq E_1$, $B_0 \geq B_1$ and $E_0 < E_1$, $B_0 > B_1$ and $E_0 \geq E_1$, $B_0 \leq B_1$ and $E_0 < E_1$).

These four categories are the resource spectrum for *C. maculatus*. (If you find that incomprehensible, it is because you are used to thinking human thoughts and unused to thinking beetle thoughts!) Then P_1 is the probability for a given egg, of its being laid on a bean of type 1 ($B_0 < B_1$ and $E_0 \geq E_1$). P_2, P_3, and P_4 are similarly defined. Let q_i be the probability of a bean encountered by the beetle being of type i; this probability is a function of the density of eggs per bean already in the universe. Let p_i be the probability of a beetle ovipositing on a bean of type i; then, from figure 6, $p_1 = 1$, $p_3 = 0$, $p_4 = 0$. Therefore $\mathbf{P}(q_1, q_2) = (q_1/(q_1 + q_2), q_2/(q_1 + q_2), 0, 0)$. For this system R_i is the expected survivorship for an egg laid on a bean of type i; R_i is a function of bean weight and of average number of eggs per bean, and could be calculated from figure 5. $\mathbf{R} = (R_1, R_2, R_3, R_4)$. Then if A is the average number of eggs per beetle, the number of surviving adults per beetle is given by $A\Sigma R_i P_i$. Multiplying this by the number of beetles per generation will give Q; the energy output O_s, does not appear explicitly in this formula for Q because it is assumed to affect Q by determining A.

We can independently calculate the potential yields of different oviposition tactics; this provides a scale against which the performance of *C. maculatus* can be judged. Random oviposition, $\mathbf{P}=(q_1,q_2,q_3,q_4)$ will give a Poisson distribution of eggs. Figure 7 shows the yields given by this random oviposition tactic, the observed tactics, and the maximum tactics described earlier. The observed tactics give ›ι 21–66 percent gain in fitness over the random tactics; this is over 70 percent of the maximum achievable gain in fitness. Output costs were not measured; but since searching costs are proportional to the fraction of the beans discovered that are rejected, and since just over half the discovered beans are rejected, searching costs for *C. maculatus* are twice the searching costs of a randomly ovipositing beetle. Thus if, for example, the energy saved by the randomly ovipositing beetle was more than enough for 21–66 percent more eggs, the observed tactics would not have been selected for. Such hypothetical beetles serve to show us what is at stake, but they do not provide experimental evidence that these oviposition tactics were selected

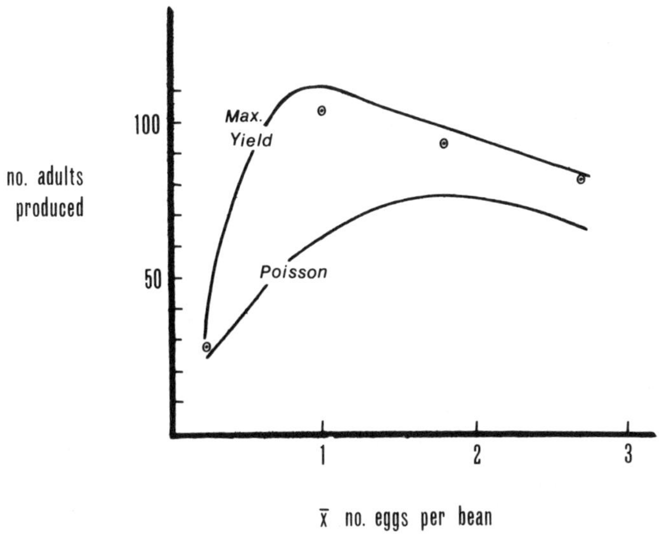

Fig. 7. Fitness response for three distributions of *Callosobruchus maculatus* eggs on mung beans. The yields were calculated from known survival functions (fig. 5) for samples of beans of known weights. In the Poisson, eggs are dispersed at random. Under maximum yield, the first eggs are put on the largest to the smallest beans in sequence and, after there is 1 egg per bean, eggs are added on the smallest to the largest. The circled dots are the observed values.

for. (Of course, no sane biologist believes that such a striking, and strikingly effective, tactic is accidental; but that is not experimental evidence.)

THE VECTORS AND THEIR SPECTRA

The strategy model is a way of characterizing the n-dimensional hypervolume that Hutchinson (1957) called the niche. A major operational problem with the niche is that there is no objective way of deciding which axes are important and when enough axes have been described. Such problems are solved in the strategy model, in which the components of the resource spectra are the niche axes, because the sum of the products of the components of **R** and **P** must be equal to the sum of the components of the energy budget vector **E**. Since **E** can usually be estimated reasonably well, the importance of the components of **P** and the adequacy of **P** in any given study can be determined.

P gives the extent of use for each component of the spectrum of resources, where the spectrum has been found through analysis. For example, in the *Notropis* study described earlier, the fact that "spatial rather than taxonomic considerations are the chief determinants of prey selection in these species" was a major conclusion of the study; similarly, in the *Callosobruchus* study, the fact that the beetles divide beans into the four types mentioned was a major conclusion of the study. In each case the conclusion was supported by tests of statistical significance. Ultimately, the spectra found by field ecologists, and tested for significance, will be the data used in determining the niche dimensions.

The range of a species will be the area in which **RP** yields sufficient energy. Most studies of endotherms, such as McNab's (1973) study of vampire bats, deal with nonreproductive adults that do not grow. If growth is zero, then experimental measures of metabolism can be extrapolated to give an estimate of the total energy budget. The estimates of metabolism show that endotherms with similar feeding habits often have similar metabolic patterns, and from this we would infer the energy budget spectra to be similar. The most interesting of these patterns is the relationship between territory, body weight, and metabolism (McNab 1963).

Both the territory size and basal metabolism of leaf-eating croppers and the seed or insect hunters are related to body weight by the same power function (McNab 1963, Schoener 1968). (In extending the analysis to birds, Schoener (1968) argued that the power function for pure carnivores was greater than for other guilds.) The food of hunters (seeds, fruits, and insects) generally yields about 25 percent more nutrients, as well as being richer than leaves in micronutrients; taking this into consideration, the relative abundances of resources in leaves versus seeds and fruit are at the level expected if the coefficients of the different guilds were determined by resource abundances that do not vary widely over major climatic zones or from community to community (table 3). (The general pattern of the log-transformed data suggests that the over-all abundance of a particular kind of resource does not vary widely.) It appears that the total amount of available nutrients is about the same for these guilds; hence it will not explain major differences in the ecology of various consumers.

TABLE 3

POTENTIAL RESOURCE BASE (NPP IN GM M^{-2} YR^{-1}) FOR FOREST AND NON-FOREST (GRASSLANDS, SAVANNAHS, STEPPES, AND SWAMPS) COMMUNITIES OF SUBBOREAL AND TROPICAL REGIONS

	SUBBOREAL		TROPICAL	
	Tree	Non-Tree	Tree	Non-Tree
Net primary production[1]				
Humid	1,240 (1,200-1,300) n=5	1,500 . . . . n=1	2,250 (1,500-3,500) n=8	. . . .
Semi-arid	. . . .	820 (70-1,300) n=16	1,600 (1,500-1,700) n=2	1,120 (700-1,500) n=5
Arid	. . . .	400 (50-260) n=9	. . . .	110 (10-400) n=7
Range of mean values	1,240	400-1,500	1,600-2,250	110-1,120
Proportion of NPP in:				
Roots, stems	.68[2] (.32-.77)	.16[3]	.70[5] (.55-.82)	.22[6]
Leaves	.30[2] (.22-.44)	.59[3]	.24[5] (.16-.40)	.53[6]
Seeds, fruits	.02[2] (.01-.07)	.25[4]	.02[5] (.02-.05)	.25[4]
Estimated resource abundances[7]				
Roots, stems	843	64-240	1,120-1,575	24-246
Leaves	372	236-885	448-630	58-594
Seeds, fruits	25	100-375	32-45	28-280

1. From Rodin, Bazilevich and Rozov (1975) with the exceptionally high NPP of bogs and floodplains excluded.
2. Based on data for both stands and individual species from Hughes (1971), Marks (1974), Whitaker et al. (1974), Whitaker and Niering (1975).
3. Based on 11 species (Golley, 1965: Harper and Ogden, 1970; Hinds, 1975; Ogden, 1974) and 2 stands (Wiegert and Evans, 1964) to get root-shoot portions.
4. Based on 8 species (Harper and Ogden, 1970; Hinds, 1975; Ogden, 1974) and used as a general estimate for grasslands.
5. John (1973).
6. Singh and Yadava (1974).
7. Calculated from the NPP and estimates of proportions.

McNab (1974) examined the relationship between, in our terminology, ΣE_i and the trait spectrum

(body size, climate in species range, food habits);

using basal metabolism as an estimate of energy expenditure and traits such as desert dwelling, arctic, and so on for the climate component, and seed-eaters, insect-hunters, and so on for the food-habit component, he pointed out a large number of cases in which the species of particular guilds shared a set of peculiar metabolic features. These extrapolations from metabolic data, which are taken under very artificial conditions, reveal quite clear patterns. The same data have also been successfully used in specific explanations of body shape and size in fossorial rodents (McNab 1966), weasels (Brown and Lasiewski 1972), and whales (Brodie 1975).

The general correlations and specific explanations based on extrapolations from basal metabolism seem likely to have a biological basis despite the fact that there are three obvious limitations to such extrapolations: First, the relation of basal metabolism to metabolism under various activity patterns is not known. Second, the extrapolations exclude data on reproducing individuals. Third, they exclude data on growing individuals. Thus these extrapolations rely on assumptions that simplify the energy spectrum from

(maintenance, growth, reproduction)

to a spectrum with basal metabolism as its sole component. These simplifying assumptions need to be reexamined to see why these factors may be ignored.

There is some evidence that the relation between activity and metabolism may be close enough that differences in basal metabolism give a fair indication of the differences in the total maintenance energy budget of various endotherms. Activity is a size-related function and Wunder's (1975) algorithm based on empirical data gives reasonably accurate estimates of activity costs from basal metabolism. More evidence is needed, but this at least suggests why it may be possible to find biologically significant similarities among species from basal metabolism alone.

Reproduction may be of little importance in this use of energy budget data for two reasons: (1) only a small fraction of the total energy budget goes into reproduction; or (2) reproduction is scheduled when resources are extremely abundant. The first of these reasons may apply to poikilotherms, in which less than 20 percent of the annual energy budget is in reproduction (table 4). In some poikilotherms energy for reproduction is accumulated over the annual cycle, and in others it is built up at the time of reproduction; in either case, the cost of reproduction is a small fraction of the energy budget and is very low compared with the differences in resource abundance over the annual cycle.

The second reason may apply to endotherms, for which the few data available suggest that reproductive costs may be from one-half to three-quarters of the *daily* energy budgets (table 5). But, as Kendeigh's (1972) diagram (figure 8) for the *annual* energy budget of a house sparrow shows, the energy need is greatest during the winter when resources are scarce and large amounts of energy are needed to keep warm; the increment for reproduction occurs when

TABLE 4

ANNUAL ENERGY BUDGETS AND THE COEFFICIENT FOR REPRODUCTIVE YIELD OVER THE ANNUAL CYCLE FOR POIKILOTHERMS

	Respiration	Growth (B_p)	Reproduction (B_y)	Yield ($B_{y/Assim}$)
Lizards (Tinkle and Hadley, 1975)				
Sceloporus jarrovi	47,040	11,768	7,261	0.11
Uta stansburiana	34,680	694	8,394	0.19
Sceloporus graciosus	26,640	710	8,630	0.24
Insects				
Philaenus spumarius (Wiegert, 1964)	109.4*	20.9	4.2	.031
Sitophilus oryzae (Singh, Campbell & Sinha,1976)	28.2	4.36	4.4	.144
Pieris rapae (Kitizawa, 1959)	477.*	217.	7.	.010
Hyalophora cecropia (Schroeder, 1972, pers. comm.)	21,900*	3,240.	1,380.	.052
Bombyx mori (Nakano and Monsi, 1968)	539.*	132.	67.	.132

*Includes all losses from the terminal larva to the adult stage.

TABLE 5

REPRODUCTIVE COSTS RELATIVE TO MAINTAINANCE METABOLISM FOR INDIVIDUAL FEMALE ENDOTHERMS

	MAINTAINANCE	REPRODUCING		YIELD ($B_{y/A}$)
		Pregnant or Ovigerous	Parental Care	
Mammals				
Clethrionomys glariolus (Kaczmarski, 1966)	17.5	23.7	39.0	0.55
Mus musculus (Grodzinski & Wunder, 1975)	....	....	....	0.78
Microtus arvalis (Grodzinski & Wunder, 1975)	....	....	....	0.82
Birds				
Taeniopygia castanotus (El-Wailly, 1966)	19.5	22.2	....	0.14
Streptopelia risoria (Brisbin, 1969)	27.0	28.4	40.5	0.50
Colinus virginianus (Case, 1972)	35.4	49.2	....	0.39

NOTE: Data are all maximum daily values.

basal metabolism is lowest and food resources most abundant. All these data are for north temperate climates, in which the animals that survive the winter will be almost certain to be able to meet the costs of reproduction in the favorable season when basal metabolism drops and resources are abundant.

A great deal of speculation in population biology is based on suppositions about energy budgets, but precious little attention has been given to the details of energy budget spectra and their relation to the resource spectrum. Extravagant extrapolations from basal metabolism do sort animals out into guilds. There is reason to suspect that these general patterns reflect the fact that the resource spectrum and energy budget spectrum are similar for the members of a guild. If this is so, then strategy models, supported by Darwinian analysis, should be useful tools in revealing the structure of these patterns and thus in showing that general patterns in natural history do exist.

DISCUSSION

The function of grand principles, such as the principle of the Hutchinsonian niche, is not to provide a program for action in the development of the science; it is to provide insight into the structure of reality. Compare the Hutchinsonian principle with the following equally famous dictum from a similar stage in the development of physics:

> An intelligence knowing all the forces acting in nature at a given instant, as well as the momentary positions of all things in the universe, would be able to comprehend [everything that ever would or ever had occurred in the universe]. (Laplace, *Théorie analytique des probabilités*)

How impractical! How *useless*! But how extraordinarily evocative of the essence of the structure of Newtonian physics! Such dicta are important because they point out what the important concepts of a part of science are, although they leave to us the task of discovering the laws that operationalize the concepts.

How can this task be accomplished? Theoretical ecologists such as Levins (1968) are attempting to accomplish it deductively. Using graphical analysis, this deductive approach seeks to discover what the niche relations may be at a community level and bases speculations on simple indices of the niche.

> The only axes, or niche dimensions, we need to consider are those variables which serve to separate species and along which species may be serially arrayed. (Cody, 1974)

The justification for this statement is not based on experimental results; Cody is simply defining the form that data must take in order to be used in a specific

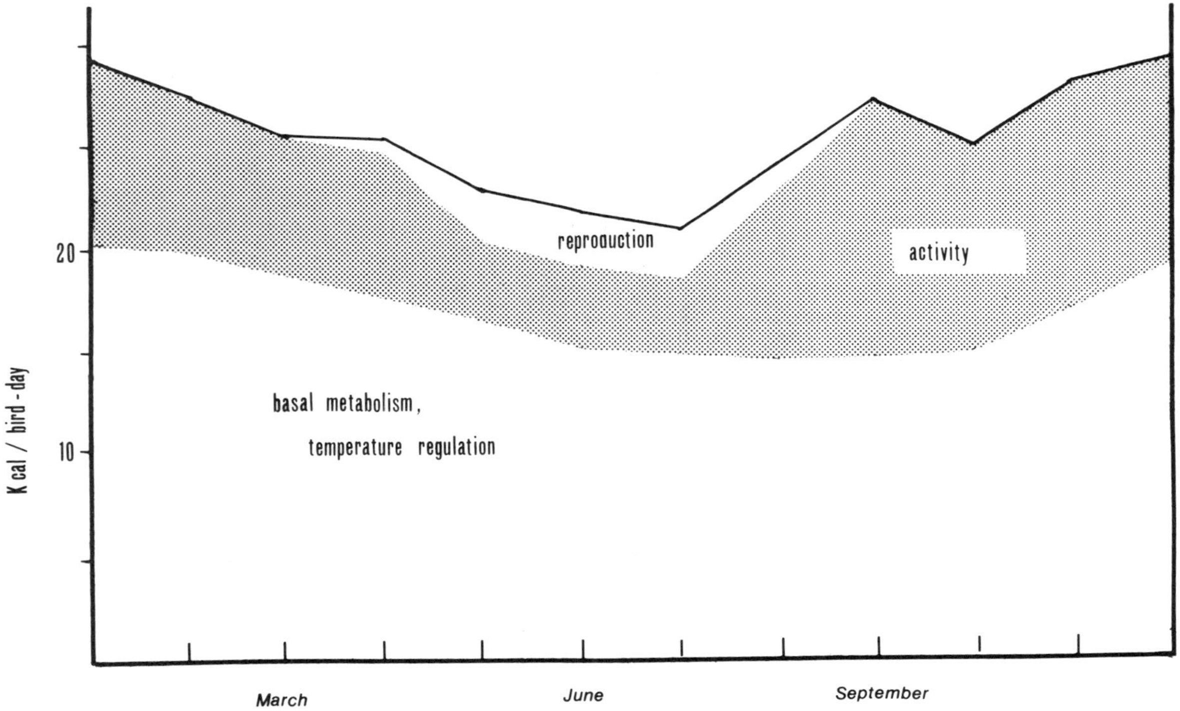

Fig. 8. Annual energy budget for *Passer domesticus*. (Modified from Kendeigh, 1972)

analytic scheme. This scheme leaves out important features of the biology of the species that are necessary to an understanding of how populations evolve. So this scheme has impoverished the niche concept; such impoverishment is an ever-present danger for the deductive approach.

Field ecologists, on the other hand, are faced with an embarrassment of riches. They are so continually faced with the richness of the interrelationships among species that they despair of finding generalizations. The niche concept inspired the development of increasingly sophisticated analyses of the role of behavioral and physiological adaptations in determining the cropping of resources. In some instances the foraging tactics are known well enough to be the basis for algorithms that will predict the cropping that occurs for a specified resource vector, and those predictions can be tested and refined by comparison with observed cropping. They have not been able to place these analyses rigorously within a more general theory, but observed patterns suggest that there are a limited set of potential outcomes in energy budget spectra. The study of these patterns through Darwinian analysis may reveal new generalizations.

Our purpose in the present paper has been to distill out the essence of the structure underlying the work of field ecologists, to formulate this structure explicitly and to make explicit the way in which this work can be rigorously related to Darwinian theory. The strategy model that we propose uses foraging tactics, data on resources, energy budgets, and so on to characterize particular parts of a species' interactions with its environment; Darwinian analysis of the strategy model applies cost-benefit analysis for well-defined alternatives for which data is available to test hypotheses about the adaptiveness of suites of traits.

ACKNOWLEDGEMENTS

The opportunity to begin this sort of work was given to RM while on an NIH Postdoctoral Fellowship in the Acarology Program at Ohio State University. In the years since then, Don Johnston has continued to be a source of encouragement and criticism. Jerry Downhower has put a great deal of time and effort into helping to clarify ideas and applications.

Work on this project was partially supported by NSF Grant GS 41878 to MBW.

REFERENCES

Brisbin, I. L. 1969. Bioenergetics of the breeding cycle of the ring dove. Auk 86:54–74.

Brodie, P. F. 1975. Cetacean energetics, an overview of intraspecific size variation. Ecol. 56:152–61.

Brown, J. H., and R. C. Lasiewski. 1972. Metabolism of Weasels: the cost of being long and thin. Ecol. 53:939–43.

Case, R. 1972. Energetic requirements for egg-laying bobwhites. pp. 205–210. *In* Morrison and J. C. Lewis (eds.), Proceedings of the First National Bobwhite Quail Symposium. Oklahoma State University.

Cody, M. L. 1974. Competition and the structure of bird communities. Monogr. Pop. Biol., Princeton University Press.

El-Wailly, A. T. 1966. Energy requirements for egg laying and incubation in the zebra finch, *Taeniopygia castanotis*. Condor 68:582–94.

Fisher, R. A. 1929. The genetical theory of natural selection. (Reprinted in 1958, Dover, New York) 291 pp.

Golley, F. 1965. Structure and function of an old farm broomsedge community. Ecol. Monogr. 35:113–37.

Grodzinski, W., and B. A. Wunder. 1975. Ecological energetics of small mammals, 5:173–204. *In* Small mammals: their productivity and population dynamics. Cambridge University Press, Cambridge.

Harper, J. L., and J. Ogden. 1970. The reproductive strategy of higher plants. I. The concept of strategy with special reference to *Senecio vulgaris* L. J. Ecol. 58:681–98.

Hinds, W. T. 1975. Energy and carbon balances in cheatgrass: an essay in autecology. Ecol. Monogr. 45:367–88.

Huey, R. B., E. R. Pianka, M. E. Eagan, and L. W. Coons. 1974. Ecological shifts in sympatry: Kalahari fossorial lizards (*Typhlosaurus*). Ecol. 55:304–16.

Hughes, M. K. 1971. Tree biocontent, net production and litter fall in a deciduous woodland. Oikos 22:62–73.

Hutchinson, G. E. 1957. Concluding remarks. Cold Spring Harbor Symp. on Quant. Biol. 22:415–27.

John, D. M. 1973. Accumulation and decay of litter and net production of forest in tropical West Africa. Oikos 24:430–35.

Kaczmarski, F. 1966. Bioenergetics of pregnancy and lactation in the bank vole. Acta Theriologica. 19:409–17.

Kendeigh, S. C. 1972. Monthly variations in the energy budget of the house sparrow throughout the year. *In* S. C. Kendeigh and J. Pinowski (eds.), Proceedings of the general meeting of the working groups on granivorous birds, IBP, PT Section. Warsaw.

Kitizawa, Y. 1959. Bio-economic study of natural populations of animals. Jap. J. Zool. 12:401–48.

Levins, R. 1968. Evolution in changing environments. Monogr. Pop. Biol., Princeton Univ. Press.

MacArthur, R. H. 1958. Population ecology of some warblers of north-eastern coniferous forests. Ecol. 39:599–619.

McNab, B. K. 1963. Bioenergetics and the determination of home range size. Amer. Nat. 97:133–40.

McNab, B. K. 1966. The metabolism of fossorial rodents: a study of convergence. Ecol. 47:712–33.

McNab, B. K. 1973. Energetics and the distribution of vampires. J. Mammal. 54:131–44.

McNab, B. K. 1974. The energetics of endotherms. Ohio J. Sci. 74:370–80.

Marks, P. L. 1974. The role of pin cherry (*Prunus pennsylvania* L.) in the maintenance of stability in northern hardwood ecosystems. Ecol. Monogr. 44:73–88.

Mendelson, J. 1975. Feeding relationships among species of *Notropis* (Pisces: Cyprinidae) in a Wisconsin stream. Ecol. Monogr. 45:199–230.

Mitchell, R. 1975. The evolution of oviposition tactics in the bean weevil, *Callosobruchus maculatus*. Ecol. 56:696–702.

Nakano, K., and M. Monsi. 1968. An experimental approach to some quantitative aspects of grazing by silkworms (*Bombyx mori*). Jap. J. Ecol. 18:217–29.

Ogden, J. 1974. The reproductive strategy of higher plants II. The reproductive strategy of *Tussilago farfara* L. J. Ecol. 62:291–324.

Rodin, L. E., N. I. Bazilevich, and N. N. Rozov. 1975. Productivity of the world's main ecosystems. *In* Productivity of world ecosystems. Pp. 13–26. Nat. Acad. Sci., Washington.

Schoener, T. W. 1968. Sizes of feeding territories among birds. Ecol. 49:124–41.

Schroeder, L. A. 1972. Energy budget of Cecropia moths, *Platysamia cecropia* (Lepidoptera: Saturniidae), fed lilac leaves. Ann. Ent. Society Amer. 65:367–72.

Singh, N. B., A. Campbell, and R. N. Sinha. 1976. An energy budget of *Sitophilus oryzae* (Coleoptera: Curculionidae). Ann. Ent. Soc. Amer. 69:503–12.

Singh, J. S., and P. S. Yadava. 1974. Seasonal variation in composition, plant biomass and net primary productivity of a tropical grassland at Kurukshetra, India. Ecol. Monogr. 44:351–76.

Slade, N. E., and D. F. Balph. 1974, Population ecology of Unita ground squirrels. Ecol. 55:989–1003.

Smigel, B. W., and M. L. Rozenzweig. 1974. Seed selection in *Dipodomys merriami* and *Perognathus penicillatus*. Ecol. 55:329–39.

Smith, J. M. N., and H. P. A. Sweatman. 1974. Food-searching behavior of titmice in patchy environments. Ecol. 55:1216–32.

Tinkle, D. W., and N. F. Hadley. 1975. Lizard reproductive effort: caloric estimates and comments on its evolution. Ecol. 56:427–34.

Whittaker, R. H., F. H. Bormann, G. E. Likens, and T. G. Siccama. 1974. The Hubbard Brook Ecosystem Study: forest biomass and production. Ecol. Monogr. 44:233–52.

Whittaker, R. H., and W. A. Niering. 1975. Vegetation of the Santa Catalina Mountains, Arizona. V. Biomass, production and diversity along the elevation gradient. Ecol. 56: 771–90.

Wiegert, R. G. 1964. Population energetics of meadow spittlebugs (*Philaenus surmarius* L.) as affected by migration and habitat. Ecol. Monogr. 34:217–41.

Wiegert, R. G., and F. C. Evans. 1964. Primary production and the disappearance of dead vegetation on an old field in southeastern Michigan. Ecol. 45:49–63.

Williamson, P. 1971. Feeding ecology of the red-eyed vireo (*Vireo olivaceus*) and associated foliage gleaning birds. Ecol. Monogr. 41:129–52.

Wunder, B. 1975. A model for estimating metabolic rate of active or resting mammals. J. Theor. Biol. 49:345–54.

D. G. EMBREE

The Ecology of Colonizing Species, with Special Emphasis on Animal Invaders

3

My purpose is to present some empirical observations on four animal colonizing species with which I have been confronted as a field biologist involved with forest protection in Canada. These are a moth, a sawfly, an aphid, and a deer. The colonizers are not usually the type that one recruits to explain ecological theories, for as examples they support existing ecological notions in some ways, and in other ways they do not. However, three are forest insect pests, and these, somewhat remotely, enact the "well-worked-over" competitive exclusion principle (Grinnell, 1904; Ito, 1971) in that the insects and forest industries (with whom I work indirectly) are each attempting to exclude the other.

In general, all species at one time or another are colonizers. And to be successful they must be able to compete for food and space, survive the climate of their new environment, and protect themselves from other organisms that would make a meal out of them or in some fashion do them in. They must have a properly adjusted biological clock so they can synchronize themselves to the seasons and to the daily turn of events (Brown, 1972). If their first adventure into a new area is within their normal range of dispersion, they can rely on successive waves of their own kind for support. However, if they have arrived by accident or have been brought by some contrived scheme of mankind, they may have only one chance at it.

In the general scheme of things, it is better to stay at home because, from what we can tell, most colonizers fail. If their new home is a barren island, the result of a volcanic spasm (Diamond, 1974), land newly bared by a receding

D. G. Embree, Dept. of the Environment, Canadian Forestry Service, Maritimes Forest Centre, P. O. Box 4000, Fredericton, New Brunswick, Canada E3B 5G4

ice cap (Worsley and Ward, 1974), or newly emerged sea bottom (Tichnelaar and Vruggink, 1975), they may stake out their claim in a suitable habitat and then fight it out with each succession of new invaders. If their home is to be in an already established environment, they must find that special place in the community called their ecological niche, where they can remain secure, at least until another invader would take their place.

If the migrant is an animal, one should ponder over the reasons for the migration in the first place, as this may govern the success of the migration and, in the end, the characteristics of the new colony. If the migration is the result of some accident or has been contrived, the condition of the migrant is most probably close to the mode of the parent population. Usually it is only when the invasion if within the normal dispersion capability of the species that the migrants might differ from the parent population (Mayr, 1965).

One would expect that invasions would be most apt to arise out of a dense or crowded population. Large numbers of a particular organism usually result from an extended period of conditions favorable to the survival of that organism. Many weak individuals that would normally have perished become possible contenders for migration. For example, the number of sluggish individuals in tent caterpillar populations increases with time, and the active vigorous individuals colonize less hospitable places. Thus if the invader is an insect, it is liable to be a formidable competitor indeed (Wellington, 1965). However, if it is a mouse, quite the reverse can be true. Recent work on mice populations (Krebs et al., 1976) has shown that thin, apparently unhealthy mice are first to migrate; fat, apparently healthy mice are last.

For plants the story is quite different, for seeds and rhizomes are neither sluggish nor aggressive. Their migration depends on passive dispersal by water, wind, and animals. If, once seeded, they grow rapidly, producing copious progeny, they will be the successful pioneers of disturbed or destroyed habitats. Most often these colonizing species will be annuals (Horn, 1974). Later migrants, long-lived, with full provisioning but fewer offspring and the ability to utilize exhaustively the resources in their new habitat, will eventually succeed the early opportunists.

The appearance of a colonizer in an established habitat eventually leads to some form of competition between species, and must initially increase the diversity of the habitat, if only a fraction. Also, chances are they may displace some indigenous species. It is therefore difficult to avoid considering two ecological concepts — the competitive exclusion principle (that different species having identical ecological niches cannot coexist for long in the same habitat) (Turnbull, 1967; Hutchinson, 1959) and the generalization that diversity begets stability (Goodman, 1975; May, 1973). These ecological notions are tested each time an invader enters a new habitat. Unfortunately the application of these concepts to actual events in a field situation is often confusing and can be easily misinterpreted. Nevertheless, in my descriptions of the fortunes of the four colonizers, I will, from time to time, touch on these concepts.

My interest in colonizing species is an indirect one and arose from the introduction into Nova Scotia of the winter moth (*Operophtera brumata* L.), which erupted in a series of continuing outbreaks in the area for about 10 years. The insect was controlled by the introduction of two parasites, a tachinid, *Cyzenis albicans* (Fall.), and an Ichneumonid, *Agrypon flaveolatum* (Grav.) (Embree, 1971). My work on the winter moth extended over a period of eighteen years at which time rather intensive population studies on a number of related forest insects were carried out, which have not been reported until now.

Nova Scotia is a unique area for studies of colonizing species. It has a seafaring history, and in early years enjoyed a favorable trade balance with Europe. Sailing ships arrived relatively empty, deposited their ballast along shore lines, loaded up with lumber, coal, and fish, and sailed away, leaving in their ballast a wealth of colonizers in the form of exotic plants and interesting invertebrates. Figure 1 from Lindroth's (1957) "Faunal connections between Europe and North America" shows clearly the extent of the introductions of invertebrates, in this case carabid beetles, into Nova Scotia.

In all, 31 additional insect pests have been introduced into the province, and 24 species of parasites and predators were released to control them. Approximately 12 parasites became established (Simmonds, 1971; McGugan and Coppel, 1962).

Colonizers are still arriving, but of course no longer in ballast. The latest record of an introduced species is the geometrid *Chloroclystis rectangulata* (L.),. which was collected in the Halifax area in 1972 (Ferguson, 1972). A particularly troublesome colonizer has been the American dog tick, which has spread slowly from the southern tip of the province.

About 368 plant species (3% of the total flora of the province) have been accidently or deliberately introduced. (Roland, 1944). A few fish (Gilhen, 1974), ten species of birds (3% of the total bird population) (Tufts, 1973), but few mammals have been introduced (Smith, 1940). The Norway rat reached Nova Scotia in 1750 and the black rat sometime previous to 1869 (Smith, 1940; Lecarbot, 1914).

THE WHITE-TAILED DEER, ODOCOILEUS VIRGINEANUS (ZIMM.)

This species was introduced into Nova Scotia in 1894 and also probably migrated into the province from neighboring New Brunswick (Dodds, 1963).

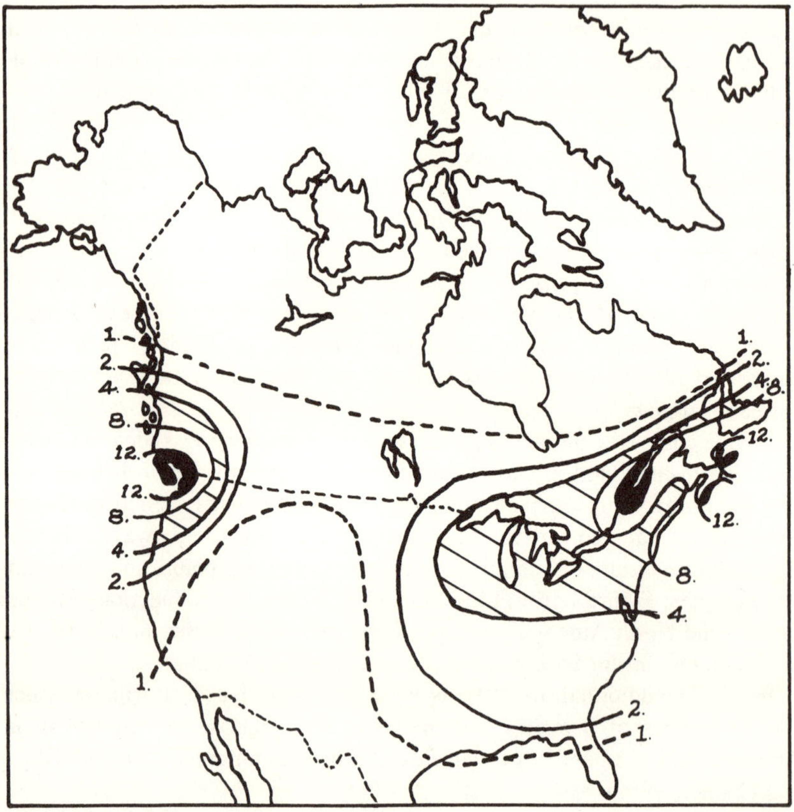

Fig. 1. "Europeization" of the North American fauna, according to the number of introduced Carabid beetles (Lindroth, 1957).

The appearance of this animal had profound repercussions on native populations of moose and woodland caribou. Deer, along with moose, had been present in Nova Scotia for at least 400 years until 1200 A.D. Significantly there were no caribou. Four hundred years later, when the first French settlers arrived, moose *Alces alces* (L.) were the major ungulates, but there were no clear records of deer (Lescarbot, 1914). In their place were caribou *Rangifer tarandus caribou* (L.). Archaeological finds (Erskine, 1957) suggest that the deer disappeared suddenly. The thirteenth century was a time of pronounced climatic change, and after several hundred years of unusually mild weather, which permitted Norse settlements and farms in Greenland, temperatures lowered drastically in the next hundred years. The Norsemen were forced to abandon

Greenland (Mowat, 1965), and perhaps the deer abandoned Nova Scotia as well, not returning for 600 years.

Archaeologists speculate that the deer were decimated because of deep snow, wolves, and hunting pressure by Indians, who had been primarily fish eaters and who were forced to abandon their coastal settlement for the more protective environment inland.

The white-tailed deer returned first to neighboring New Brunswick in the mid-1800s and were well established in the southern part of that province by 1900. By 1915, virtually all the caribou in New Brunswick were gone.

Some said that the deer drove out the caribou, others that hunting pressure was the cause. Some suspected that a disease killed the caribou, but no one knew. Caribou persisted for some years on Bald Mountain in New Brunswick but eventually disappeared. A few years later, a forest fire burned through the area revealing caribou bones strewn everywhere. An account of this (Johnson 1945), stated that "all the bones were of the same age, as though a mighty scythe had swept through the herds and slaughtered the whole of them in one fell swoop. What was it? Some epizootic infection, maybe caribou influenza or caribou pneumonia or some animal disease that no scientist had ever heard of? Any way the caribou were gone." Actually the disease "that no scientist had ever heard of" was almost certainly a meningeal worm harbored by the deer. The infective larvae were carried by terrestrial gastropods feeding on deer feces (Anderson 1965). The snails were ingested accidentally by caribou, which are wide-ranging grazers. Entire herds of caribou could be infected when crossing through areas occupied by deer.

The nematode was not detected until 1963, when, in attempts to explain declining moose populations, the mystery of the disappearance of the caribou was finally solved. Reacting differently in moose and caribou than in deer, the worm causes fatal lesions in the brain of these magnificent animals, and they die within three months to one year after being infected.

There is a strong suggestion that hunting pressure was a contributing factor. Caribou surviving the nematode were decimated by increased hunting pressure on a diminishing resource. This is similar to a collapse of insect populations that are regulated by virus diseases. High population levels of insects permit the build-up of parasite populations. When the host population succumbs to the virus disease, these parasites kill virtually every survivor. In a similar fashion large numbers of hunters inspired by previously high caribou populations killed most of the few caribou surviving the nematode. Even so, because the caribou and the deer were not competing for the same resource, the exclusion of the caribou is ecologically rather unique. I tend to categorize it as one of nature's dirty tricks.

Moose populations were not so drastically affected because moose are sedentary individuals often being confined to a few acres.

Deer are generally meandering munchers frequenting edge areas in the forest, disturbed forest areas, or croplands. Deer cannot usually function in snow depths of over two feet and during the winter gradually retreat to lowland areas where they yard in softwood thickets. Moose overwintering near deer yards almost invariably are stricken with the nematode. Consequently, the only moose that survive are those that overwinter in highland areas where snow depths often reach four feet. (Moose can travel easily through snow depths of two feet and do not begin to congregate in yards until snow depths approach three feet [Kelsall and Prescott, 1971].)

What has happened is that the moose has changed its habitat and survived. Interestingly, moose and deer at times compete with one another, but not with the caribou, for the same food resource. Yet it is the caribou that has been excluded.

The innate inability of the parasite, which itself is a colonizer, to adapt to the moose and caribou is regrettable. Nova Scotia, as a result, has a much less diversified population of ungulates than it otherwise would.

Several attempts have been made to reintroduce caribou into areas frequented by deer, in Wisconsin, Maine, and Georgian Bay, Ontario. In each case the caribou succumbed to the nematode. Recent introductions of caribou into Cape Breton Island yielded similar results (Dauphine, 1975). Of the 55 caribou that were released, none were found because of the dense forest on the island's plateau. One sighting of a caribou with typical "moose disease" sickness and no records of hunting kills suggested that the nematode was the cause of the disappearance.

As a colonizer, the white-tailed deer, which inhabits a wide variety of sites and feeds on 61 species of woody plants as well as many farm crops, can be classified as a generalist. Extremely successful, its populations reached over 22 deer per square mile, with some estimates being as high as 50 per square mile, 50 years after its first entry into the province. By 1950 on the mainland, and 1968 in Cape Breton, deer populations had declined drastically after frequently depleting the winter-carrying capacity of their range. Figure 2 shows estimates of deer and moose populations in Nova Scotia, 1925–70. After each severe winter die-off, more time is required for the herd to come back (Dodds, 1963).

THE BALSAM WOOLLY APHID, ADELGES PICEA (RATZ.)

This colonizer can be termed a specialist. It is a parthenogenetic insect whose feeding habitat is restricted to the phloem of true firs, including balsam fir.

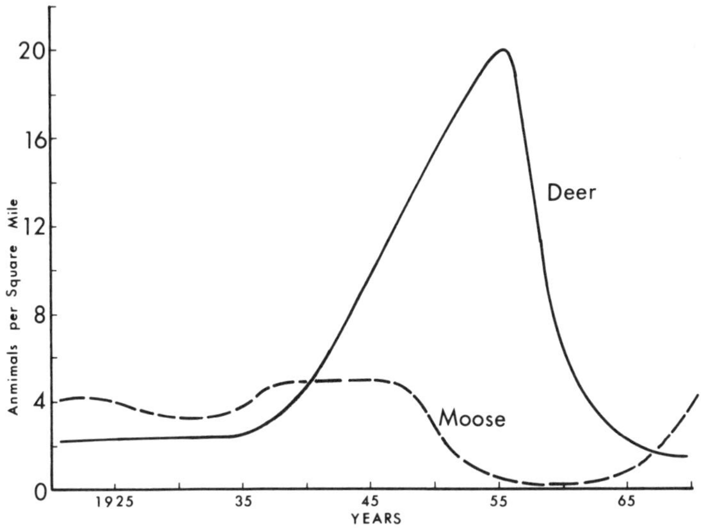

Fig. 2. Approximations of moose and deer populations in Nova Scotia over a 40-year period.

On arrival from Europe, where it was relatively harmless, it began killing trees. There were two types of attack, a twig attack, which caused gout-like swellings but did not kill the tree, and a stem attack, which was usually fatal. A large proportion of mature fir in Nova Scotia were killed or stunted and eventually died by the mid-1930s, and trees that were not killed failed to produce valuable wood volume (Balch, 1952).

The northward spread was restricted by cold temperatures. Only the first-instar nymphs can withstand freezing temperatures, but these succumb to temperatures ranging from −16 to −25°F. Nymphs preconditioned to cold can survive an additional 5° drop in temperature.

Although today the aphid is present throughout Nova Scotia, it is concentrated in coastal regions, where temperatures are mild. The aphid still persists everywhere on the twigs. As stands mature and the crown canopy closes, the aphids migrate down the stems in response to increasing humidity, eventually killing the trees (Greenbank, 1970). As a species, fir has survived by prolific regeneration, but its age-class distribution has been appreciably altered. The aphid has, therefore, changed the nature of the fir forests but not the species composition of those forests.

With the advent of just two invaders, the fir forest has been altered and the caribou has disappeared. The moose, which now has a limited range in Nova Scotia, has become a colonizer. Usually it is an unsuccessful one at that, which in periods of high population attempts to recolonize its former range.

The ecologies of the aphid and the moose are therefore somewhat similar. Each is restricted primarily from particular areas, the aphid because of low temperatures, and the moose because of the presence of the nematode.

EUROPEAN SPRUCE SAWFLY, DIPRION HERCYNIAE (HARTIG)

Like the balsam woolly aphid, the European spruce sawfly is parthenogenic; but unlike the aphid and the deer, it moved into a fully occupied habitat. It also is a specialist in that it feeds on only the old foliage of spruce. Spruce has a wide variety of approximately eighty defoliators. The sawfly was introduced from Europe into Nova Scotia by way of Quebec in the mid-1930s. The effect on the forest was devastating (in eastern Canada more than 11 million cords of wood were lost through tree mortality alone). The accidental introduction of a virus disease plus the deliberate introduction of parasites caused the collapse of the outbreak, and the sawfly populations have remained low ever since. In 1960 an indication of the degree of balance that exists between the sawfly and its enemies was demonstrated when a study area was sprayed with DDT against the spruce budworm. During the spray years, sawfly populations were reduced to the lowest levels ever recorded, and the parasites and the virus were virtually eliminated from the area. However, after the spraying was terminated, sawfly populations began to increase and within five generations (2½ years) reached the levels of the latter years of the original outbreak. After eight generations the virus and the parasites, which had reappeared, suppressed the sawfly population to its former low level (Neilson and Morris, 1964). The colonizer had an initial impact on its new environment; but once its natural enemies were introduced and its population reduced, it successfully fitted itself into the spruce habitat, where it has survived ever since.

THE WINTER MOTH, OPEROPTHERA BRUMATA (L.)

The winter moth is a generalist, in that it feeds on a number of deciduous trees. However, its principal host is red oak. This species is the only forest species of oak in Nova Scotia, and most of the stands are the result of forest fires. Burned areas regenerate to red oak and white pine, with lesser amounts of grey birch, aspen, and red maple. Stands that are burnt more than once have the highest proportion of oak. Most of these stands are concentrated along the south shore of the province and are not continuous but consist of small patches. These average about 20 acres with the largest areas being no more than 2,000 acres.

Aside from the winter moth, at least 38 species of Lepidoptera, 1 sawfly, 10 gall, and 2 aphid species have at one time or another fed on the foliage of oak in Nova Scotia. However, during the entire study period, only two species of Lepidoptera aside from the winter moth were present in appreciable numbers. These were the fall cankerworm *Alsophila pometaria* Harr. and the leaf roller *Pseudexentera cressoniana* Clem. These three defoliators are strikingly unique.

Of the three species, the winter moth and the fall cankerworm are the most competitive. Both are loopers, similar in size, and both defoliate red oak at the same time each year. Both pupate in the soil after forming earth-covered cocoons. Females are wingless, and in November and December crawl up tree trunks to oviposit. Moisture and temperature requirements are similar. In short, here is a classic, but relatively rare, situation where two sympatric species compete for the same resource, and, according to some ecological theory at least one should go.

Initial winter moth populations were free of parasites, and native parasites of associated insects did not attack them. Once established, populations remained in a general outbreak situation.

Winter moth egg-hatching, unlike that of the cankerworm, was not synchronized with the host species, red oak, and only a portion of the larvae survived. Populations fluctuated considerably as a result of weather conditions. However, where the winter moth and the cankerworm were competing for food, the winter moth survived because the larvae, hatched earlier than the cankerworm, completed feeding earlier and were less subject to starvation. The cankerworm, in effect, had to feed on what the winter moth left. Moreover, a difference in behavior pattern of the larvae was quite critical. Cankerworms are extremely active, spinning down readily when disturbed. Shake an oak tree and a cascade of cankerworms spins down! By contrast, winter moth larvae curl the leaves slightly and cling to the foliage. Coupled with their earlier development time, this was enough to ensure that winter moth larvae survived and cankerworm larvae did not.

Slight differences in adult behavior were also critical. Cankerworm adults emerge about two hours before dusk and are therefore more vulnerable to bird predation than are the winter moths, which emerge just at dusk. Moreover, cankerworm eggs are laid in an exposed egg mass and are much more visible and susceptible to insect parasites than winter moth eggs, which are sequestered singly in crevices and under lichens. Not uncommonly, however, a winter moth commits a tactical error and deposits its eggs in an empty cankerworm egg mass.

Cankerworm outbreaks have persisted in Nova Scotia presumably as long as there had been an oak forest there. As outbreaks develop, an egg parasite population (*Telenonus alsophilae* Veireck.) gradually builds up behind the cankerworm. When cankerworm populations eventually run out of food and space, large numbers of larvae starve. The parasite build-up from the previous year is by this time large enough to parasitize virtually all of the eggs, and the cankerworm population collapses.

The process usually occurs over a four-year period, beginning from the first year of heavy defoliation. Presence of the winter moth in cankerworm populations increases the rate of cankerworm starvation. This results in an even more favorable, to the parasite, ratio of parasites to cankerworm eggs than during a typical cankerworm build-up. However, the successful introduction of winter moth parasites removed the advantage of the winter moth over the cankerworm (fig. 3). Coupled with its poor host synchronization, the winter moth virtually disappeared from the oak forest. It has not gone completely, however, and persists in apple orchards and in shade tree species whose phenologies are synchronized with its egg hatch. Through all of the fluctuations of cankerworm and winter moth, the third species, the leaf roller, was always present on a few trees. With the demise of the winter moth, leaf-roller populations began to increase (fig. 4).

The leaf roller is also similar to the winter moth and the cankerworm. It has a similar life cycle, pupates in the soil, and lays its eggs on the bark. It has two principal parasites, a tachinid, which is a species of *Cyzenis*, and an ichneumonid. Thus its parasite complex is extremely close to that of the winter moth. However, the insect is smaller than either of its competitors, the females are winged and they appear in the spring instead of the fall.

Only one population graduation of this species has been measured. The collapse was apparently due to parasites and occurred before the food supply was depleted. This was followed by a series of outbreaks of the cankerworm; such outbreaks, although harmful, are cyclical, and in the absence of effective parasitism do not have the destructive potential of the more persistent winter moth outbreaks.

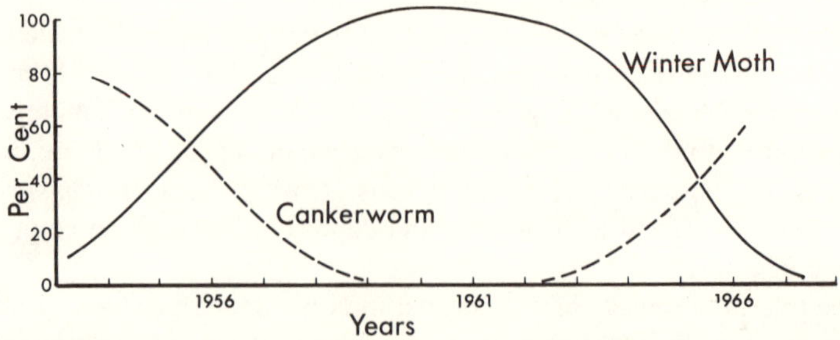

Fig. 3. Winter moth, leaf roller (*Pseudexeutera cressoniana* Clem.) and fall cankerworm populations in Nova Scotia over a 24-year period.

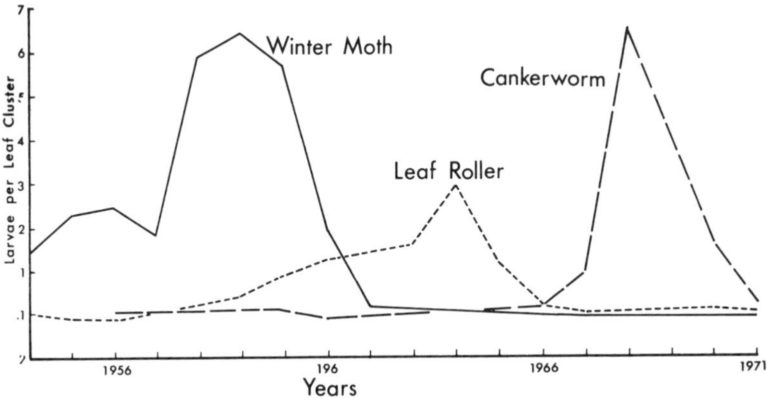

Fig. 4. Relative populations of winter moth and fall cankerworm over a 14-year period.

The fortunes of the three insect colonizers can be summarized as follows. In the beginning, the defoliators, the winter moth and the sawfly, invaded relatively diverse communities and decreased the stability of the forest habitat by killing large numbers of trees. The insect populations were also unstable because of extreme fluctuations in their numbers. With the introduction of the natural enemies of the two defoliators, the insect-host complex returned to a much more stable situation. The two insects are persisting at low levels in the forest, being no more numerous than the many other defoliators with which they compete. Interestingly, both species are controlled by two parasites and a virus disease. The importance of the disease is reversed in the two species, the disease being the major means of control of the sawfly, and the parasites the major control of the winter moth. In Europe the sawfly is an innocuous insect, whereas the winter moth, in spite of its having at least 63 parasites on that continent, is often troublesome.

Similarly, the initial introduction of the aphid decreased stability, but the introduction of natural enemies did little, if anything, to improve it. However, the initial onslaught on the forest has been somewhat reduced, probably because of the depletion of susceptible trees. An equilibrium now exists between host and insect, but at a much higher level than in Europe. Trees are still dying in response to changes in aphid populations. Does diversity increase stability? No conclusions are possible from these examples. In some cases, perhaps, diversity increases stability; in other cases it does not.

Also, no change in the species composition resulted. In spite of the introduction of 400 species of plants and animals into Nova Scotia, the only species that are known to have disappeared have been three species of large mammals and one bird. Aside from the caribou, the others are the almost legendary cougar (reported seen many times but never shot, photographed, or found dead), the wolf, and the passenger pigeon, which doubtlessly succumbed to human predation. Admittedly, it is virtually impossible to determine if very rare species of smaller animals and plants have actually disappeared.

THE BRUCE SPANWORM, OPEROPHTERA BRUCEATA

Further insight into the performance of the winter moth can be gained by examining a similar but indigenous insect, the bruce spanworm. This insect is present on northern hardwoods at high elevations. Morphologically the two species are almost identical, but there are critical differences in life cycle and behavior. The spanworm feeds principally on the northern hardwoods, beech, birch, and sugar maple, and is synchronized with its host. The spanworm actually accomplishes this by staying in the egg longer than the winter moth; consequently spanworm eggs are larger, and there are four rather than five instars. Its fecundity is approximately half that of the winter moth. Its parasite complex is virtually identical to the winter moth, each being parasitized by a species of *Cyzenis* and *Agrypon*, and by a similar, but far more virulent, virus disease. However, the spanworm is rarely associated with the fall cankerworm, the outbreaks of which in Nova Scotia are usually confined to hardwood stands of predominately red oak in the southern portion of the province. The spanworm and the winter moth most probably arose from a common ancestor. Perhaps the spanworm is what the winter moth would have been had the winter moth originally inhabited the North American continent. Its ability to synchronize itself with its host is improved, but at the expense of a drop in fecundity. This lower fecundity may also be the reason the spanworm has been excluded from those areas normally occupied by the cankerworm.

We can expect from the example of the European spruce sawfly that the winter moth, and the cankerworm will coexist in oak forests of southern Nova Scotia, at least on the short term. In the long term, my guess is that the winter moth will probably remain an obscure defoliator as are most of the other 38 species of defoliators now found on oak; or will it evolve in the direction of the bruce spanworm?

An interesting corollary to the winter moth species complex in Nova Scotia is found in Scandinavia (Tenow, 1972). Here, the winter moth and a second species of *Operophtera, O. fagata* Scharf., and a geometrid, *Operinia autamnata* Bkn., have been problem insects for over 100 years. Since 1862, twelve synchronized outbreaks of these insects have occurred. Although the distributions of these insects overlapped in some areas, the focal point for each species was in a separate area. *O. autamata* was confined to highland plateaus, *O. fagata* to seaward slopes, and *O. brumata* to lower elevations. (Admittedly, some confusion existed between *O. fagata* and *O. brumata* because of difficulties in verifying early records.)

One explanation of the separation of species in Scandinavia is the competitive exclusion principle. The synchronization of the outbreaks is added support for this principle, since competition always occurs when the food resource is being depleted in areas common to all three species. Consequently, selection pressure on individuals adapted to such areas would be most severe. However, in most forest situations, outbreaks of more than one species of defoliator are not usually synchronized. The habitat is so vast that residual members of each species survive to recolonize after outbreaks subside. This, in my view, is the most logical outcome of the winter moth colonization of the oak forest of Nova Scotia, where oak forests are not continuous. The situation is akin to the plant climax concept, where, although a climax occurs, it varies from place to place in a continuous fashion.

I would like to close with a few comments on the response that colonizers elicit from one particular species, mankind. This response is usually one of panic. Witness the number of horror movies, plays, or books on strange animals from the past or outer space, invasions by bats, birds, and bushes, or the real world of fire ants, African bees, and swine influenza.

As for my own studies on the winter moth, I have often wondered what the outcome would have been if the initial outbreak of the winter moth had followed an outbreak of the bruce spanworm, instead of the other way around. The winter moth almost certainly would have been misidentified as the bruce spanworm, and we would have resigned ourselves to the problem rather than being alerted to it. As it was, the winter moth was in fact first misidentified as the spring cankerworm. It was only when the realization came that here was an introduced species that there was concern. And this really brings the ecology of colonizing species down to cases. Hard experience tells us that successful colonizing species always make a profound impact on the environment, and, more often than not, continue to do so in spite of our efforts to control (or to stabilize) them.

LITERATURE CITED

Anderson, R. C. 1965. *Cerebrospinae nematodiasis* in North American cervids. North Am. Wildl. Conf. Trans. 30:156–67.

Balch, R. E. 1952. Studies of the balsam woolly aphid, *Adelges picea* (Ratz.) and its effect on balsam fir, *Abies balsamea* (L.) Mill. Can. Dept. Agric. Publ. 867, 76 pp.

Brown, F. A. Jr. 1972. The "clocks" timing biological rhythms. Am. Sci. 60(6):756–66.

Dauphine, T. C., Jr. 1975. The disappearance of caribou reintroduced to Cape Breton highlands park. Can. Field Nat. 89(3):299–310.

Diamond, J. M. 1974. Colonization of exploded volcanic islands by birds, the supertramp strategy. Science 184(4138):803–6.

Dodds, D. C. 1963. The white-tail deer in Nova Scotia. Dept. Lands and Forests, Halifax. 30 pp.

Embree, D. G. 1971. *In* Biological Control, Huffaker Plenum Press, New York.

Erskine, J. S. 1957. Shell trap archaeology of southerwestern Nova Scotia. Proc. N.S. Inst. Sci. 24(4):339–75.

Ferguson, D. G. 1972. The occurrence of *Chloroclystis rectangulata* (L). in North America. J. Lepidopterists's Soc. 26(4):220–21.

Gilhen, J. 1974. The fishes of Nova Scotia's lakes and streams. Nova Scotia Museum, Halifax, N.S.

Goodman, Daniel. 1975. The theory of diversity-stability relationships in ecology. Quart. Rev. Biol. 50(3):237–66.

Greenbank, D. O. 1970. Climate and the ecology of the balsam woolly aphid. Can. Ent. 102:546–78.

Grinnel, J. 1904. The origin and distribution of the chestnut-backed chickadee. Auk 21:364–82.

Horn, H. S. 1974. The ecology of secondary succession. Ann. Rev. Ecol. Syst. 5:25–37.

Hutchinson, G. E. 1959. Homage to Santa Rosalia or why are there so many kinds of animals? Am. Nat. 93:145–59.

Ito, Yosiaki. 1971. Some notes on the competitive exclusion principle. Researches on Population Ecology 8(1):46–54.

Johnson, G. B. 1945. Miramichi Woodsman. Wittet and Shepardson, Richmond, Virginia.

Kelsall, J. P., and W. H. Prescott. 1971. Moose and deer behavior in snow. Can. Wildl. Serv. Rep. Series 15, Inf. Can. 27 pp.

Krebs, C. J., I. Wingate, J. LaDuc, J. A. Redfield, M. Taitt, and R. Hilborne. 1976. *Microtus* population biology: dispersal in fluctuating population of *M. towsendii*. Can. J. Zool. 54:79–95.

Lescarbot, Marc. 1914. The history of New France. Grant and Biggs, The Champlain Society.

Lindroth, C. H. 1957. The faunal connections between Europe and North America. John Wiley and Sons, New York.

May, R. M. 1973. Stability and complexity in model ecosystems. Princeton University Press, Princeton, N.J.

Mayr, Ernest. 1965. *In*: The genetics of colonizing species. Academic Press, New York and London.

McGugan, B. M. and H. C. Coppel. 1962. A review of biological control attempts against insects and weeds in Canada. Part 2. Tech. Commun. No. 2, Commonw. Inst. Bio. Contr.

Mowat, Farley. 1965. West Viking: the ancient Norse in Greenland and North America. McClelland and Stewart, Toronto.

Neilson, M. M., and R. F. Morris. 1964. The regulation of European spruce sawfly numbers in the Maritime Provinces of Canada from 1937 to 1963. Can. Ent. 96:773–84.

Roland, A. E. 1944. The flora of Nova Scotia. Nova Scotia Museum, Halifax, N.S.

Simmonds, F. J. 1971. Biological control programmes against insects and weeds in Canada 1959–1968. Tech. Commun. No. 4 Commonw. Inst. Bio. Contr., Trinidad.

Smith, R. W. 1940. The land mammals of Nova Scotia. Am. Mid. Nat. 24:213–41.

Tenow, O. 1972. The outbreaks of *Oporinia autumnata* Bkh. and *Operophthera* spp. (Lep., Geometridae) in the Scandanavian mountain chain and Northern Finland 1862–1968. Zoological contributions from Uppsala, Almquist, and Wiksells, Sweden.

Tichnelaar, G. M., and H. Vruggink. 1975. A survey of fungi and actinomycetes colonizing newly reclaimed soil of the polder zuidelijk flevoland. Plant Soil 42:241–53.

Tufts, R. W. 1973. The birds of Nova Scotia. Nova Scotia Museum, Halifax, N.S.

Turnbull, A. L. 1967. 1967 Population dynamics of exotic insects. Bull. Ent. Soc. Am. 13(4):333–37.

Wellington, W. G. 1965. Qualitative changes in populations in unstable environments. Can. Ent. 96:436–51.

Worsley, P., and M. R. Ward. 1974. Plant colonization of recent annual morraine ridges at Austre o Kstindbreen North Norway. Arc. Alp. Res. 6(2):217–36.

ERIC R. PIANKA, RAYMOND B. HUEY, AND
LAWRENCE R. LAWLOR

Niche Segregation in Desert Lizards

4

Ecologists have long been fascinated by the possibility of an upper limit on the degree of ecological similarity between coexisting species. Such thinking has generated a rich literature and has led to the notion of "competitive communities," in which species have reached an equilibrium with their resources and rates of resource consumption are exactly balanced by their respective renewal rates. This appealing view of community structure has produced numerous useful concepts, including the principle of competitive exclusion (Gause, 1934; Cole, 1960; Hardin, 1960; Patten, 1961; DeBach, 1966), character displacement (Brown and Wilson, 1956; Hutchinson, 1959; Schoener, 1965; Grant, 1972), limiting similarity (MacArthur and Levins, 1967; Fenchel and Christiansen, 1976), species packing (MacArthur, 1969, 1970, 1972), and maximal tolerable niche overlap (Pianka, 1972; May and MacArthur, 1972; May, 1974).

The term *niche* has been, and doubtlessly will continue to be, used in a wide variety of ways (see, e.g., Whittaker and Levin, 1975). The niche concept has gradually become closely linked to the phenomenon of interspecific competition, however, and has recently been increasingly identified with resource utilization patterns (Levins, 1968; MacArthur, 1968, 1970, 1972; Schoener and Gorman, 1968; Pianka, 1969, 1973, 1974, 1975; Colwell and Futuyma, 1971; Roughgarden, 1972, 1976; Vandermeer, 1972; Pielou, 1972; May and MacArthur, 1972; May, 1974, 1975; Cody, 1974; Schoener, 1968,

Eric R. Pianka and Lawrence R. Lawlor, Department of Zoology, University of Texas, Austin, Texas 78712.

Raymond B. Huey, Department of Zoology, University of Washington, Seattle, Washington 98195.

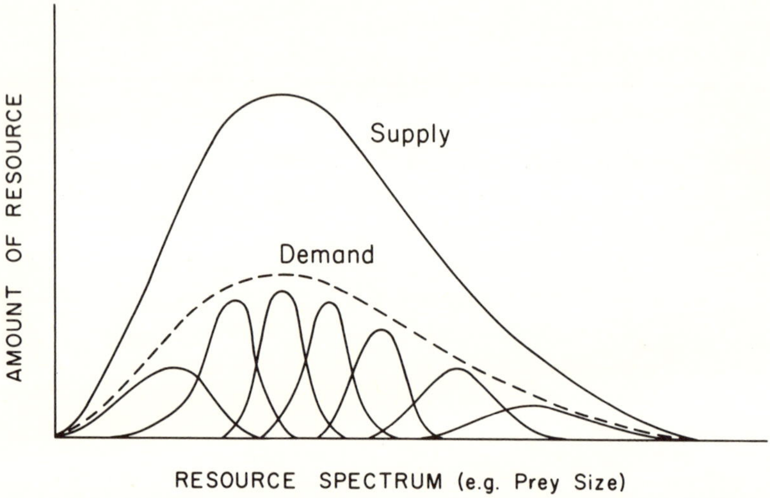

Fig. 1. The vertical axis can be expressed either as the amount of resource available at an instant or as the rate of resource consumption and/or renewal along a spectrum of resources. A positive correlation between resource abundance and turnover rate is assumed for convenience. The uppermost curve represents supply as measured by standing crop along the resource continuum, and the lower curves represent utilization by 7 hypothetical species. Species toward the tails of the spectrum have broader utilization curves because their resources are less abundant. The sum of the species utilization curves is the over-all demand (dashed curve). At equilibrium, resources are consumed at the same rate they are renewed along the entire continuum. Moreover, minimization of interspecific competition results in a relatively constant ratio of demand to supply along any given resource continuum as shown. Analogous arguments can be made for discrete resource categories.

1975a, 1975b). Thus niche relationships among potentially competing species are often visualized and modeled with bell-shaped utilization curves along a resource gradient, such as height above ground or prey size (fig. 1). The sum of the utilization curves of component species represents the total demand along the resource spectrum. When expressed as rates, this demand curve is equal to the rate of resource consumption and/or renewal. The supply curve, on the other hand, is a measure of the instantaneous availability of various resources as measured by their standing crops. Along any given resource continuum, pressures (both ecological and evolutionary) leading to the avoidance of interspecific competition dictate that the ratio of demand to supply must be constant. Hence, all resources should be used in proportion to their availability, and the intensity of competition does not vary among resources. See MacArthur (1969, 1970) for an alternative but related treatment.

Although this emphasis upon resource use is operationally very convenient and will be employed here, it does neglect reproductive tactics and reproductive success, which clearly must eventually be integrated into niche theory (Pianka, 1976a).

In any case, we focus here on differential use of resources among members of desert saurofaunas. We search for answers to some fundamental questions about community structure, such as the following:

1. *How* are available resources divided among members of an ecological community? Is this segregation nonrandom? If so, how?
2. How much do pairs of species overlap in their use of resources, and how does maximal overlap vary with factors such as environmental variability, intensity of competition, and number of species?
3. Are species evenly dispersed in niche space, or are there clusters of functionally similar species with similar ecologies ("guilds") separated from other such groups by lower overlaps? If so, why? How are such guilds structured?
4. How many, and which, niche dimensions are important in separating species (and hence reducing interspecific competition)? Why?

Competition is the conceptual backbone of much current ecological thought; nonetheless, it remains surprisingly elusive to study and hence is poorly understood (probably because avoidance of competition is always advantageous when possible). Nevertheless, precise mechanisms by which available resources are divided among members of a community must be known before determinants of species diversity and community structure can be fully understood. For these reasons, resource-partitioning among coexisting species, or niche segregation, has attracted considerable interest (for recent reviews, see Lack, 1971; MacArthur, 1972; Cody, 1974; Schoener, 1974; Pianka, 1976b).

We do not attempt to review the now extensive body of theory on niche breadth and overlap (for an entry to this literature, see Whittaker and Levin, 1975); rather, we simply outline methods of quantifying niche overlap and indicate some difficulties that are of particular concern in the present context.

The basic raw data for analysis of niche overlap is the resource matrix, which is simply an m by n matrix indicating the amount (or rate of consumption) of each of m resource states utilized by each of n different species. From this matrix, one generates an n by n matrix of overlap with ones on the diagonal and values less than unity as off-diagonal elements. Overlap is sometimes equated with competition coefficients ("alphas") because overlap is much easier to measure. However, the caveat is often issued that overlap need not result in competition unless resources are in short supply. Indeed, extensive overlap may well be possible when there is a surplus of resources (low demand/supply), whereas maximal tolerable overlap may be much less in more saturated environments. Because the ratio of demand over supply must be constant along any *particular* resource gradient, intensity of competition should be directly proportional to the actual overlap observed along that resource spectrum. But caution must be exercised in comparing patterns of niche overlap along different resource axes or between different communities.

Throughout this paper, we quantify niche breadth with the diversity index of Simpson (1949) and niche overlap with the symmetrical overlap index of Pianka (1973, 1974). (The precise formula used to measure overlap is somewhat arbitrary since all overlap indices generate values ranging between zero and one with various degrees of departure at intermediate values.)

Niche overlap theory is usually framed in terms of a single niche dimension. As such, each species has only two neighbors in niche space, and overlap matrices contain many zeros and only two positive entries on the off-diagonal per row. Real plants and animals, however, differ in their use of just one resource only infrequently; rather, pairs of species often show moderate niche overlap along two or more niche dimensions. Complementarity of niche dimensions also occurs, with pairs that have high overlap along one niche dimension overlapping little on another dimension and vice versa (Schoener, 1974, 1975). As the effective number of niche dimensions rises, the number of neighbors in niche space increases more or less geometrically. Moreover, overlap matrices contain fewer off-diagonal elements of zero, and the variance in observed overlap usually falls, both within rows and over the entire matrix. Niche dimensionality also strongly affects the potential for "diffuse" competition arising from the total competitive effect of all interspecific competitors (MacArthur, 1972). Note that the over-all effect of relatively low competitive inhibition per species summed over many other species could well be as strong or even stronger than much more intense competitive inhibition (per species) by fewer competing species. Thus an increased number of niche dimensions, by generating a greater potential for immediate neighbors in niche space, can intensify diffuse competition.

We now discuss some problems and consequences of multidimensional niche relationships. Imagine that height above ground and prey size are the two critical niche dimensions that species use differentially and thus avoid or reduce interspecific competition (fig. 2). Analysis of resource utilization and niche separation along more than a single niche dimension should proceed through estimation of proportional *simultaneous* utilization of all resources along each separate niche dimension. These define a three-dimensional resource matrix, with each entry representing the probability of capture of a prey item of a given size at a particular height by each of the species present. Obtaining such multidimensional utilization data is extremely difficult, however, because animals move and integrate over both space and time. Accurate

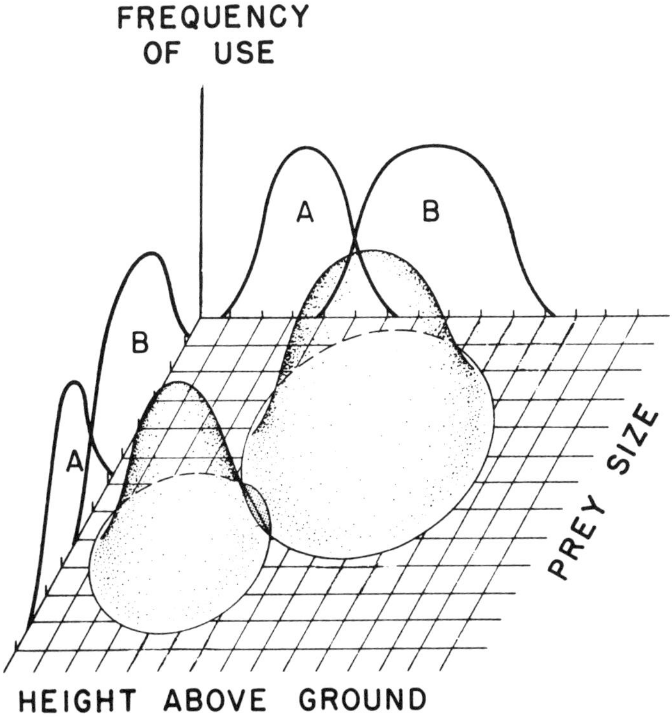

Fig. 2. Resource utilization surfaces for two hypothetical species, A and B, along two different niche dimensions. Silhouettes of these peaks on either resource axis alone overlap, but true multidimensional overlap is slight (see text).

estimates of an animal's true use of a multidimensional niche space could be obtained only by monitoring continually an individual's use of all resources (even then, the degree to which prey individuals move between microhabitats will affect competition in obscure but vitally important ways!). Because such continual observation is often extremely tedious or even impossible, one usually approximates from separate unidimensional utilization distributions (fig. 2 and 3). Just as the three-dimensional shape of a mountain cannot be accurately determined from two of its silhouettes as viewed at right angles, these "shadows" do not allow inference of the true multidimensional utilization. The question of the degree of dependence or independence of dimensions becomes critical. Provided that niche dimensions are truly independent, with prey of any size being equally likely to be captured at any height, over-all multidimensional utilization is simply the product of the separate unidimensional utilization functions (May, 1975). Under perfect indepen-

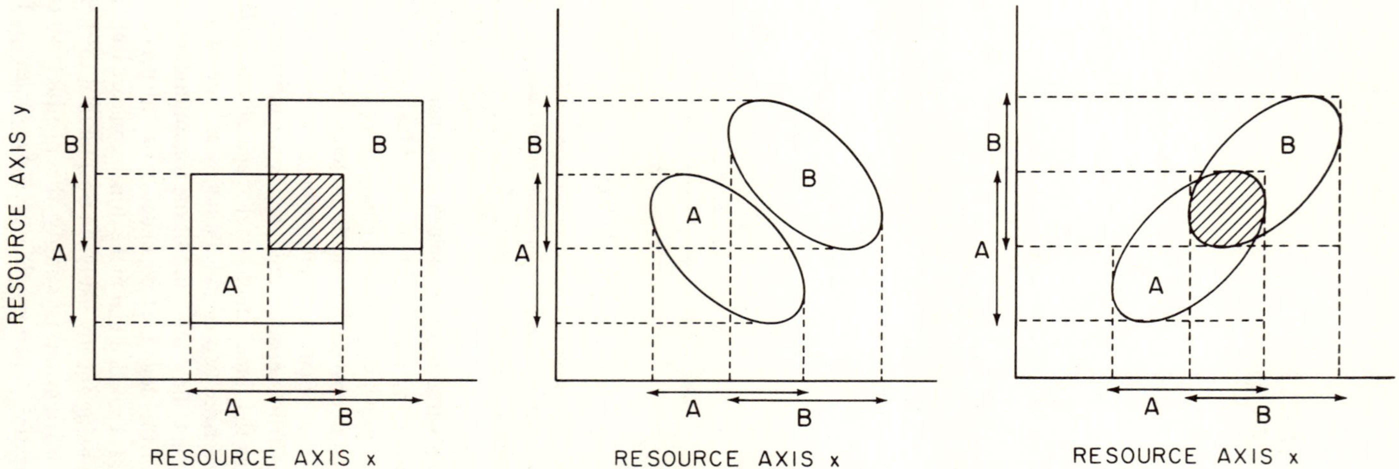

Fig. 3. Three different cases of niche overlap on two resource dimensions with identical unidimensional projections. In the leftmost plot, niche dimensions do not interact and unidimensional projections indicate multidimensional conditions accurately (dimensions are truly independent and niche axes are orthogonal). In the center and right plots, however, niche dimensions are partially interdependent and their unidimensional shadows can be misleading.

dence, the probability of capture of prey item i in microhabitat j is then equal to the probability of capture of item i times the probability of being in microhabitat j. Unidimensional estimates of various niche parameters (including overlap) along component niche dimensions may then simply be multiplied to obtain multidimensional estimates. However, should niche dimensions be partially interdependent (fig. 3), there is no substitute for knowledge of the true multidimensional utilization. True multidimensional overlap can vary greatly depending upon the exact form of this dependence (see fig. 3). In the extreme case of complete dependence (if, for example, prey of each size are found only at one height), appearances to the contrary, there is actually only a single niche dimension and a simple average provides the best estimate of true utilization. Moreover, the arithmetic average of estimates of undimensional niche overlap obtained from two or more separate unidimensional patterns of resource use actually constitutes an *upper bound* on the true multidimensional overlap (May, 1975).

In most lizard species, it would be virtually impossible to evaluate the degree of interdependence of niche dimensions. However, in a relatively sedentary subterranean skink (*Typhlosaurus lineatus*) from the Kalahari desert that specializes on fairly sedentary prey (termites), one can assess the degree to which foods eaten are influenced by the microhabitat of collection (Huey et al., 1974). Most species and castes of termites are eaten in similar proportions by lizards taken from different microhabitats (table 1), indicating that these two niche dimensions are largely independent in *Typhlosaurus lineatus*.

In reviewing major factors leading to ecological isolation among birds, Lack (1971) concluded that the most important were differences in geographic range, habitat, and foods eaten. Schoener (1974) recently reviewed patterns of resource-partitioning in some eighty-odd natural communities ranging from simple organisms such as slime molds through various mollusks, crustaceans,

TABLE 1

PERCENT AND TOTAL NUMBER OF PREY ITEMS EATEN BY TYPHLOSAURUS LINEATUS UNDER SPECIFIC MICROHABITATS

	LOGS		LEAF LITTER		CROTALARIA		GRASS
	Sandplain Areas	Sandridge Areas	Sandplain Areas	Sandridge Areas	Sandplain Areas	Sandridge Areas	Sandridge Areas
Allodontermes (schultzei?)							
Minor workers	32.3	34.2	34.4	21.9	29.2	30.2	40.7
Major workers	24.5	52.7	48.0	74.1	42.3	64.5	54.8
Psammotermes allocerus							
Workers	36.5	7.1	7.3	1.8	24.6	1.9	1.6
Soldiers	1.0	3.9	3.0	1.4	3.2	1.4	1.8
Hodotermes mossambicus	5.5	1.8	6.0	0.6	0.7	1.1	0.7
Other termites	0.3	0.3	1.4	0.1	---	0.9	0.4
Total number of termites	2,385	4,214	2,500	4,746	1,037	3,386	2,620

insects, and other arthropods to various members of the five classes of vertebrates including lizards. He identified and attempted to rank five resource dimensions by their degree of importance in niche segregation: macrohabitat, microhabitat, food type, time of day, and seasonality of activity. As will be seen below, desert lizards use all five of these dimensions in niche segregation.

Schoener concluded that habitat dimensions are generally more important in separating niches than food-type dimensions, which in turn tend to be important more often than temporal dimensions. Moreover, he concludes that terrestrial poikilotherms partition food by being active at different times of day relatively often compared with other animals.[1] He also suggests that predators partition resources by diurnal differences in time of activity more than other groups and that vertebrates segregate less by seasonal activity differences than do lower animals. Lastly, Schoener asserts that segregation by food type is more important for animals feeding on large foods relative to their own size than it is among animals that feed on relatively small items. We examine how well Schoener's first generalization (above) holds up among desert lizards below.

THE NULL HYPOTHESIS: RANDOM PARTITIONING OF RESOURCES

Observed estimates of niche overlap are often compared with values of limiting similarity predicted from theoretical arguments, such as those of MacArthur and Levins (1967; see, e.g., Orians and Horn, 1969) or May and MacArthur (1972; see May, 1974). Because such "magic" numbers depend strongly upon specific assumptions of models concerned (Abrams, 1975), we follow a somewhat different approach here.

How much would niches overlap if resources were simply allocated randomly among members of a community? Such "randomized" communities are useful for comparison with real communities to detect nonrandom patterns of niche segregation and community structure (Lawlor and Pianka, in prep.).[2] Various such randomized communities can be constructed with differing degrees of semblance to observed communities to elucidate the structure of the latter. Perhaps the most random way to construct an artificial community is to assign the utilization coefficients in the resource matrix with a random number generator. Lawlor and Pianka do this by sampling m values from a uniform random distribution over the interval zero to one for each "species." Dividing these values by their sum gives each of n species its utilization rate for resource j. Overlap matrices computed from such randomized resource matrices (randomization treatment 1) have remarkably similar distributions of niche overlap for a wide variety of artificial communities (table 2 and fig. 4).

TABLE 2

Overlap Statistics Computed from 100 Randomly Constructed Communities with Different Numbers of Equally Abundant Resources and Species As Indicated

Number of Species	Number of Resource States	Average Overlap	Average Maximal Overlap	Mean Variance in Overlap
5	5	.778	.940	.016
5	15	.756	.863	.005
5	25	.755	.843	.003
10	5	.774	.975	.018
10	15	.755	.903	.006
10	25	.755	.872	.004
15	5	.771	.985	.019
15	15	.759	.918	.006
15	25	.756	.885	.004
40	5	.774	.994	.019
40	15	.753	.943	.006
40	25	.752	.914	.004

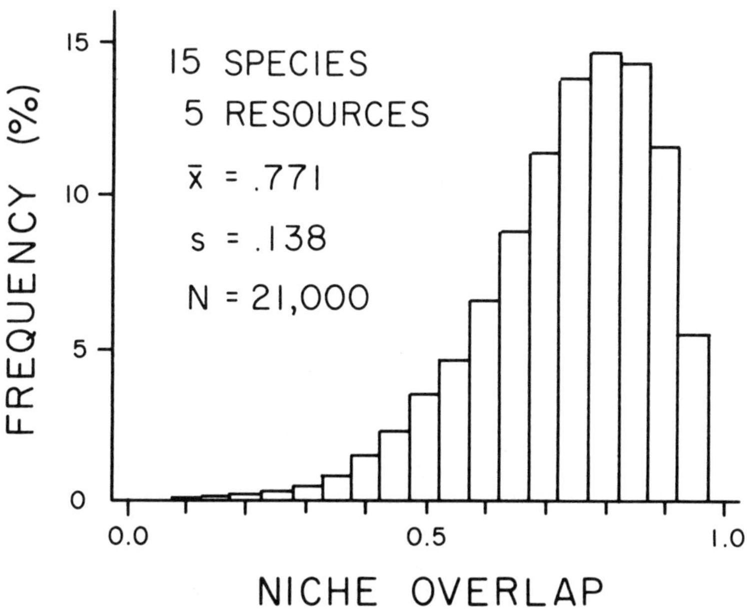

Fig. 4. Distribution of niche overlaps in 100 randomly constructed communities with 15 species and 5 equally abundant resource states. Five numbers were drawn from a uniform random distribution for each species and used to assign utilization coefficients in the resource matrix. Communities with different numbers of species and resource states have very similar distributions of overlap (see table 2).

Using a computer, 100 such randomized communities were constructed for all combinations of 5, 10, 15, and 40 species and 5, 15, and 25 resource states. Average overlap in these communities varied only from .752 to .778, increasing slightly with both increased number of species and with decreased number of resource states (table 2). Variance in overlap is quite low in these randomized communities, increasing slightly with increased number of species and with a reduced number of resource states (table 2). In a more realistic randomization routine (treatment 2), Lawlor and Pianka incorporate some of the structure of the real community by retaining zero utilization coefficients for resources that are not used by observed species and simply repartitioning randomly those that are actually used among each consumer. Still another randomization process is to interchange observed utilization rates among the actual resource states used by each species. Comparisons of the distributions of overlap in such randomly constructed communities with observed overlap reveal intriguing patterns in the structure of real communities (below).

NICHE RELATIONSHIPS AMONG SELECTED LIZARD GUILDS

In the remainder of this paper, we examine a series of cases of niche segregation among desert lizards, beginning with a brief but previously unpublished analysis of niche relationships and possible competitive interactions among seven species of lacertids. Niche relations among such guilds, or clusters of species with very similar ecolgies, are of greatest interest because these presumbly represent the arenas of most intense interspecific competition.

Kalahari Lacertids

During an investigation into the ecology of Kalahari desert lizards (Pianka, 1971), we encountered and studied seven species of lacertids on a series of ten study areas varying in physiography, topography, climate, and vegetative structure (fig. 5).[3] From three to five of these seven species were sympatric on different sites (table 3). Of the twenty-one possible different pairs of species, only two were never found in sympatry, probably as a direct result of non-overlapping geographic distributions arising from habitat differences. We found *Nucras tessellata* only in the sandridge part of the Kalahari, whereas both *N. intertexta* and *Ichnotropis squamulosa* were collected only on non-sandridge sites. The other four species (*Eremias lineo-ocellata, E. lugubris, E. namaquensis*, and *Meroles suborbitalis*) are found on both sandridge and non-sandridge areas (table 3) and are usually sympatric.

On our sandridge study areas, *E. lineo-ocellata* and *E. lugubris* frequent habitats with red sands and large shrubs such as *Acacia mellifera*, often on the slopes or crests of sandridges. In contrast, *E. namaquensis* and *Meroles*

TABLE 3

LACERTIDS FOUND ON TEN KALAHARI DESERT STUDY SITES
VARYING IN CLIMATE, VEGETATION, AND TOPOGRAPHY

	SANDRIDGE AREAS						NON-SANDRIDGE AREAS					
LIZARD SPECIES	L	K	M	B	A	X	G	D	R	T_s*	T_f*	T
Eremias lineo-ocellata	x	x	x	x	x	x	x	x	x	x	x	x
Eremias namaquensis	x	x		x		x	x		x		x	x
Eremias lugubris	x	x	x		x	x	x	x	x	x	x	x
Ichnotropis squamulosa							x			x		x
Meroles suborbitalis	x	x	x	x	x	x	x	x	x			
Nucras tessellata	x			x	x							
Nucras intertexta										x	x	x
Total number of lacertid species	5	4	3	4	4	4	5	3	4	4	4	5
Total number of lizard species	16	15	14	17	15	16	13	13	11	14	13	16

*T_s and T_f are sub-areas of T.

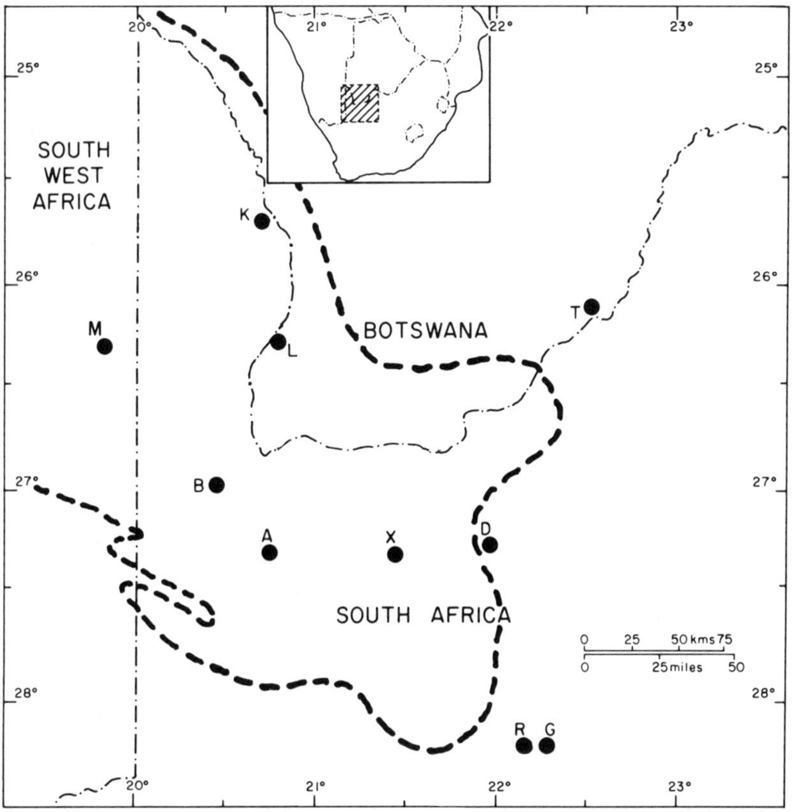

Fig. 5. Positions of ten study areas in the Kalahari desert of southern Africa. Six sites lie within the sandridge area, the approximate boundary of which is delineated by the dashed line following Leistner (1967).

suborbitalis are usually found in interdunal flats with either white or light-brown sands, generally with smaller shrubs such as *Rhigozum trichotomum* and often near rodent burrows (table 4). Although this habitat separation is certainly not complete, differences between the two red-sand species and the two white-sand species are fairly striking (tables 4 and 15). The fifth sand-ridge study area species, *N. tessellata*, occurs largely on red sandy flats but tends to be active somewhat later in the day than the preceding species (below). Habitat niche breadth is low in *E. lugubris* and *E. namaquensis*, intermediate in *E. lineo-ocellata* and *Meroles suborbitalis*, and high in *Nu-*

TABLE 4

HABITAT RELATIONSHIPS OF FIVE SPECIES OF LACERTIDS IN THE SANDRIDGE REGION OF THE SOUTHERN KALAHARI (PERCENTAGES)

	Eremias lineo-ocellata	Eremias lugubris	Eremias namaquensis	Meroles suborbitalis	Nucras tessellata
Habitats					
Dune streets					
Red sands					
with rodent burrows	8.9			21.1	14.3
without rodent burrows	46.1	6.3	3.3	9.4	28.5
White or light brown sands					
with rodent burrows	2.1	6.3	50.0	42.5	14.3
without rodent burrows	11.9	12.5	23.3	20.4	28.5
Slopes and crests of sandridges					
Red sands					
with rodent burrows	2.4	12.5		1.3	14.3
without rodent burrows	24.9	56.3	3.3	1.3	
White or light brown sands					
with rodent burrows	0.3		16.7	1.7	
without rodent burrows	3.4	6.3	3.3	2.3	
Sample size (N)	293	16	30	299	7
Habitat niche breadth	.42	.35	.37	.45	.56
Dune streets					
Red sands	46.4	13.5	6.4	33.6	68.8
White or light brown sands	16.8	13.5	58.7	59.6	21.9
Slopes and crests of sandridges					
Red sands	31.9	67.6	11.1	2.7	9.4
White or light brown sands	4.9	5.4	23.8	4.1	0.0
Sample size (N)	392	37	63	369	32
All red sands	78.3	81.1	21.2	36.3	78.1
All white sands	21.7	18.9	78.9	63.7	21.9
Sample size (N)	392	37	63	369	32
All dune streets	62.8	23.9	70.7	92.8	88.9
All slopes and crests	37.2	76.1	29.3	7.2	11.1
Sample size (N)	468	71	116	416	54

cras tessellata. Overlaps in habitat among these species are given in table 15 (below).

Microhabitats used by the various species also differ, with *Meroles* and *E. namaquensis* occurring in the open sun more often than other species; as a

TABLE 5

Percentages of Occurrence, Sample Sizes, and Microhabitat Niche Breadths of Seven Species of Lacertid Lizards.

	Eremias lineo-ocellata	Eremias lugubris	Eremias namaquensis	Ichnotropis squamulosa	Meroles suborbitalis	Nucras tessellata	Nucras intertexta
Michrohabitat							
Open sun	29.3	32.1	54.8	18.3	56.7	25.4	4.6
Grass sun	11.7	7.9	1.2	17.2	3.4	4.2	4.6
Bush sun	25.5	33.7	24.9	22.0	24.1	34.8	40.9
Tree sun	0.5	1.6	0.6	1.1			4.6
Other sun					0.1	0.9	4.6
Low sun	0.6				0.3		
Open shade	1.4	1.1	4.5	1.1	1.8	1.7	4.6
Grass shade	6.9	2.6	1.2	16.1	1.8	7.6	4.6
Bush shade	23.7	19.7	12.9	16.7	11.6	24.6	27.3
Tree shade	0.3	1.3		7.5			4.6
Other shade					0.1	0.9	
Low shade					0.1		
All sun	67.7	75.3	81.3	58.6	84.7	65.2	59.3
All open	30.7	33.2	59.3	19.4	58.5	27.1	9.2
Sample size (N)	635	190	167	93	482	59	11
Michrohabitat niche breadth	.369	.317	.219	.487	.211	.328	.325

TABLE 6

Time Budget Data on the Foraging Tactics of Seven Species of Lacertid Lizards (Means plus or minus one standard error)

Species	N	Total Time Observed (Minutes)	Meters Moved per Minute	Distance per Move (Meters)	Number of Moves per Minute	Proportion of Time Spent Moving
Eremias lineo-ocellata	15	152.5	1.23±0.37	0.76±0.07	1.54±0.42	0.143±0.030
Eremias lugubris	15	72.1	5.24±0.40	1.93±0.19	2.97±0.28	0.574±0.038
Eremias namaquensis	25	131.3	4.67±0.52	1.79±0.11	2.78±0.31	0.535±0.052
Ichnotropis squamulosa	5	20.9	3.20±0.41	1.12±0.18	3.10±0.14	0.546±0.079
Meroles suborbitalis	15	122.8	1.02±0.14	0.58±0.06	1.83±0.19	0.135±0.016
Nucras tessellata	11	59.7	7.71±1.23	2.23±0.33	2.90±0.37	0.502±0.052
Nucras intertexta	3	8.9	8.93±2.46	2.53±0.87	3.69±0.27	0.649±0.039

result of this heavy usage of a single microhabitat category, these two white-sand species have relatively narrow microhabitat niche breadths (table 5).

Even within the genus *Eremias*, foraging tactics differ rather strikingly among these lacertids (table 6). Both *Meroles suborbitalis* and *Eremias lineo-ocellanta* are conspicuously sit-and-wait predators, whereas the remaining five species forage more widely, moving not only more frequently but also over greater distances and a greater percentage of the time. In the white-sand versus red-sand habitat separation documented above (table 4), note that a widely foraging species is paired with a sit-and-wait species in each habitat.

TABLE 7

OBSERVED DIETS OF SEVEN SPECIES OF LACERTIDS (PROPORTIONS BY VOLUME)

Prey Category	Eremias lineo-ocellata	Eremias namaquensis	Eremias lugubris	Nucras tesellata	Nucras intertexta	Meroles suborbitalis	Ichnotropis squamulosa
Centipedes	. . . .	. . . .	. . . .	. . . .	. . . .	0.001	0.001
Spiders	0.086	0.078	0.008	0.095	0.010	0.045	0.040
Scorpions & Solpugids	0.018	0.018	0.009	0.532	0.010	0.026	0.007
Ants	0.023	0.039	0.005	0.002	0.004	0.050	0.001
Other Hymenoptera	0.020	0.007	0.001	. . . .	. . . .	0.014	. . . .
Locustidae	0.112	0.046	0.009	0.171	0.360	0.066	0.116
Blattidae	0.003	0.001	0.003	. . . .	0.022	. . . .	0.002
Beetles	0.152	0.023	0.009	0.010	0.131	0.147	. . . .
Termites							
Allondotermes (unidentified)	. . . .	. . . .	. . . .	0.001	. . . .	0.003	0.015
Allondotermes major workers	0.109	0.075	0.530	0.002	0.077	0.085	0.494
Allondotermes minor workers	0.018	0.007	0.071	. . . .	. . . .	0.022	0.047
Psammotermes allocerus soldiers	0.002	0.005	0.013	. . . .	. . . .	0.002	0.001
Psammotermes allocerus workers	0.001	0.003	0.021	. . . .	. . . .	0.003	. . . .
Hodotermes mossambicus	0.142	0.190	0.139	0.031	0.076	0.180	0.175
Trinervitermes spp.	0.123	0.288	0.120	0.006	0.009	0.177	0.034
Termite species "M"	0.001	0.004	. . . .	. . . .	. . . .	0.002	0.006
Termite species "N"	0.023	0.061	0.010	. . . .	. . . .	0.010	. . . .
Termite species "O"	0.003	. . . .	0.019	. . . .	. . . .	0.003	. . . .
Other unidentified termites	0.013	0.033	0.005	0.007	0.004	0.028	0.019
Homoptera & Hemiptera	0.041	0.007	0.005	0.0	. . . .	0.022	. . . .
Diptera	0.014	0.002	0.001	0.001	. . . .	0.018	. . . .
Lepidoptera	0.003	. . . .	0.002	0.009	0.144	0.003	. . . .
Larvae							
Coleoptera	0.023	0.008	0.002	0.001	. . . .	0.035	0.031
Lepidoptera	0.002	0.023	. . . .	. . . .	. . . .	0.001	. . . .
Neuroptera	0.002	0.001	0.001	0.064	. . . .	0.007	. . . .
Unidentified larvae	0.017	0.015	0.008	. . . .	. . . .	0.016	. . . .
Vertebrates	0.005	0.008	. . . .	. . . .	. . . .	0.002	. . . .
Plant Material	0.004	0.002	. . . .	. . . .	. . . .	0.003	. . . .
Miscellaneous	0.038	0.054	0.007	0.068	0.153	0.024	0.009
Total volume (cc)	111.04	25.18	58.97	10.58	2.29	92.95	23.69
Number of stomachs	1,135	218	238	79	6	780	112
Food Niche Breadth	0.357	0.239	0.107	0.104	0.170	0.323	0.117

TABLE 8

PERCENTAGES OF TERMITES, BY VOLUME, IN THE STOMACHS OF VARIOUS SPECIES OF KALAHARI LACERTID LIZARDS.

Species	Percentage of Termites
Eremias lineo-ocellata	43.5
Eremias lugubris	92.8
Eremias namaquensis	66.6
Meroles suborbitalis	51.5
Ichnotropis squamulosa	79.1
Nucras tessellata	6.4*
Nucras intertexta	16.6†

*Eats 53% scorpions and 17% grasshoppers.
†Eats 36% grasshoppers.

Not unexpectedly, certain dietary differences are associated with this dichotomy in foraging tactics (tables 7 and 8). One would expect sit-and-wait predators to rely largely on moving prey and to take a fairly wide variety of prey types, whereas widely foraging predators should encounter and consume non-moving types of prey more frequently and might well be able to be somewhat more selective as to which prey items they choose to consume, leading to narrower food niche breaths. For the sit-and-wait tactic to pay off, prey must be relatively mobile and prey density must be high (or predator energy requirements low). One might predict, then, that the sit-and-wait tactic would be less prevalent during periods of prey scarcity. The success of the widely foraging tactic also depends upon prey mobility and prey density and predator's energetic requirements (which should usually be higher than those of sit-and-wait predators because the latter expend less energy in search), but the spatial distribution of the prey and the searching abilities of the predator now assume substantial importance.

Diets of these lacertids fit theoretical expectations reasonably well. The two sit-and-wait foragers have broader food niches than the widely foraging species (table 7). Two widely foraging species of *Nucras* specialize on relatively large generally nonmoving prey, whereas the other widely foraging species consume more termites (also relatively sedentary prey for diurnal lizards) than do the two sit-and-wait species (table 8). Prey size distributions are given in table 9.

Times of activity vary markedly among these lacertids, on both a daily and a seasonal basis. An instructive seasonal comparison can be made between the two sit-and-wait species, *Meroles suborbitalis* and *Eremias lineo-ocellata*, both of which are active throughout the entire year (*Nucras, E. lugubris*, and *E. namaquensis* are generally inactive during the winter). However, *Meroles* are adult (average snout-vent length nearly 66 mm) during winter (May-June) when mating occurs, whereas *E. lineo-ocellata* average only 30–40 mm during this same period (table 10). Ovigerous female *Meroles* were collected

TABLE 9

OBSERVED DISTRIBUTIONS OF PREY SIZE (IN CUBIC CENTIMETERS)

Prey Size Category	Eremias lineo-ocellata	Eremias namaquensis	Eremias lugubris	Nucras tesellata	Nucras intertexta	Meroles suborbitalis	Ichnotropis squamulosa
Under .001	0.6	0.3	. . . .	. . . .	. . . .	12.4	0.4
.001–.005	66.0	58.9	91.8	13.4	69.3	49.5	92.6
.006–.014	14.6	23.4	2.9	26.1	5.3	18.6	. . . .
.015–.024	10.6	11.8	3.6	11.8	9.3	13.9	1.5
.025–.034	4.3	3.4	1.2	12.6	1.3	2.7	2.3
.035–.044	1.3	1.0	0.3	2.5	. . . .	0.9	0.3
.045–.054	1.2	0.5	0.2	10.1	1.3	1.0	0.3
.055–.064	0.1	0.0	0.001	2.5	5.3	0.1	0.2
.065–.074	0.1	0.3	0.001	0.8	. . . .	0.1	. . . .
.075–.084	0.3	0.1	. . . .	1.7	. . . .	0.3	0.1
.085–.094	0.02	. . . .	. . . .	1.7	1.3	0.01	. . . .
.095–.104	0.3	0.1	0.002	4.2	4.0	0.2	0.05
.105–.174	0.1	0.05	0.001	3.4	. . . .	0.1	0.02
.175–.224	0.2	0.1	. . . .	1.7	. . . .	0.04	0.02
.225–.274	0.04	. . . .	. . . .	. . . .	1.3	0.1	. . . .
.275–.324	0.05	0.05	. . . .	1.7	. . . .	0.01	0.02
.325–.374	0.01	. . . .	. . . .	. . . .	. . . .	. . . .	0.05
.375–.424	0.02	. . . .	. . . .	0.8	. . . .	. . . .	0.05
.425–.474	. . . .	. . . .	0.001	0.8	. . . .	. . . .	. . . .
Over .475	0.03	0.05	. . . .	4.1	1.3	. . . .	. . . .
N (lizards)	980	195	215	48	6	702	105
N (total number of prey items)	10,947	2,574	12,221	119	75	9,626	4,362
Mean head length	11.53	10.87	12.24	11.74	14.41	12.23	12.24
Mean prey volume	.009	.009	.005	.074	.026	.009	.005

over a near five-month period that includes winter, from 21 May through 13 October (with a single record on 12 December); in contrast, ovigerous female *E. lineo-ocellata* were found from 24 November to 12 March. Juvenile *Meroles* emerged earlier and at a larger size than *E. lineo-ocellata*, resulting in a clear separation by size of the juveniles of the two species as they grow (fig. 6 and table 10). This size difference could well serve to reduce interspecific competition between juveniles of the two species (see also Broadley, 1969 and Schoener, 1976).

Distinct temporal patterns of activity occur within the day as well. For example, like many diurnal desert lizards, both *Meroles suborbitalis* and *E. lineo-ocellata* typically have a unimodal daily period of activity during the cold winter months, but this gradually gives way to a bimodal daily activity pattern during warmer months (figs. 7 and 8). Still other species appear to have only a single peak of activity each day. Some species, notably the two species of *Nucras* under study here, emerge and are active considerably later than the other lacertid species (table 11). Differences in thermal relationships were correlated with, and doubtlessly causally related to, these observed differences in temporal patterns of activity. Thus, *Meroles* have relatively low mean air and body temperatures, whereas these same statistics for *Nucras* are

TABLE 10

STATISTICS ON SNOUT-VENT LENGTHS (IN MM) OF TWO SPECIES OF LACERTIDS BY MONTHS (SEE TEXT AND FIGURE 6)

	Jan	Feb	Mar	Apr	May	July	Aug	Sept	Oct	Nov	Dec
Eremias lineo-ocellata											
Juveniles											
$\overline{X}$	24.89	33.72	21.50	42.95	36.35	36.42	40.16	45.06	48.61	....	....
S	3.18	6.19	2.12	13.48	10.26	5.16	4.94	7.05	6.81	....	....
SE	0.52	0.91	1.50	1.29	1.38	0.70	0.52	0.63	0.59	....	....
N	38	46	2	109	55	55	91	126	131	....	....
Range	19–33	21–46	20–23	21–66	20–61	27–58	29–60	34–70	37–69	....	....
Adults											
$\overline{X}$	59.97	60.31	59.27	....	....	....	....	....	....	55.00	55.00
S	3.45	2.57	3.81	....	....	....	....	....	....	3.99	4.65
SE	0.27	0.21	0.57	....	....	....	....	....	....	0.94	0.43
N	160	153	45	....	....	....	....	....	....	18	115
Range	48–68	54–66	50–67	....	....	....	....	....	....	47–62	41–63
Meroles suborbitalis											
Juveniles											
$\overline{X}$	38.73	46.06	52.19	....	....	....	....	....	....	....	32.97
S	5.26	6.12	5.73	....	....	....	....	....	....	....	3.57
SE	0.62	0.87	0.76	....	....	....	....	....	....	....	0.46
N	71	49	57	....	....	....	....	....	....	....	61
Range	26–49	28–53	38–64	....	....	....	....	....	....	....	26–43
Adults											
$\overline{X}$	62.38	62.08	....	56.22	57.03	58.41	58.30	58.49	58.73	57.20	58.74
S	3.38	2.75	....	3.34	6.11	3.83	3.03	4.22	3.62	2.17	3.14
SE	0.85	0.76	....	0.39	0.57	0.50	0.38	0.50	0.49	0.97	0.72
N	16	13	....	72	115	58	63	70	55	5	19
Range	57–66	59–70	....	46–64	30–71	48–63	50.68	42–69	48–67	55–60	54–64

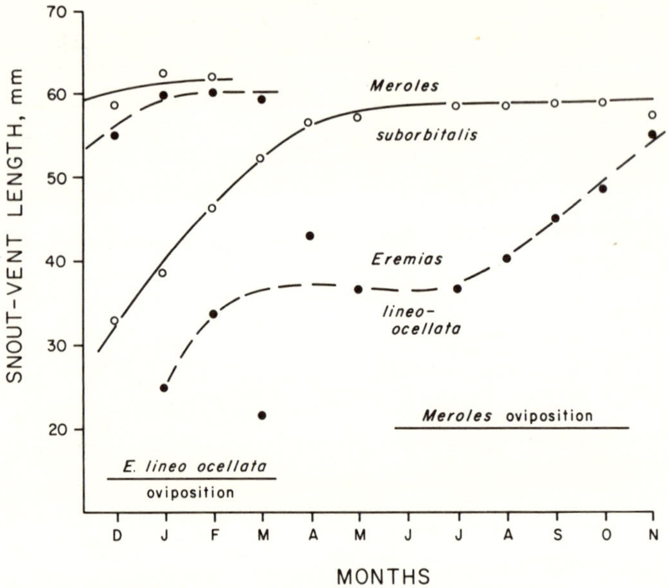

Fig. 6. Mean snout-vent lengths, by months, of two sit-and-wait lacertid species showing temporal separation by size. Statistics are given in table 10.

TABLE 11

Numbers of Lacertid Lizards Observed Active above Ground during Hourly Intervals at Various Seasons

Species and Season	Time Categories											
	7	8	9	10	11	12	13	14	15	16	17	18
Eremias lineo-ocellata												
Spring	. .	. .	19	33	36	45	43	32	31	27	9	1
Summer	10	58	108	83	32	3	10	14	16	18	67	57
Fall	. .	3	16	37	56	31	5	11	14	6	8	4
Winter	. .	. .	. . .	3	20	28	24	43	19	3	. .	. .
Eremias lugubris												
Spring	. .	. .	3	10		1	1	3	. .	. .	. .	. .
Summer	. .	25	58	32	15	1	12	6	6	4	2	1
Fall	. .	. .	12	19	14	7	2	4	. .	. .	1	. .
Winter	. .	. .	. .	. .	2	1	2	2	. .	. .	. .	. .
Eremias namaquensis												
Spring	. .	. .	3	10	5	5	3	3	5	5	0	3
Summer	. .	3	16	11	8	5	0	4	7	9	14	26
Fall	. .	. .	14	12	7	11	8	5	5	2	2	2
Winter	. .	. .	. .	. .	1	1	1	2	. .	. .	. .	. .
Meroles suborbitalis												
Spring	. .	. .	9	12	21	20	18	14	12	16	6	0
Summer	4	28	35	37	18	1	3	3	11	9	31	44
Fall	. .	1	6	30	41	27	45	32	19	14	5	4
Winter	. .	. .	. .	. .	15	27	28	46	16	2	. .	. .
Nucras tessellata												
Spring	. .	. .	. .	. .	. .	3	1		1	. .	1	. .
Summer	1	4	20	27	11	2	1	3	3	2	1	1
Fall	. .	. .	. .	. .	4	2	. .	1	. .	. .	1	. .

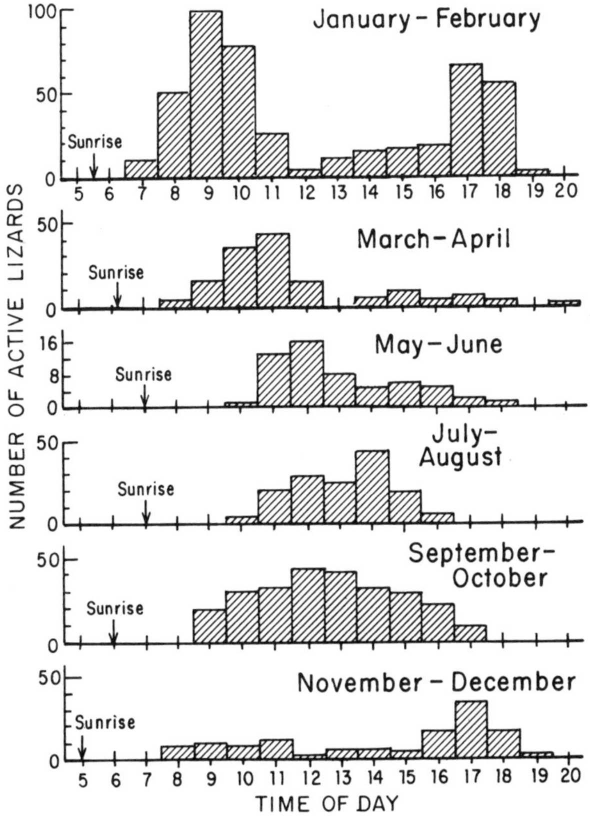

Fig. 7. Histograms of time of activity of *Eremias lineo-ocellata* at bimonthly intervals. Activity is bimodal during hot summer months but unimodal during midday in cooler winter months.

TABLE 12

STATISTICS ON BODY AND AIR TEMPERATURES OF SEVEN SPECIES OF LACERTID LIZARDS

Species	Air Temperature				Body Temperature				r_{ATxBT}
	$\overline{X}$	S	SE	N	$\overline{X}$	S	SE	N	
Eremias lineo-ocellata	28.87	5.06	.171	875	36.84	2.33	.092	636	.579
Eremias lugubris	29.08	3.29	.221	222	37.69	2.54	.192	176	.453
Eremias namaquensis	29.83	3.52	.263	179	37.70	1.92	.167	133	.330
Ichnotropis squamulosa	31.22	2.43	.230	112	36.33	1.76	.184	92	.307
Meroles suborbitalis	26.44	5.60	.218	661	35.54	2.16	.098	485	.553
Nucras tessellata	31.55	3.55	.394	81	39.39	3.19	.481	44	.669
Nucras intertexta	34.03	0.53	.270	4	38.87	1.52	.878	3	.864

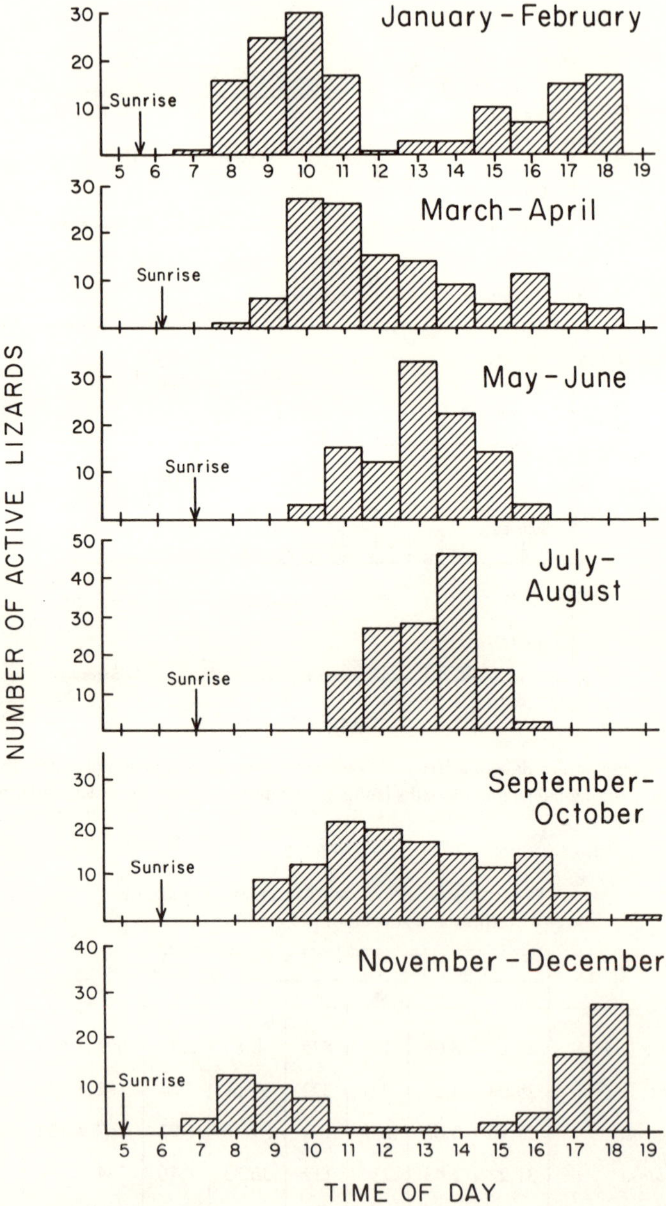

Fig. 8. Frequency distributions of time of activity of *Meroles suborbitalis* at bimonthly intervals.

TABLE 13

OVERLAP MATRICES FOR THE LACERTID GUILD ALONG THREE NICHE DIMENSIONS

	Eremias lugubris	Eremias namaquensis	Ichnotropis squamulosa	Meroles suborbitalis	Nucras intertexta	Nucras tessellata
			Time of Activity (hours since sunrise)			
Eremias lineo-ocellata	.887	.954	.851	.972	.403	.903
Eremias lugubris	....	.790	.956	.803	.251	.905
Eremias namaquensis	....	....	.799	.933	.513	.859
Ichnotropis squamulosa	....	....	....	.754	.208	.863
Meroles suborbitalis	....	....	....	....	.431	.856
Nucras intertexta	....	....	....	....	....	.466
			15 Basic Microhabitats			
Eremias lineo-ocellata	.976	.902	.921	.901	.442	.967
Eremias lugubris	....	.945	.856	.935	.386	.977
Eremias namaquensis	....	....	.708	.996	.232	.895
Ichnotropis squamulosa	....	....	....	.710	.703	.886
Meroles suborbitalis	....	....	....	....	.233	.883
Nucras intertexta	....	....	....	....	....	.472
			19 Prey Categories by Volume			
Eremias lineo-ocellata	.899	.938	.926	.985	.532	.203
Eremias lugubris	....	.986	.990	.945	.271	.095
Eremias namaquensis	....	....	.990	.967	.346	.145
Ichnotropis squamulosa	....	....	....	.953	.374	.136
Meroles suborbitalis	....	....	....	....	.433	.151
Nucras intertexta	....	....	....	....	....	.331

comparatively high (table 12). Thermal relations of Kalahari lizards are considered in greater detail elsewhere (Huey and Pianka, 1977b).

Overlap matrices among these lacertides along various niche dimensions are presented in tables 13, 14, and 15. Table 14 gives multiplicative and summation estimates of over-all overlap, the latter representing an upper bound on the true multidimensional niche overlap (May, 1975) for the entire assemblage of seven species. Among the five species of lacertids in the

TABLE 14

Estimates of Over-all Niche Overlap among Seven Species of Lacertids along the Three Niche Dimensions of Table 13

	Eremias lineo-ocellata	Eremias lugubris	Eremias namaquensis	Ichnotropis squamulosa	Meroles suborbitalis	Nucras intertexta	Nucras tessellata
Eremias lineo-ocellata	. . . .	.779	.808	.726	.863	.095	.177
Eremias lugubris	.921	. . . .	.736	.810	.710	.026	.084
Eremias namaquensis	.931	.907	. . . .	.559	.899	.041	.112
Ichnotropis squamulosa	.899	.934	.832	. . . .	.511	.055	.104
Meroles suborbitalis	.953	.894	.965	.806	. . . .	.044	.114
Nucras intertexta	.459	.303	.364	.428	.366	. . . .	.073
Nucras tessellata	.691	.659	.633	.628	.630	.423	. . . .

Note: Product values above the diagonal and summation values below (the latter represent upper bounds on true multidimentional niche overlap [see May, 1975]).

TABLE 15

Overlap in Habitat among Five Species of Sandridge Desert Lacertids

	Eremias lugubris	Eremias namaquensis	Meroles suborbitalis	Nucras tessellata
Eremias lineo-ocellata	0.581	0.200	0.346	0.838
Eremias lugubris	. . . .	0.240	0.217	0.504
Eremias namaquensis	. . . .	. . . .	0.877	0.555
Meroles suborbitalis	. . . .	. . . .	. . . .	0.714

Note: Habitat categories are those of table 4.

sandridge area of the Kalahari, dietary overlap is essentially either all or none, whereas overlap in habitat varies more continuously, with values ranging from 0.2 to 0.88 (Table 15). Only two sandridge pairs overlap extensively in both diet and habitat (fig. 9). Interestingly enough, these two pairs with high over-all niche overlap (*Eremias lineo-ocellata* versus *E. lugubris* and *E. namaquensis* versus *Meroles suborbitalis*) both represent a pairing of a widely foraging with a sit-and-wait species.

Average overlap in microhabitat and time of activity in a 7-species lacertid guild randomized by treatment 2 (.70 and .65, respectively) did not differ significantly from the observed mean overlaps (.76 and .73). However, the observed overlap in diet among all seven species ($\bar{x} = .60$) *is* significantly

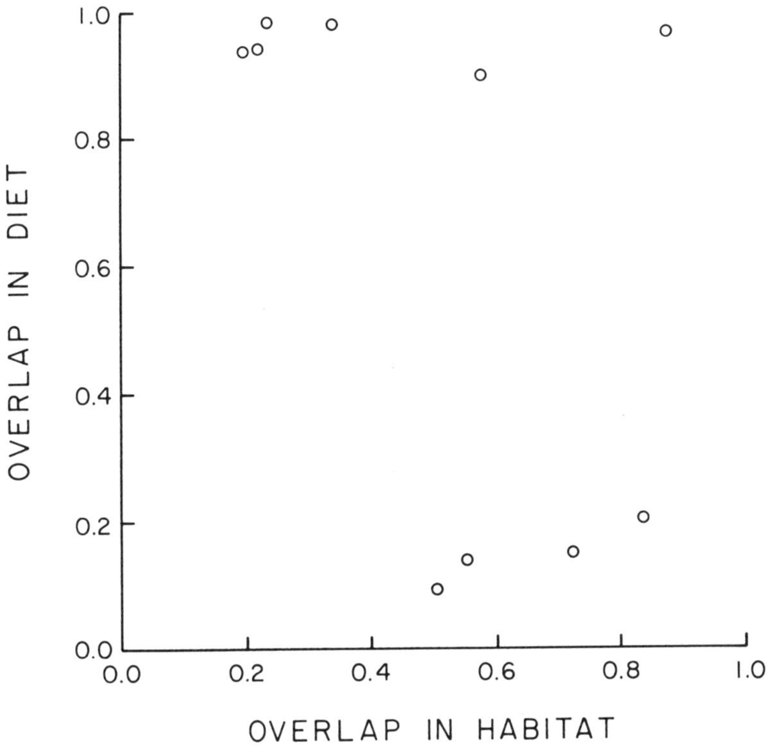

Fig. 9. Dietary overlap plotted against overlap in habitat among the five species of lacertids in sandridge areas. The two pairs with the highest over-all overlap include a sit-and-wait forager and a widely foraging species.

lower than that in communities randomized by treatment 2 ($\bar{x} = .84$, range .64 to .99 in 100 randomized communities). This nonrandomness is due to inclusion of the two species of *Nucras*, whose diets overlap relatively little with the remainder of the lacertid species (table 13): observed overlap in diet among the remaining five species averages .96, not significantly different from average overlaps under randomization treatment 2 ($\bar{x} = .85$, range .45 to .99). Among the five sandridge species, observed overlap in habitat based on the eight categories in table 4 (tables 15 and 20) is significantly lower than in guilds randomized by treatment 2 (observed $\bar{x} = .47$, randomized communities $\bar{x} = .73$ with a range of .49 to .89). Habitat separation may well be the primary factor allowing coexistence of the five lacertids in the sandridge area of the Kalahari.

TABLE 16

ESTIMATES OF OVER-ALL OVERLAP ALONG THREE NICHE DIMENSIONS
(FOOD, MICROHABITAT, AND TIME OF ACTIVITY)
AMONG SEVEN SPECIES OF SYNTOPIC CTENOTUS

	Ctenotus						
Ctenotus	*calurus*	*grandis*	*helenae*	*pantherinus*	*piankai*	*quattuor-decimlineatus*	*schomburgkii*
calurus	. . .	.30	.37	.22	.04	.25	.60
grandis	.69	. . .	.79	.87	.02	.35	.37
helanae	.73	.93	. . .	.80	.03	.44	.43
pantherinus	.64	.95	.93	. . .	.02	.31	.36
piankai	.39	.49	.49	.45	. . .	.42	.02
quattuorde-cimlineatus	.63	.73	.79	.71	.77	. . .	.23
schomburgkii	.86	.73	.77	.73	.32	.62	. . .

NOTE: Multiplicative values above the diagonal and summation values below (the latter represent upper bounds on true multidimensional overlap [see May, 1975]).

A Genus of Australian Skinks

The scincid genus *Ctenotus* contains perhaps the most diverse known assemblage of coexisting congeneric species of lizards, with as many as eleven species occurring together on a heterogeneous study area (Pianka, 1969). Up to seven species are truly syntropic, occurring side by side in the same habitats in true ecological sympatry. These skinks differ in at least three distinct niche dimensions: place (habitat and microhabitat), food (prey size and certain qualitative prey categories weighted by volume), and time of activity (both daily and seasonal). Two species frequently forage in the open (*C. calurus* and *C. schomburgkii*), whereas the remaining five species are usually found either within or near spinifex grass tussocks, resulting in a bimodality in microhabitat overlaps. Among the five species of spinifex foragers, there are also two subguilds, one termite-eating group (*C. grandis, C. helenae,* and *C. pantherinus*) and another pair of species that consume considerably fewer termites (*C. piankai* and *C. quattuordecimlineatus*). Among the assemblage of seven species, dietary overlap varies from 0.06 to 0.99 (mean 0.64), whereas overlap in microhabitat is higher, varying only between 0.60 and 0.98 (mean 0.82). Over-all overlap is extremely high among several pairs of species of *Ctenotus* (table 16 and fig. 10). Somewhat surprisingly (and probably because of the subguild structure of the *Ctenotus* guild), mean observed overlap does not differ significantly from those generated by randomization treatment 2 for any of the three niche dimensions.

Australian Geckos

From four to nine species of nocturnal gekkonid lizards, including up to five congeneric *Diplodactylus*, occur in sympatry in the Great Victoria Desert

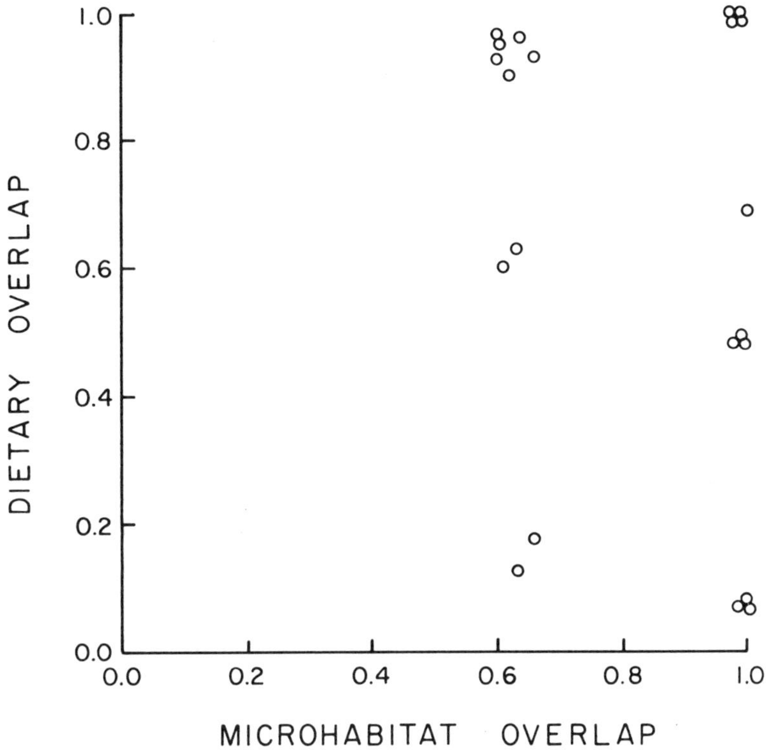

Fig. 10. Dietary overlap versus overlap in microhabitat among a guild of seven syntopic species of *Ctenotus* skinks consisting of three subguilds. Several pairs have exceedingly high over-all niche overlap.

of Western Australia (Pianka and Pianka, 1976). Differences in temporal patterns of activity are slight or nonexistent, but foods eaten as well as habitats and microhabitats exploited differ strikingly among these gekkonids. Larger species tend to eat larger prey than smaller species. Three species are food specialists, eating essentially nothing but termites. Three others are distinctly arboreal, with the majority of specimens first sighted off the ground. Two species are semiarboreal, one of which is apparently restricted almost entirely to *Triodia* grass tussocks. Due to clear-cut differences in habitat requirements, three large terrestrial species of *Nephurus* are always allopatric: one is restricted to sandridges, another to sandplain-*Triodia* habitats, and still another to shrub-*Acacia* habitats.

Observed overlap in microhabitat and dietary overlap both vary greatly among these geckos, ranging respectively from 0.03 to 0.99 and from zero to 0.85 (plus two exceptionally high dietary overlaps of unity). Matrices of

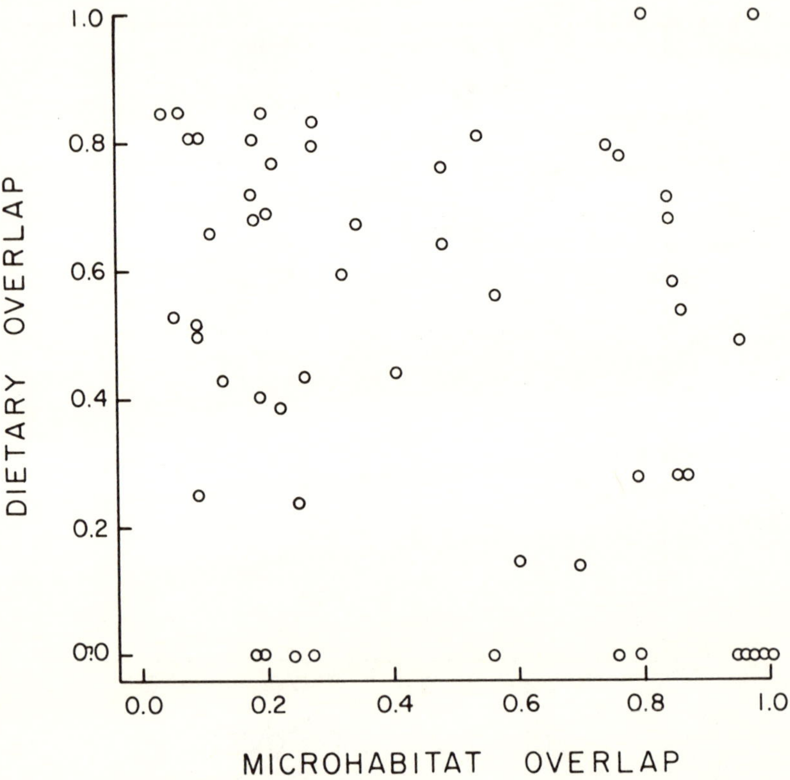

Fig. 11. Overlap in diet plotted against microhabitat overlap among sympatric species of Australian geckos. Some complementarity is suggested with relatively few pairs having high overlap along both niche dimensions.

estimates of overlap in diet and microhabitat are presented by Pianka and Pianka (1976). Exept for two pairs of termite specialists that overlap extensively in both diet and microhabitat, most sympatric species pairs with high dietary overlap tend to overlap relatively little in microhabitat and vice versa (fig. 11). Average overlaps actually observed in both diet and microhabitat among these geckos (means of .46 and .52, respectively) are significantly lower than those in guilds randomized by treatment 2 (respective means are .54 with a range of from .47 to .59 and .77 with a range of from .62 and .93 in 100 trials). Hence the observed differences in resource utilization promoting niche segregation are significantly nonrandom.

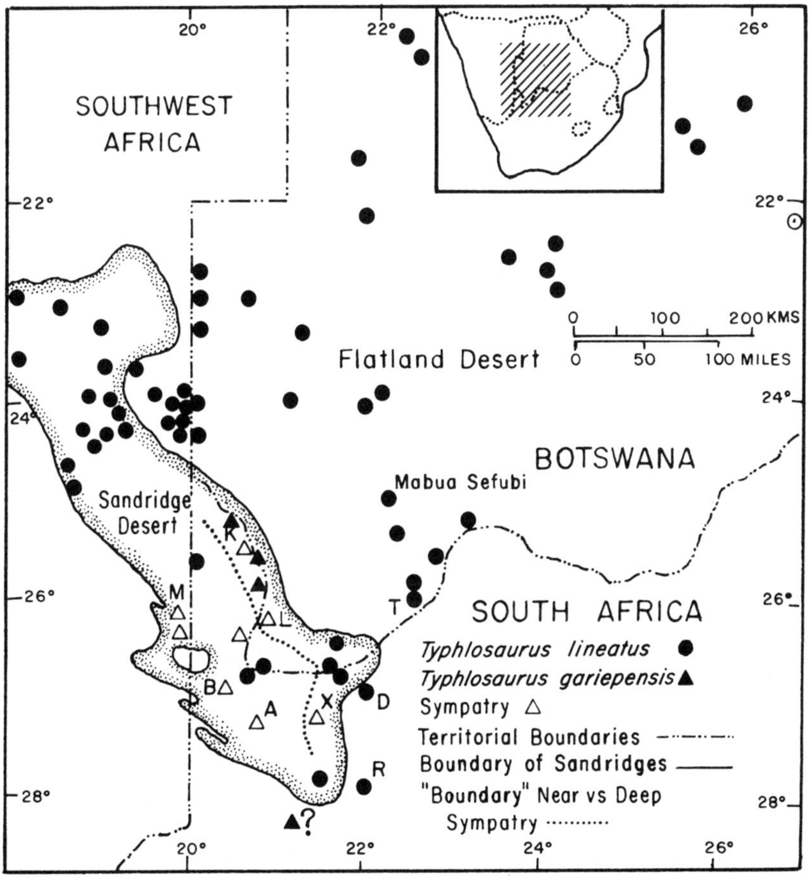

Fig. 12. Localities of collection of *Typhlosaurus lineatus* and *T. gariepensis* in central southern Africa.

Kalahari Fossorial Lizards: Character Displacement

Two species of legless, termite-eating, subterranean skinks are partially sympatric in the Kalahari sandveld. *Typhlosaurus lineatus*, the larger species, inhabits both flatland and sandridge provinces of the Kalahari (fig. 12) and *T. gariepensis* is restricted to southern sandridges. Snout-vent lengths (as a step function, not as a cline; fig. 13), head dimensions, proportional head lengths, and prey size (fig. 14) of *T. lineatus* are larger in sympatry than in allopatry (Huey et al., 1974; Huey and Pianka, 1974). These data support the hypothesis that morphological and behavioral character displacement of *T. lineatus* has occurred that reduces dietary overlap with *T. gariepensis*.

Morphological shifts of *T. lineatus* might alternatively relate to some physical difference in the subterranean environment between flatland and sandridge habitats rather than to competition with *T. gariepensis*. The inability to separate this possible environmental influence from that of competition was a major weakness in our study, as it has been for most cases of character displacement (Grant, 1972).

For *Typhlosaurus*, however, we discovered a crude test that separates out the physical environmental from the competitive influences on body size

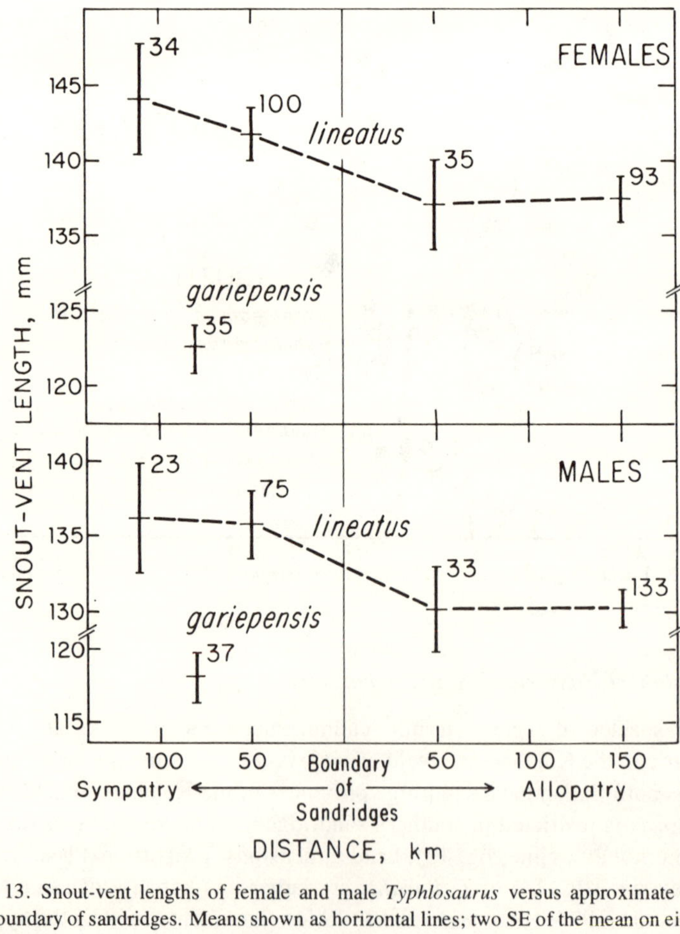

Fig. 13. Snout-vent lengths of female and male *Typhlosaurus* versus approximate distance from boundary of sandridges. Means shown as horizontal lines; two SE of the mean on either side by vertical lines. Numbers inside graph are number of lizards. (From Huey et al., 1974).

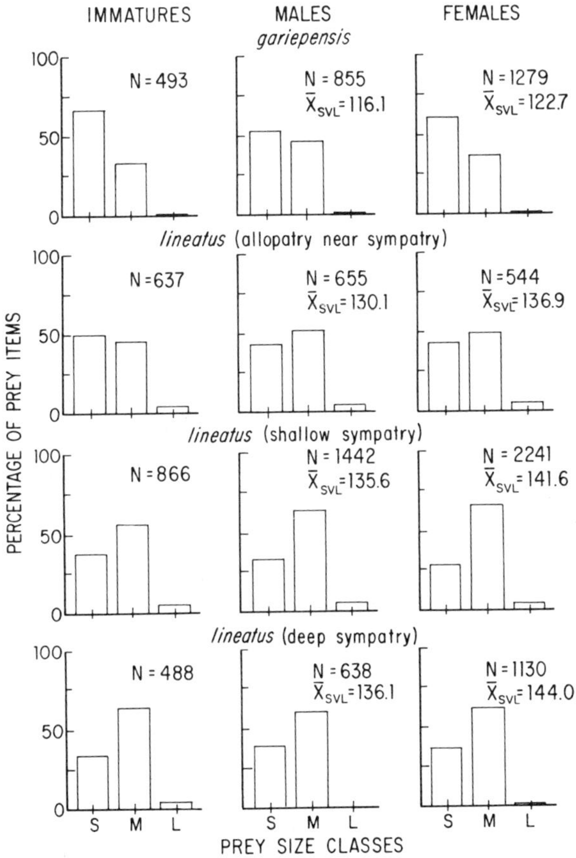

Fig. 14. Percentage of prey items in each of three prey volume categories (S = small [0 – 2.5 μl], M = medium [3 – 11 μl], L = large [over 11 μl]). Total number of prey items and mean snout-vent lengths included for adults. Sympatric *T. lineatus* eat more larger prey than allopatric *T. lineatus*. (From Huey et al., 1974)

(Huey and Pianka, 1974). Noting that *T. lineatus* is seemingly allopatric in the northern sandridge areas (fig. 12), we proposed comparing SVL's from a series from this region and from northern flatland localities. If *T. lineatus* from these two areas are similar in size, the hypothesis of character displacement is supported; if, however, *T. lineatus* are larger in nothern sandridges, the hypothesis of a response to the dune environment is favored. Alternatively, we could compare only northern and southern sandridge populations of *T. lineatus*. However, a possible inverse relationship between latitude and SVL in this species (Broadley, 1968) could complicate interpretations.

TABLE 17

STATISTICS ON SNOUT-VENT LENGTHS OF TYPHLOSAURUS LINEATUS

	Males		Females	
Allopatric *T. lineatus*				
South flatlands	130.5±0.59	166	137.2±0.68	128
North flatlands	133.2±1.25	51	141.6±1.03	56
North dunes	134.7±1.41	59	143.4±1.47	54
Sympatric *T. lineatus*				
South dunes	135.7±0.98	98	142.2±0.82	135
Sympatric *T. gariepensis*				
South dunes	116.1±0.78	37	122.9±0.77	35

To conduct this test, we returned to the Kalahari in the summer of 1976 and measured series of *T. lineatus* from northern sandridges and adjacent flatlands. Snout-vent lengths of both adult males and adult females average slightly larger in the dunes than in the flatlands (table 17), but these differences are insignificant (P's > .2). Males and females in the northern flatlands are significantly longer than those in the southern flatlands (P's < .05), documenting the inverse relationship between latitude and SVL in *Typhlosaurs* (Broadley, 1968).

"Tests" like this one must be interpreted with caution. The apparent absence of *T. gariepensis* from the north suggests that northern and southern sandridges do differ in some way, rendering any comparison suspect. *T. garienpensis* are strongly associated with a particular type of bunch grass (Huey et al., 1974), which is sporadically abundant in the south but rare or absent from northern sandridges. Since *T. lineatus* rarely use bunch grasses but instead are found chiefly under logs, leaf litter, and fallen *Crotalaria*, all of which are abundant both in north and in south, our test may have some validity.

These new data are consistent with the hypothesis that the larger size of *T. lineatus* where sympatric with *T. gariepensis* primarily reflects morphological character displacement, not interaction with the physical environment. The latter may have some influence, as suggested by the slightly larger size of males and females in the northern dunes than in the northern flatlands. However, the lack of significance for these differences suggests a minimal direct environmental influence.

Opportunities for natural experiments are invariably imperfect. Our assertions rest on the existence of an inverse relationship between latitude and SVL (Broadley, 1968; above) and on the assumption that the northern sandridges are effectively similar to the southern sandridges, despite the restriction of *T. gariepensis* to the south. We therefore argue only that our new data are apparently consistent with the hypothesis of character displacement. The

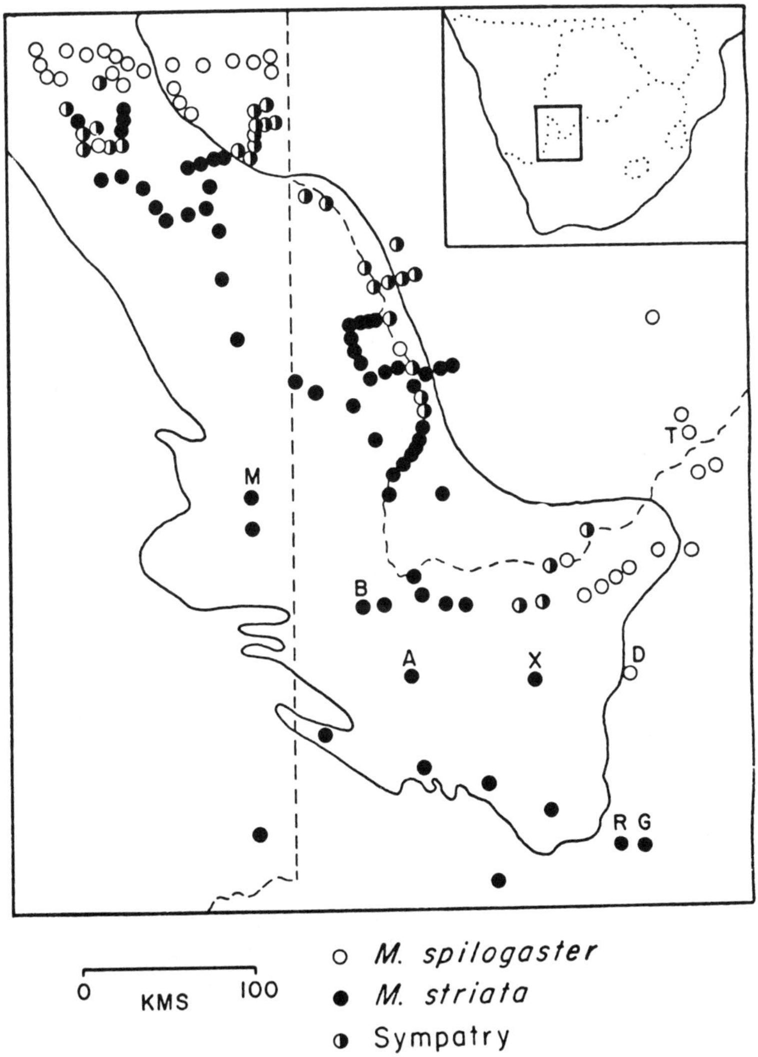

Fig. 15. Map showing known localities of collection of *Mabuya striata* and *M. spilogaster* in the southern Kalahari. The boundary of the sandridges is also shown as well as detailed distributions along several transects through the zone of parapatry. (From Huey and Pianka, 1977a).

strongest evidence in favor of this hypothesis remains the abrupt shift in SVL (fig. 13), the change in proportional head length, and the shift to larger termites (fig. 14).

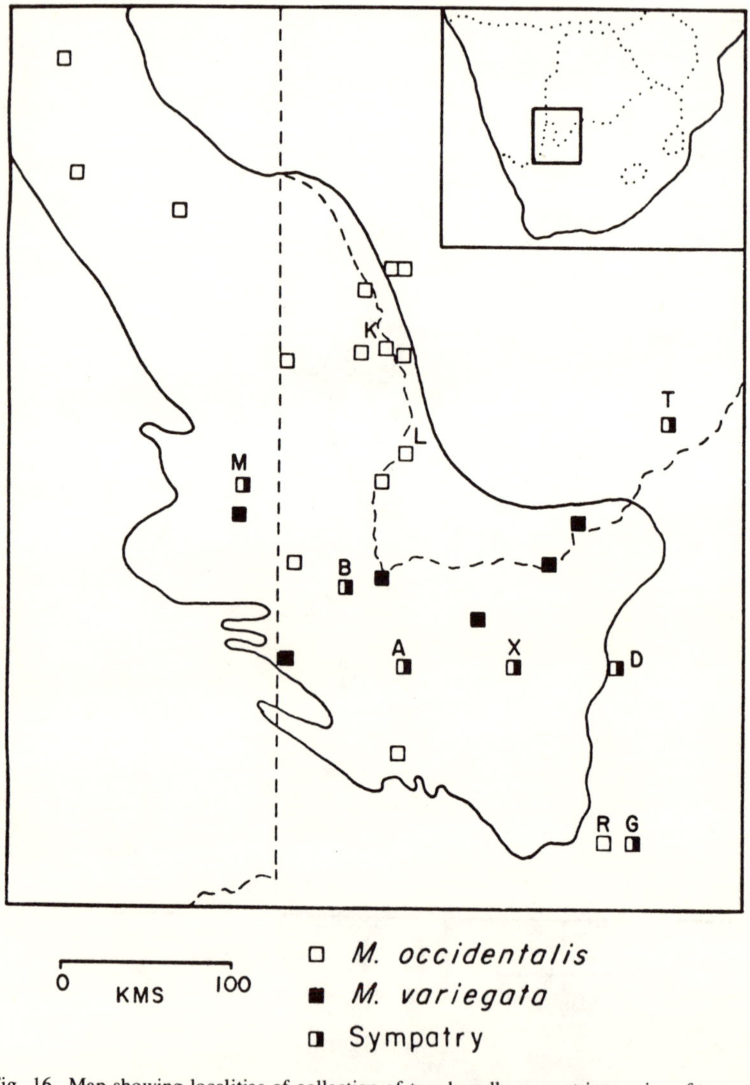

Fig. 16. Map showing localities of collection of two broadly sympatric species of terrestrial *Mabuya*, *M. occidentalis* and *M. variegata* in the southern Kalahari. Compare with figure 15.

Sympatry versus Parapatry among Kalahari Skinks

Four species of skinks of the genus *Mabuya* are common in the southern Kalahari. Distributions of the two semiarboreal species, *M. striata sparsa* and *M. spilogaster*, are parapatric and coincide crudely with the sandridge and flatland provinces, respectively (figs. 5 and 15). Two terrestrial species, *M. occidentalis* and *M. variegata punctulata*, occur in both provinces and thus are broadly sympatric with each other and with the two semiarboreal species (compare figs. 15 and 16). We attempted to determine whether these sympatric and parapatric distributional patterns were a function of differing degrees of adaptation to the habitat provinces or of differing degrees of interspecific competition (Huey and Pianka, 1977a).

Parapatry of the semiarboreal species could result from their adaptations to aspects of the physical environment that abut geographically. Necessarily, *M. occidentalis* and *M. variegata* either do not share these adaptations or are not restricted by them. Distributions of the former pair do parallel sandridge and flatland habitats; and these contiguous habitat zones seemingly affect distributions of Kalahari subterranean lizards (*Typhlosaurus*, Huey and Pianka, 1974) and terrestrial geckos (*Colopus*, Haacke, 1976), and may be important to certain terrestrial lacertid lizards as well (*Nucras*, Broadley, 1972). However, distributions of these *Mabuya* are only similar to, but not entirely congruent with, these major habitats. For example, *M. spilogaster*, though normally associated with flatland habitats, replaces *M. striata* in the northern dune areas of South West Africa. Moreover, *M. striata* occurs in some flatland localities further south (fig. 15).

Changes in other environmental factors occur near the junction of the sandridge and flatland habitats. Rainfall isohyets are roughly coincident (Leistner, 1967). Not surprisingly, some vegetation changes, though not dramatic can be observed. For example, *Acacia giraffee* is the dominant large acacia in *M. striata* areas, and *A. reficiens* (?), a darker-barked tree, seemingly replaces it in areas inhabited primarily by *M. spilogaster*. These changes are general, not absolute. Nonetheless, given the number of environmental changes that seemingly occur near the zone of parapatry and given the complex pattern of that zone, we believe that the basic distributional patterns of the semiarboreal *Mabuya* result from adaptations to different habitats.

Why is the boundary between *M. striata* and *M. spilogaster* relatively narrow? Environmental transitions are too gradual and too patchy to provide a simple answer. Interspecific competition might play a role in restricting the zone of sympatry. *M. occidentalis* and *M. variegata* overlap little with each other and with other species (tables 18 and 19). *M. striata* and *M. spilogaster* are, however, nearly identical in all niche parameters examined (position on sandridges, microhabitat associations, time of activity, body temperatures, and types and sizes of prey). Thus the parapatric species pair has high niche overlap, and broadly sympatric species pairs have much lower overlaps. This pattern is consistent with the hypothesis that competition between *M. striata*

TABLE 18

MATRIX OF OVERLAP VALUES BETWEEN PAIRS OF MABUYA IN THE KALAHARI

	spilogaster	occidentalis	variegata
striata	.990	.994	.982
	.975	.225	.056
	.970	.529	.833
	.919	.898	.499
	.915	.542	.575
	.960	.995	.992
spilogaster		.983	.977
		.290	.035
		.809	.936
		.951	.450
		.760	.404
		.984	.966
occidentalis			.952
			.261
			.941
			.495
			.643
			.993

NOTE: From top to bottom, overlap values represent microgeography, microhabitat, season of activity, time of activity in summer, prey taxa, and prey size (from Huey and Pianka, 1977a).

TABLE 19

ESTIMATES OF OVER-ALL OVERLAP ALONG THE SIX NICHE DIMENSIONS OF TABLE 18

	striata	spilogaster	occidentalis	variegata
striata	1.000	0.755	0.065	0.013
spilogaster	0.955	1.000	0.164	0.006
occidentalis	0.702	0.796	1.000	0.074
variegata	0.656	0.628	0.715	1.000

NOTE: Multiplicative values above the diagonal and summation values below (the latter represent upper bounds on true multidimensional niche overlap [see May, 1975]). (From Huey and Pianka, 1977a.)

and *M. spilogaster* is relatively intense and might be sufficient to restrict the zone of sympatry.

Cody (1974) and Nevo et al. (1975) have noted that parapatric species are often interspecifically territorial or aggressive. Given a stable environment, selection would favor such aggression only if competition is potentially severe. However, several individuals of both semiarboreal species commonly occur on the same log or tree, and we noted little or no interspecific aggression between species in either indirect observations or tethering experiments.

The apparent lack of interspecific aggression does not necessarily invalidate the hypothesis that intense competition results in parapatry. These *Mabuya*, particularly *M. striata*, are social lizards and frequently occur in large numbers of the same log or tree. Evolution of interspecific aggressive behavior might be difficult or impossible in such gregarious species, even in the absence of gene flow with nearby allopatric populations.

Average overlaps in *Mabuya* communities randomized by treatment 2 do not differ from those actually observed in either diet or time of activity, but observed overlap in microhabitat ($\bar{x} = .30$) is significantly lower than in communities randomized by treatment 2 ($\bar{x} = .54$, range .38 to .67). However, this nonrandomness in microhabitat utilization stems from low overlaps with *M. occidentalis* and *M. variegata* since the observed overlap between *M. striata* and *M. spilogaster* is nearly complete (.98).

Conclusions: Comparisons among Guilds

Three guilds in addition to those discussed above were similarly analyzed. These include six species of nocturnal Kalahari geckos, five species of Australian *Varanus*, and seven species of Australian *Amphibolurus*. Patterns of niche overlap observed in seven guilds are summarized in table 20, which, for comparisons, also includes results of the treatment 2 randomization process. Note that, in all but one guild (*Ctenotus*), resource utilization is nonrandom along at least one or more niche dimensions, reducing overlap and promoting niche segregation. Overlap in time of activity is relatively high in most guilds ($\bar{x} = .78$; in all cases, observed mean overlaps either do not differ significantly from treatment 2 means or, if they do differ, observed overlap is significantly *greater* than in comparable randomized assemblages (that is, species are "packed in" on particular times of activity, which may simply reflect limited opportunities for ectotherms to exploit rapidly changing thermal conditions of desert environments). Hence differences in time of activity do not appear to contribute to niche segregation among these guilds of desert lizards. In contrast, many guilds show significantly nonrandom utilization of foods and microhabitats leading to reduced niche overlap. For example, dietary overlap among these guilds are usually lower ($\bar{x} = .67$) than overlap in time of activity ($\bar{x} = .78$); moreover, in 3 of the 7 guilds utilization of foods in significantly nonrandom *and* reduces overlap from that observed in treatment 2 randomized assemblages. Similarly, observed overlap in microhabitat ($\bar{x} = .59$) is low and resource utilization is significantly nonrandom (table 20).

SOME ANALYSES OF ENTIRE SAUROFAUNAS

Niche relationships and community structure of the complete lizard faunas on twenty-eight desert study sites were examined by Pianka (1973, 1974, 1975). Ten North American areas support from four to ten species of lizards, whereas ten southern African Kalahari sites had from eleven to seventeen species and eight areas in Western Australia supported from eighteen to forty

TABLE 20

SUMMARY OF NICHE OVERLAP STATISTICS IN OBSERVED GUILDS AND IN 100 COMPARABLE ASSEMBLAGES RANDOMIZED UNDER TREATMENT 2 (SEE TEXT) ALONG THREE NICHE DIMENSIONS

GUILD	MICROHABITAT				FOOD				TIME OF ACTIVITY			
	observed		Treatment 2		Observed		Treatment 2		Observed		Treatment 2	
	$\bar{x}$	var.	$\bar{x}$	var.	$\bar{x}$	var.	$\bar{x}$	var.	$\bar{x}$	var.	$\bar{x}$	var.
Kalahari Lacertids (all 7 species)	.76	.06	.70[NS]	.07	.60	.13	.84***	.03	.73	.06	.65[NS]	.05
Kalahari Lacertids (5 sandridge species)	.94	.002	.81[NS]	.02	.63	.16	.87**	.02	.87	.003	.78***	.01
Habitat overlap for above 5 lacertids	.47	.06	.73***	.02								
Australian *Ctenotus*	.81	.03	.85[NS]	.03	.64	.12	.69[NS]	.13	.59	.04	.67[NS]	.03
Australian Geckos	.52	.11	.77***	.05	.46	.11	.54***	.09	.95	.002	.83***	.03
Australian *Varanus*	.30	.07	.80***	.03	.88	.005	.84[NS]	.04	.53	.03	.56[NS]	.04
Kalahari *Mabuya*	.30	.12	.54***	.06	.85	.007	.79[NS]	.02	.84	.01	.77[NS]	.01
Kalahari Geckos	.50	.12	.79***	.03	.76	.07	.82[NS]	.03	.96	.001	.80***	.03
Australian *Amphibolurus*	.74	.02	.74[NS]	.05	.57	.05	.75***	.02	.74	.03	.67*	.03

NS = not significantly different from observed
* = 5 or fewer out of 100 treatment means overlapped observed
** = only 1 out of 100 treatment means overlapped the observed mean
*** = none of 100 treatment means overlapped observed mean

species (Pianka, 1967, 1969, 1971). Two parallel analyses were performed on these entire saurofaunas. In the first (Pianka, 1973), to reduce the sampling error in characterizing resource utilization of component species, all lizards in each continental desert-lizard system were lumped into a single merged "community." An area-by-area analysis (Pianka, 1974, 1975) gave basically comparable results, outlined briefly below (for example, compare the niche overlap distributions in figure 2 of Pianka (1973) with those in figures 18, 19, and 20 of this paper). Details of methodology, and so forth, can be found in Pianka's papers.

Flatland Desert Lizards of North America

Partly because the number of lizard species coexisting on a given study area in the flatland deserts of western North America is low, guilds are not easily identifiable (indeed, they may not exist). Lizards are easily separated out as in the "ecological key" shown in table 21. Overlaps along the two niche dimensions that separate species in this merged North American lizard community are plotted in figure 17. Note that microhabitat overlap is low between many pairs and that only a few species pairs overlap extensively in both diet and microhabitat. Resource matrices for this assemblage of nine species[4] were randomized by treatment 2. Average observed overlap in time of activity ($\bar{x} = .58$) did not differ significantly from the mean in 100 randomized com-

TABLE 21

SUMMARY OF THE WAYS IN WHICH RESOURCES ARE PARTITIONED AMONG TWELVE SPECIES OF NORTH AMERICAN FLATLAND DESERT LIZARDS

Habits	Species
I. *Diurnal species*	
A. Widely-foraging	*Cnemidophorus tigris*
B. Sit-and-wait foragers	
1. Underneath bushes	
a. Small prey items	*Uta stansburiana*
b. Large prey items	*Crotaphytus wislizeni*
2. Open spaces between shrubs	
a. Generalized diet of large insects	*Callisaurus draconoides*
b. Diet consists largely of ants	*Phrynosoma platyrhinos*
3. Arboreal (Trees, large bushes)	
a. Trunks and ground around base	*Sceloporus magister*
b. Primarily in outer smaller branches	*Urosaurus graciosus*
4. Sand Dunes (primarily open spaces)	*Uma scoparia*
C. Herbivorous (both floral & vegetative parts)	*Dipsosaurus dorsalis*
II. *Nocturnal Species*	
A. Open-foraging	*Coleonyx variegatus*
B. In Joshua Tree rubble	*Xantusia vigilis*
C. Capture prey by olfaction and digging (eats eggs and young of various vertebrates)	*Heloderma suspectum*

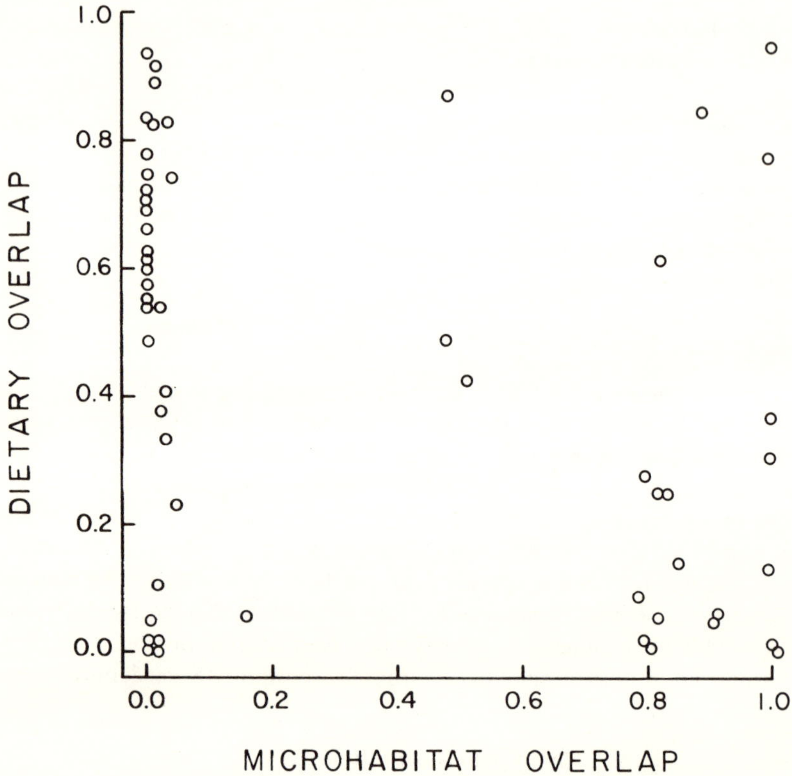

Fig. 17. Dietary overlap plotted against overlap in microhabitat among flatland desert lizards of western North America. Overlap in microhabitat tends to be either all or none, whereas dietary overlap values are distributed more evenly. Some complementarity of niche dimensions is suggested in that relatively few pairs of species overlap extensively along both axes. Compare with figure 11.

munities ($\bar{x}=.52$). Observed mean overlaps in diet (.43) and microhabitat (.34) were, however, significantly lower than in their respective randomized communities ($\bar{x}=.65$, range .55 to .75, and $\bar{x}=.47$, range .40 to .56), promoting niche segregation.

Niche Overlap Distributions: Evidence for Nonrandom Niche Segregation

Distributions of observed niche overlap along three dimensions were presented for the merged communities by Pianka (1973, figure 2). Figures 18, 19, and 20 show comparable distributions from an area-by-area analysis. Both the degree of overlap and the importance of various niche dimensions in separat-

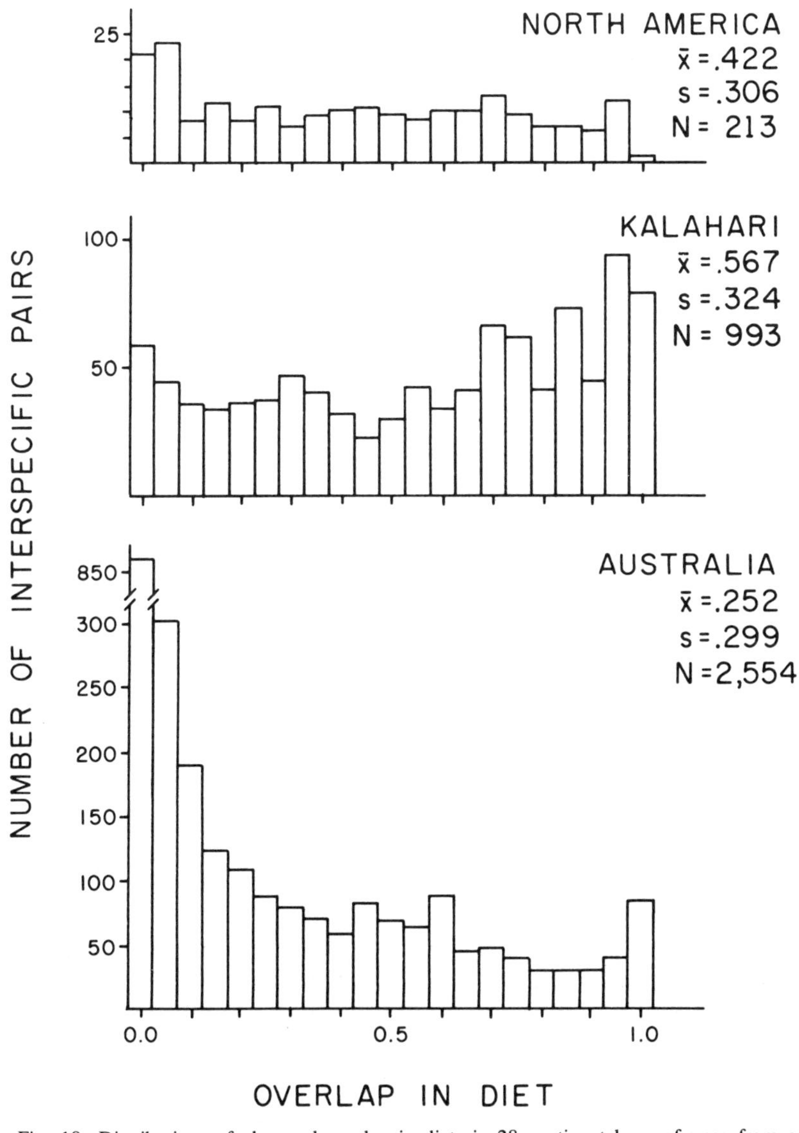

Fig. 18. Distributions of observed overlap in diets in 28 continental saurofaunas from an area-by-area analysis. Dietary overlap is considerable in Kalahari lizards, where termites comprise 41% of the diet. Compare with distribution in figure 4.

ing species clearly differs between continents. For example, dietary overlap is quite high in the Kalahari ($\bar{x} = .65$) where termites comprise some 41 percent of the lizard diet by volume (in contrast, termites represent only about 16

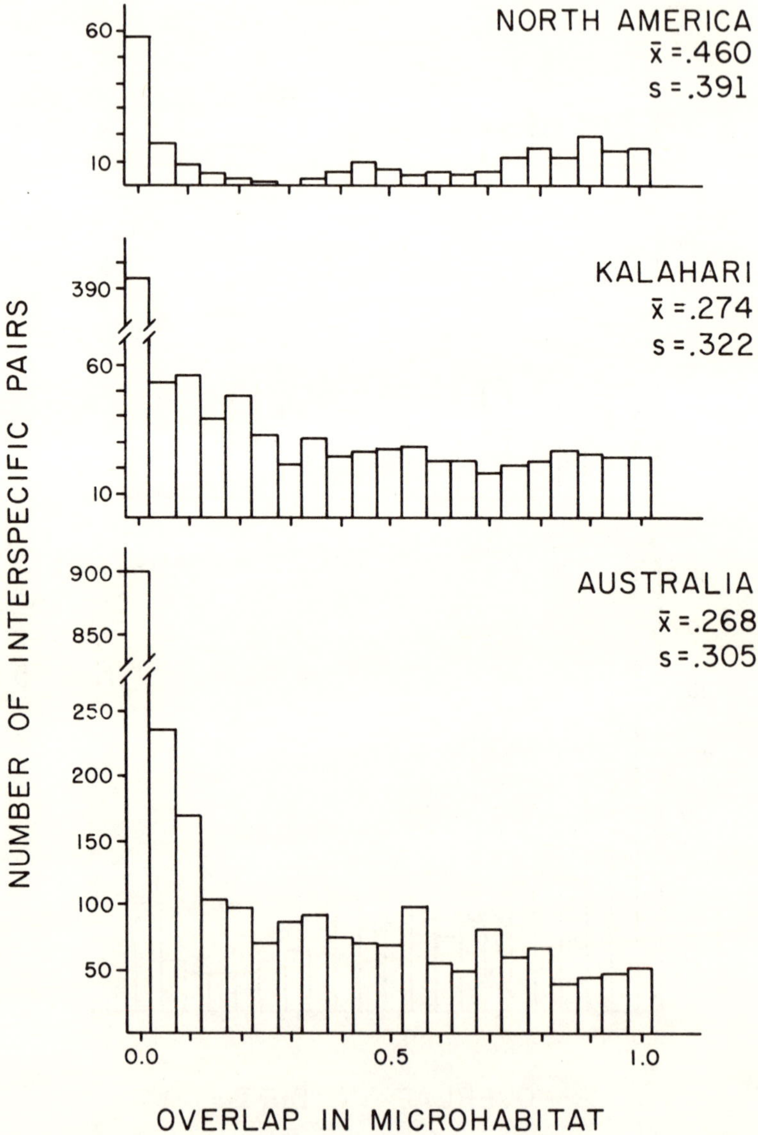

Fig. 19. Histograms of observed overlap in microhabitat among lizards on the 28 sites in an area-by-area analysis. Compare with distribution in figure 4. Excepting the large zero overlap category, these distributions are more similar to those of randomly constructed communities (fig. 4) than are the distributions of overlap in diet and microhabitat (figs. 18–19).

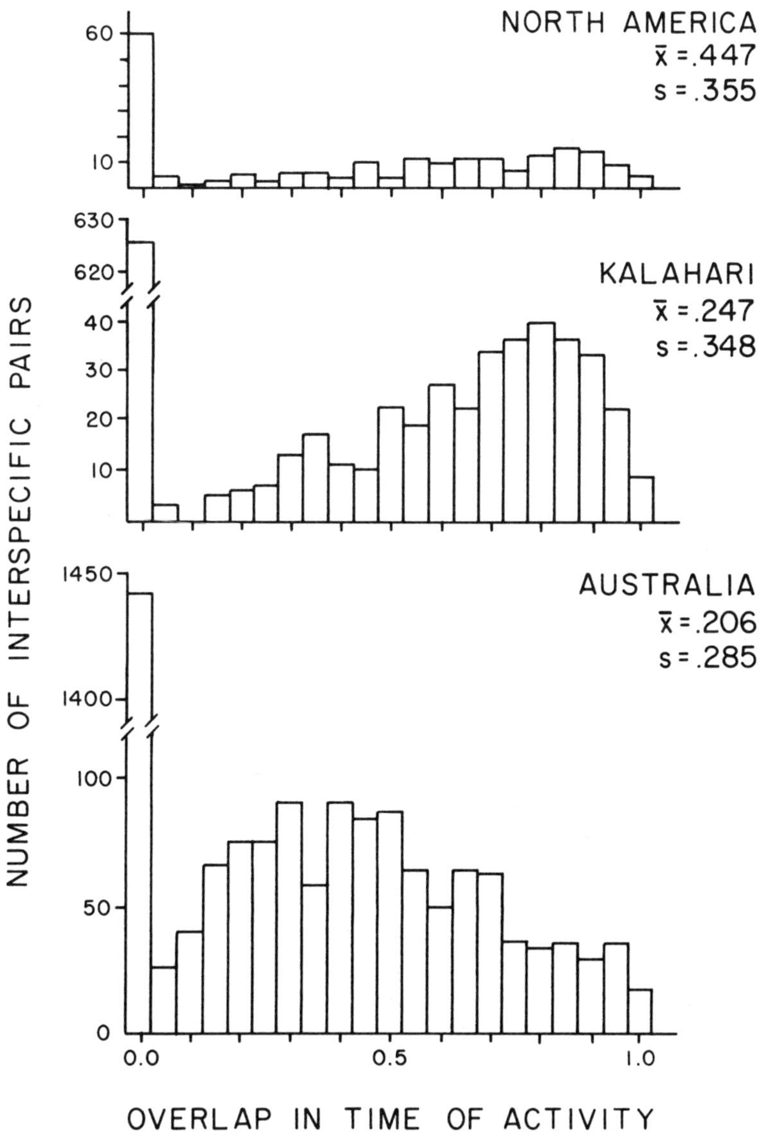

Fig. 20. Observed frequency distributions of overlap in time of activity in the 28 continental saurofaunas (area-by-area).

percent of the foods eaten by lizards on the other two continents). Even so, observed dietary overlap among all Kalahari lizards averages significantly less than in 100 communities randomized by treatment 2 ($\bar{x}$ = .81, range .69 to .89). Observed average overlap in microhabitat is similarly lower than in randomized Kalahari communities (table 22). Overlap in time of activity among Kalahari lizards, however, averages significantly *less* in randomized communities than in the observed. Comparisons of observed average overlap with randomized communities for Australian desert lizards also show greater niche separation than expected under the null hypothesis of random utilization for the food and microhabitat niche dimensions. Mean overlap in observed times of activity of Australian lizards does not differ, however, from those of randomized communities (table 22).

It is instructive to compare overlap statistics from lizard guilds (table 20) with those from the entire saurofaunas in which they occur (table 22). As would be expected, average overlap within guilds is generally considerably greater than in entire faunas (but note that dietary overlap among all Kalahari lizards averages .65, whereas the same mean among the seven lacertids is only .60).

Recall that Schoener (1974) asserted that habitat dimensions tend to be more important in separating niches than food dimensions, which in turn are generally important more often than temporal niche dimensions. In conformity with Schoener's very plausible generalization, mean overlap in time of activity observed in three saurofaunas (table 22) as well as seven guilds (table 20) either does not differ from that of comparable assemblages randomized by treatment 2 (six of the ten cases) or the average of observed overlaps in time of activity is actually *higher* than in 100 randomized treatments (four cases—the entire Kalahari saurofauna, both the Australian and Kalahari gecko guilds, and perhaps Australian *Amphibolurus*). As indicated above, this may well be a result of limited opportunities for ectothermic activity in the rapidly changing thermal environment of deserts. The same sort of analyses of the food niche dimension suggest that nonrandom resource utilization facilitates niche segregation in three guilds (Kalahari lacertids and both *Amphibolurus* and geckos) as well as in all three continental saurofaunas. Similarly, observed overlap in microhabitat averages lower than expected under the hypothesis of random utilization in all three saurofaunas (table 22) and in all guilds except the lacertids, *Ctenotus*, and Australian *Amphibolurus* (table 20). In addition, nonrandom utilization of habitats separates the five species of lacertids in the sandridge region of the Kalahari. Thus these studies of niche segregation among desert lizards certainly support Schoener's generalization that habitat, microhabitat, and food niche dimensions are more important in niche segregation than temporal niche dimension, both in guilds and in entire saurofaunas.

TABLE 22

SUMMARY OF NICHE OVERLAP STATISTICS AMONG ENTIRE SAUROFAUNAS AND FOR 100 COMPARABLE ASSEMBLAGES RANDOMIZED UNDER TTREATMENT 2 (SEE TEXT) ALONG THREE NICHE DIMENSIONS

(Compare with Table 20)

Saurofauna	Microhabitat				Food				Time of Activity			
	Observed		Treatment 2		Observed		Treatment 2		Observed		Treatment 2	
	$\bar{x}$	var.	$\bar{x}$	var.	$\bar{x}$	var.	$\bar{x}$	var.	$\bar{x}$	var.	$\bar{x}$	var.
North America†	.34	.18	.47***	.15	.43	.11	.65***	.04	.58	.17	.52^{NS}	.15
Kalahari†	.29	.10	.45***	.11	.65	.08	.81***	.03	.35	.17	.32***	.14
Australia†	.29	.09	.43***	.10	.32	.09	.43***	.09	.30	.11	.30^{NS}	.11

†Saurofauna means may differ slightly from those presented by Pianka (1973) due to corrections and addition of new data for some species.
NS = not statistically significant.
*** = range of means observed in 100 trials did not overlap observed.

Niche Overlap and Diffuse Competition

Various estimates of over-all multidimensional niche overlap can be generated from unidimensional patterns of resource utilization. As indicated above, the product of several unidimensional estimates of niche overlap usually (though not always) underestimates true multidimensional overlap unless niche dimensions are independent (orthogonal). In contrast, a simple arithmetic average of several unidimensional estimates of overlap almost always[5] overestimates true overlap; indeed, such an average provides the *upper bound* on true multidimensional niche overlap (May, 1975). When averaged among all pairs within each of the twenty-eight saurofaunas, both these estimates of over-all overlap are strongly intercorrelated ($r = .97$, $P << .001$) so that statements to follow concerning upper bound estimates (summation overlap) also apply to multiplicative estimates. Both measures of over-all overlap vary inversely with number of lizard species (fig. 21) and with the standard deviation in annual precipitation (fig. 22), a measure of environmental variability (see also Pianka, 1974, 1975). The latter correlation vanishes when the number of lizard species is held constant by partial correlation, but the former remains significant ($r_{xy.z} = -.56$ $P < .01$), suggesting that the extent of tolerable overlap is not a function of environmental variation but rather that it

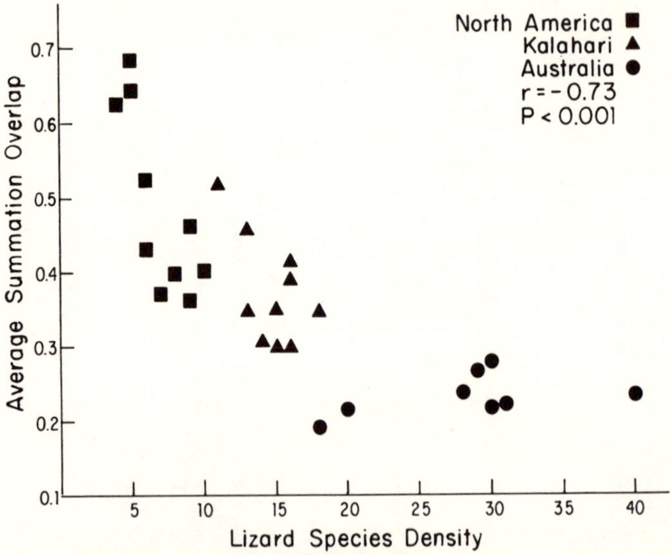

Fig. 21. Average niche overlap (over-all summation) plotted against estimated numbers of lizard species (continents coded by shape). Similar inverse correlations exist with multiplicative over-all overlap and with several estimates of maximal tolerable niche overlap.

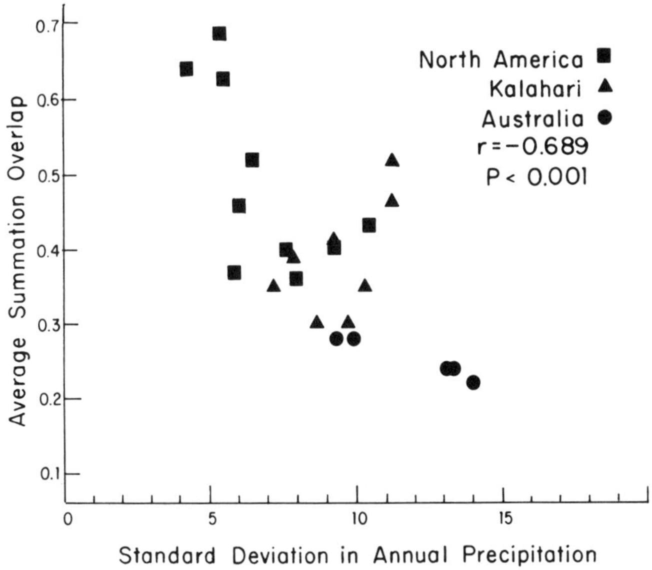

Fig. 22. Average over-all summation niche overlap plotted against the standard deviation in annual precipitation for 23 desert study areas (continents coded by shape). This correlation disappears when the number of lizard species is held constant by partial correlation.

is related to the number of competing species and the intensity of diffuse competition. Pianka (1974, 1975) interprets this pattern as indicating that competitor-species density influences tolerable-niche overlap, with a greater number of competing species (more intense diffuse competition) demanding greater average niche separation among coexisting species. Extensive niche overlap with fewer competitors could be roughly equivalent to lower overlaps between pairs of species but summed over more competitors.

SUMMARY

Nonrandom utilization of resources reduces overlap among desert lizards, thereby facilitating resource partitioning; in three continental saurofaunas and all but one of seven lizard guilds examined, niche segregation was demonstrated along at least one niche dimension. The relative importance of different niche dimensions in separating species differs among guilds and among continental saurofaunas. Spatial dimensions (habitat and microhabitat) separate lizard species more often than food dimensions, although both are important. No temporal niche segregation was discernible in this analysis; desert lizards examined are "packed" into short time periods, presumably because of severe constraints imposed on ectothermic activity by the rapidly changing thermal environment of deserts. Niche overlap varies inversely with the number of species, probably due to the impact of diffuse competition in diverse saurofaunas. Thus a greater number of competing species results in stronger diffuse competition and demands greater average niche separation among coexisting species.

ACKNOWLEDGMENTS

Persons who have contributed greatly to various aspects of the development of research reported here include L. W. Coons, V. Denniston, M. E. Egan, P. R. Grant, W. D. Haacke, H. S. Horn, G. Kaufman, P. Levin, A. R. Main, R. M. May, R. H. MacArthur, W. S. Parker, H. D. Pianka, N. A. Pianka, J. J. Schall, T. W. Schoener, W. Shaneyfelt, R. C. Snyder, G. M. Storr, and M. Thomas. Work was supported by the National Institute of Health, the National Science Foundation, the National Geographic Society, the Miller Institute for Basic Research in Science, the Museum of Vertebrate Zoology (University of California, Berkely), the Los Angeles County Museum of Natural History Foundation, and the University of Texas Research Institute. We are very grateful to all.

1. The use of time of activity as a niche dimension can be justified in several ways (MacArthur and Levins, 1967; Pianka, 1969, 1973, 1975; Schoener, 1974, 1975). If resources are rapidly renewed, exploitative competition cannot occur unless individuals are active a fairly short time of each other: otherwise, resources may be replenished during the interval separating species. Moreover, when microhabitat and food categories are crude and temporally changing, differences in time of activity may in fact be associated with differential use of space and food resources that would not be reflected without much greater precision in recognition of appropriate nontemporal resource states. Thus, the microhabitat "open sun" clearly changes with the daily march of temperature; similarly a given crude prey category such as "ants" may usually lump a series of prey species with temporal segregation. Hence it is often both appropriate and useful to treat time as a niche dimension.

2. We offer only a preliminary and abbreviated treatment of Lawlor and Pianka's analysis here.

3. For descriptions of these study areas and their funas, as well as details of our methods, see Pianka, 1971, and Pianka and Huey, 1971.

4. Three species with very restricted distributions, *Heloderma, Xantusia,* and *Uma,* were omitted.

5. If all niche dimensions are perfectly dependent upon one another (in which case there is really one a single niche dimension), the average of observed unidimensional values equals the true multidimensional value.

LITERATURE CITED

Abrams, P. 1975. Limiting similarity and the form of the competition coefficient. Theoretical Population Biology 8:356–75.

Broadley, D. G. 1968. A revision of the African genus *Typhlosaurus* Weigmann (Sauria: Scincidae). Arnoldia (Rhodesia) 3:1–20.

Broadley, D. G. 1969. Two sympatric species of the *Mabuya striata* complex (Sauria: Scincidae) in the southern Kalahari. Koedoe 12:11–14.

Broadley, D.G. 1972. A review of the *Nucras tessellata* group (Sauria: Lacertidae). Arnoldia (Rhodesia) 5:1–36.

Brown. W. L., and E. O. Wilson. 1956. Character displacement. Systematic Zoology 5:49–64.

Cody, M. L. 1974. Competition and the structure of bird communities. Princeton University Press.

Cole, L. C. 1960. Competitive exclusion. Science 132:348–49.

Colwell, R. K., and D. J. Futuyma. 1971. On the measurement of niche breadth and overlap. Ecology 52:567–76.

DeBach, P. 1966. The competitive displacement and coexistence principles. Ann. Rev. Entomol. 11:183–212.

Fenchel, T., and F. B. Christiansen. 1976. The niche and limiting similarity. Springer-Verlag.

Gause, G. F. 1934. The struggle for existence. Hafner reprint, 1964.

Grant, P. R. 1972. Convergent and divergent character displacement. Biol. J. Linn. Soc. 4:39–68.

Haacke, W. D. 1976. The burrowing geckos of southern Africa, 3 (Reptilia: Gekkonidae). Ann. Transvaal Mus. 30:29–39.

Hardin, G. 1960. The competitive exclusion principle. Science 162:1243–48.

Huey, R. B., E. R. Pianka, M. E. Egan, and L. W. Coons. 1974. Ecological shifts in sympatry: Kalahari fossorial lizards *(Typhlosaurus)*. Ecology 55:304–16.

Huey, R. B., and E. R. Pianka. 1974. Ecological character displacement in a lizard. American Zoologist 14:1127–36.

Huey, R. B., and E. R. Pianka. 1977a. Patterns of niche overlap among broadly sympatric *versus* narrowly sympatric Kalahari lizards (Scincidae: *Mabuya*). Ecology 58:119–28.

Huey, R. B., and E. R. Pianka. 1977b. Seasonal variation in thermoregulatory behavior and body temperature of dirunal Kalahari lizards. Ecology 58:1066–75.

Hutchinson, G. E. 1959. Homage to Santa Rosalia, or why are there so many kinds of animals? American Naturalist 93:145–59.

Lack, D. 1971. Ecological isolation in birds. Blackwell. 404 pp.

Lawlor, L. R., and E. R. Pianka. 1976. Niche overlap and community structure. In preparation for Theoretical Population Biology.

Leistner, O. A. 1967. The plant ecology of the southern Kalahari. Bot. Surv. South Africa, Mem. No. 38:1–172.

Levins, R. 1968. Evolution in changing environments. Princeton University Press.

MacArthur, R. H. 1968. The theory of the niche. *In* R. C. Lewontin (ed.), Population Biology and Evolution. Pp. 159–176. Syracuse University Press.

MacArthur, R. H. 1969. Species packing and what interspecific competition minimizes. Proc. Nat. Acad. Sci. USA 64:1369–71.

MacArthur, R. H. 1970. Species packing and competitive equilibrium for many species. Theoretical Population Biology 1:1–11.

MacArthur, R. H. 1972. Geographical ecology. Harper and Row, New York. 269 pp.

MacArthur, R. H., and R. Levins. 1967. The limiting similarity, convergence, and divergence of coexisting species. American Naturalist 101:377–85.

May, R. 1974. On the theory of niche overlap. Theoretical Population Biology 5:297–332.

May, R. 1975. Some notes on estimating the competition matrix, α. Ecology 56:737–41.

May, R., and R. H. MacArthur. 1972. Niche overlap as a function of environmental variability. Proc. Nat. Acad. Sci. USA 69:1109–1113.

Nevo, E., G. Naftali, and R. Guttman. 1975. Aggression patterns and speciation. Proc. Nat. Acad. Sci. USA 72:3250–54.

Orians, G. H., and H. S. Horn. 1969. Overlap in foods and foraging of four species of blackbirds in the potholes of central Washington. Ecology 50:930–38.

Patten, B. C. 1961. Competitive exclusion. Science 134:1599–1601.

Pianka, E. R. 1966. Convexity, desert lizards, and spatial heterogeneity. Ecology 47:1055–59.

Pianka, E. R. 1967. On lizard species diversity: North American flatland deserts. Ecology 48:333–51.

Pianka, E. R. 1969a. Habitat specificity, speciation, and species density in Australian desert lizards. Ecology 50:498–502.

Pianka, E. R. 1969b. Sympatry of desert lizards (*Ctenotus*) in Western Australia. Ecology 50:1012–30.

Pianka, E. R. 1971. Lizard species density in the Kalahari desert. Ecology 52:1024–29.

Pianka, E. R. 1972. *r* and *K* selection or *b* and *d* selection? American Naturalist 106:581–88.

Pianka, E. R. 1973. The structure of lizard communities. Ann. Rev. Ecol. Syst. 4:53–74.

Pianka, E. R. 1974. Niche overlap and diffuse competition. Proc. Nat. Acad. Sci. USA 71:2141–45.

Pianka, E. R. 1975. Niche relations of desert lizards. In Ecology and evolution of communities, M. Cody and J. Diamond (eds.). Pp. 292–314. Harvard University Press.

Pianka, E. R. 1976a. Competition and niche theory. *In* Theoretical ecology: principles and applications, R. M. May (ed.). Pp. 114–41. Blackwell.

Pianka, E. R. 1976b. Natural selection of optimal reproductive tactics. Amer. Zoologist 16:755–84.

Pianka, E. R., and R. B. Huey. 1971. Bird species density in the Kalahari and the Australian deserts. Koedoe 14:123–30.

Pianka, E. R., and H. D. Pianka. 1976. Comparative ecology of twelve species of nocturnal lizards (Gekkonidae) in the Western Australian desert. Copeia 1976:125–42.

Pielou, E. C. 1972. Niche width and niche overlap: a method for measuring them. Ecology 53:687–92.

Roughgarden, J. 1972. Evolution of niche width. American Naturalist 106:683–718.

Schoener, T. W. 1965. The evolution of bill size differences among sympatric congeneric species of birds. Evolution 19:189–213.

Schoener, T. W. 1968. The *Anolis* lizards of Bimini: resource partitioning in a complex fauna. Ecology 49:704–26.

Schoener, T. W. 1974a. Resource partitioning in ecological communities. Science 185:27–39.

Schoener, T. W. 1974b. Competition and the form of habitat shift. Theoretical Population Biology 5:265–307.

Schoener, T. W. 1976. Competition and the niche. *In* Biology of the Reptilia, D. W. Tinkle and C. Gans (eds.). Academic Press.

Schoener, T. W., and G. C. Gorman. 1968. Some niche differences among three species of Lesser Antillean anoles. Ecology 49:819–30.

Vandermeer, J. H. 1972. Niche theory. Ann. Rev. Ecol. Syst. 3:107–32.

Whittaker, R. H., and S. A. Levin, eds. 1975. Niche: theory and application. Dowden, Hutchinson & Ross.

LAWRENCE E. GILBERT

Development of Theory in the Analysis of Insect-Plant Interactions

5

INTRODUCTION

The editors of this volume have asked that I provide an overview of research in the area of insect-plant interactions with a focus on areas of particular interest to me. Since there has been an entire volume devoted to this extremely diffuse subject (Van Emden and Way, 1973), I cannot hope, nor do I need, to develop an exhaustive review within the limits of a single paper. The most notable attempt in this direction is the paper by Southwood (1973), which, taken with the still excellent chapters (3–5) of Brues's (1946) Insect Dietary, now reprinted under a different title (Brues, 1972), provides a solid introduction to the empirical literature on insect-plant interaction.

In this paper I will be most concerned with discussing some recent attempts to develop theory in the study and analysis of insect-plant interactions. However, for the benefit of the general reader, I first provide a brief discussion of the diversity of insects and plants, review the categories of interactions known to occur between them, describe some of the approaches that have been taken in the study of insect-plant relationships, and point to some of the more interesting evolutionary and ecological consequences of such interactions. In this way I hope to provide an abbreviated but useful index to the recent research and review literature. In the final section I will use examples from my own work as well as from that of close colleagues in interpreting and evaluating current thought on plant-herbivore (parasite) interactions.

THE DIVERSITY OF INSECTS AND PLANTS

Insects account for well over one-half of the million-plus described animal species on the earth. A conservative estimate for the number of insects that

Lawrence E. Gilbert, Department of Zoology, University of Texas, Austin, Texas 78712

rely on plants for at least part of their resource needs would be about 350,000 named species. My guess is that if we suddenly knew everything about largely undescribed tropical insect faunas, the number of insect species known to interact directly with plants would more than double. This is a safe guess, because even in California, a well-studied region entomologically, Opler (1974) found that 80 percent of leaf-mining moth species (N=116) reared from oak were previously undescribed. Moreover, since poorly described groups of insects, such as microlepidoterans and small to minute coleopterans, dipterans, and hymenopterans, make up the vast bulk of plant feeders, current estimates of herbivorous insect diversity must be very conservative.

In contrast to insects, most vascular land plants are large, conspicuous, and relatively well-collected, so that we can be more satisfied with over-all estimates of plant diversity. Angiosperms, with some 235,000 species, are by far the dominant group of plants, and contribute the bulk of terrestrial biomass. Fossil evidence, although sketchy for both insects and plants (see Smart and Hughes, 1973, and references therein) is consistent with the view that angiosperms appeared in the lower Cretaceous, well after several insect orders had established relationships with plants, but that the primarily angiosperm-feeding order, Lepidoptera (ca. 120,000 species), arose later (see also Opler, 1973). It is thus reasonable to think that the parallel diversifications of both angiosperms and phytophagous insects[1] is largely the product of their interaction as Ehrlich and Raven (1965) suggested. They asserted that any attempts to explain over-all organic diversity cannot ignore insect-plant interaction (note that angiosperms and their insect faunae represent at least 60 percent of the earth's total eucaryotic species diversity if the numbers and projections given above are reasonable estimates).

Most phytophagous insect species utilize more than one host species in nature. This is true even of relatively specialized species. If we further consider that insects with complex life cycles often utilize different hosts as larvae and adult (e.g., butterflies), then the number of possible insect-plant interactions is much higher than the number of plants or plant-feeding insects. For example, for 66 species of heliconiine butterflies, there are 387 known pairwise interactions with 123 larval host species of the family Passifloraceae (Benson, Brown, and Gilbert, 1976).

Thus it is clear that biologists could spend several hundred additional years completely occupied with describing new insect species and discovering novel interactions between plants and insects without exhausting the supply. It is not clear, however, that further base-level exploration would provide many new ecological or evolutionary insights, or that additional categories of interactions would be found which fall outside those kinds that so far have been described.

AN ECOLOGICAL CLASSIFICATION OF INSECT-PLANT INTERACTIONS

Southwood (1973) has summarized the basic interactions between insects and plants under the categories of food, shelter, and transport. Subsequently Hocking (1975) elaborated this scheme slightly. Although these categories, if interpreted broadly, cover all possible interactions, they are too inclusive to be useful in ecological analysis. For example, ''transport'' would lump dispersal of pollen and seeds, two very different processes that would have to be separated in any evolutionary or ecological model of a real system. Likewise the ''food'' category lumps leaf-feeding insects with pollinators, two ecologically distinct types of insect from a plant point of view.

TABLE 1

WELL-KNOWN CATEGORIES OF INSECT-PLANT INTERACTION

Insects	Plants
Predators, parasitoids	Prey
Parasites	Host
Pollinators, seed dispersal agents	Provide a reward
Allelopathic agents, anti-herbivore defense force	Proof food reward and/or domicile
Nutrient-gatherers	Provide domicile
Host	Parasite
Prey	Predator

During the last decade there has been an increasing trend, stimulated greatly by the work and writing of D. H. Janzen, to consider insect-plant interaction in terms of the ecological relationships represented. In table 1, I have summarized some known non-neutral categories of insect-plant interaction. These include:

1. Insects as Predator, Plants as Prey

Individual plants, typically seeds or small seedlings, are killed by externally feeding insects such as harvester ants (Tevis, 1958), seed-feeding bugs (Janzen, 1972), or caterpillars (Breedlove and Ehrlich, 1971).

2. Insects as Parasitoids, Plants as Host

Insects ranging from tiny chalcid wasps to bruchid beetles undergo larval development within seeds and typically kill them. Although these insects might be considered predators (Janzen, 1971b), their internal feeding habits increase the need for host specificity and impose life-cycle constraints (see

Center and Johnson, 1974) that alter details of functional responses to host densities and change the rules of discovery for their own predators and hyperparasitoids. For purposes of ecological models the extreme examples of predator and parasitoid should be viewed separately.

3. Insects as Parasites, Plants as Hosts

Parasitic insects, though they typically do not directly kill their host plants, may nevertheless increase mortality due to secondary factors. Chewing insects, such as sawfly larvae and lepidopteran caterpillars, are well-known defoliaters of forest trees, and defoliation can be related to increased tree mortality (Kulman, 1971). Defoliation can also reduce the reproductive output of trees (Rockwood, 1973), but in *Rumex* single defoliations can actually increase seed set at least in the short term (Cavers, 1973). Sucking insects, such as leaf hoppers (DeLong, 1971), specialize on the sap of plants and are vectors for plant viruses (Smith, 1965). On the other hand, bark beetle larvae rely on fungal infection under host bark, and the adult beetles have special chambers that house the fungal propagules (Barras, 1975). Dutch elm disease is a striking indication of the drastic consequence that beetle-borne fungal disease may have for an entire tree species. Note that selection for resistance to an insect-borne pathogen might act at the level of preventing the vector's attack, rather than on the pathogen directly. Category 3 includes the largest number and the greatest diversity of insect-plant interactions. Whereas seeds and seedlings are relatively uniform resources, individual adult plants are sufficiently diverse in structure to provide an incredible variety of adaptive zones for parasitic insects. Within sucking insects, for instance, aphids have radiated on new growth of many plant groups (Eastop, 1973). Other homoptera, such as jassids, subdivide the "sap" resource, with some species specializing on vascular bundles and others tapping inside mesophyll cells (Saxena, 1954). I propose that such feeding tactics circumvent an important kind of chemical defense to which chewing insects are exposed: the production of toxins when compartmentalized glycosides are mixed with glycosidases as the leaf tissue is crushed (e.g., Jones, 1973).

Chewing insects are similarly diverse with respect to what part of the plant is used and how it is used. Both Lepidoptera and Coleoptera include "borers" (stems, roots), leaf-feeders and "leaf-miners." Leaf-mining insects are the subject of an entire volume (Hering, 1951) as are another major group of internal parasites, gall insects (Mani, 1964). The ecological impact of miners and gall-makers is probably less than that of other plant parasites, but I know of no attempts to estimate their relative effects on plant growth and reproduction.

Perhaps the most unique insect parasites of plants are the attine, or fungus ants (Wheeler, 1973), which cut leaves of a large variety of trees, vines, and herbs for use as substrate for their fungus cultures. Both in the rules that govern the host choice of *Atta* (Rockwood, 1976) and in the manner of processing the collected leaves, these ants most resemble browsing vertebrates, which also seem to prefer variety but possess gut microfloras instead of fungus gardens (see Freeland and Janzen, 1974). *Atta* colonies apparently conserve their host plants by not completely defoliating preferred species (Cherret, 1968), a policy possible only if colonies are strictly territorial (they are). Other suggested instances of host conservation by parasitic insects (Monro, 1967) are more difficult to envisage without group selection having been a factor.

4. Insects as Transporters of Plant Gametophytes, Plants as Providers of Food Reward

Historically, pollination of plants by insects has received more attention than other aspects of insect-plant relations. Much has been written on the evolution of angiosperm flowers in the context of insect pollination (see Meeuse, 1973), or on the mechanism involved in floral allurement and insect behavioral response (Yeo, 1973). Yet aside from applied studies of crop pollination (Free, 1970), the ecological aspects of this relationship have been analyzed only recently. Adequate ecological analysis requires specification of the rewards given to the insects (see Baker and Baker, 1975), estimation of pollen transfer efficiency (e.g., Levin and Berube, 1972) and physiological and behavioral significance of rewards to the insect (Gilbert, 1972; Heinrich, 1975a, 1975b; Pliske, 1975), and integration of information on the population biologies of pollinators and plants (Heinrich, 1975c; Gilbert, 1975). Although much emphasis has been given to highly specific pollination systems, many plants interact with taxonomically diverse assemblages of pollinators (Beattie, 1971), so that ecological analysis is more complex.

5. Insects as Transporters of Plant Propagules, Plants as Providers of Food Reward

Relatively little is known about seed dispersal by insects, possibly because it is inconspicuous when compared with insect pollination or seed dispersal by vertebrates (see McKey, 1975). Nevertheless, it is important for some groups of plants such as violets (see Beattie and Lyons, 1975, for a review). Of greater interest is the fact that in Australia, insects (ants) are a major non-passive mode of seed dispersal (Berg, 1975), as determined by the presence of elaiosomes (food bodies) on seeds of many species.

6. Insects Prune or Kill Competing Non-host Plants, Plant Provides Food and Domicile

Insects acting as allelopathic agents of plants are known for neotropical ant-acacias (Janzen, 1966) and for *Cecropia* (Janzen, 1969), and are a potential benefit to any plant that harbors ant colonies, although known cases are few. In both of these examples, plants provide all the necessities of life for the ants involved.

7. Insects Defend Plant against Herbivores, Plant Provides Food and/or Domicile

Much attention has been given to the more spectacular obligate cases of ant-plant mutualism (Wettstein, 1889; Hocking, 1975) in which anti-herbivore activity has been shown (Janzen, 1966) to be a major contribution of the insect partner. However, many plants, especially in tropical zones, produce extra-floral nectar that attracts and supports (?) a variety of predaceous ants, social wasps, and minute parasitic wasps, some of which are known to be significant mortality factors for the herbivorous insects that attack particular plants (Ehrlich and Gilbert, 1973). Such insects, which I will term "mutualistic defenders" of plants, have been shown to be significant in reduction of herbivory (Bentley, 1976). As yet, little is known of the precise composition of nectar rewards offered to this class of plant visitors, nor has the portion of any insect's energy budget due to such food been accurately estimated, although Hocking (1970) made progress in this direction.

Both from an individual plant point of view and at the ecosystem level, the impact of insects predaceous or parasitic upon eggs or early stage larvae of herbivorous insects is vast relative to their own biomass or relative to the cost of maintaining them (Gilbert, 1977).

8. Insects May Gather Nutrients for Plants, which in Turn Provide Domicile or Food

In nutrient-poor forests of the Asian tropics, certain ant species cache dead insects and other nitrogen-rich debris within the epiphytes that house them. The ants can be viewed as an extended root system for the plants (Janzen, 1974a). Fungus ants are similar in providing energy and nutrients for their Basidiomycete cultures (see Martin, 1974), but differ in providing shelter for their "plants."

9. Insect as "Host," Plant as Parasite

Plants can be viewed as parasites on the time and energy of insects when insects are induced to perform a service, such as pollen dispersal, for which no reward is provided. Such parasitism is known only in plants that deceive insects by tactics such as mimicry of legitimate floral signals (e.g., Mosquin, 1971) or mimicry of the odor and appearance of a female insect, which elicits copulatory behavior by males (see Yeo, 1973, for a brief review). This kind of exploitation of insects is a common tactic of the most diverse family of angiosperms, the Orchidaceae.

10. Insects as Prey, Plants as Predators

A small fraction of angiosperms (e.g., pitcher plants) attract, then trap and kill insects. Undoubtedly various nutrients such as nitrogen are the insect ingredients of primary value to these plants and thus nutrient-poor environments are the most likely areas for such plants to be conspicuous elements of the flora.

There are obvious overlaps and inconsistencies among these ten categories of interaction. However, the reader will at least understand the incredible diversity of phenomena encompassed by the simple phrase "insect-plant interaction." Furthermore, some of these categories and subcategories are ubiquitous and important worldwide, while others are geographically or ecologically restricted. For example, Hocking (1975) points out that although 15/17 *Acacia* species in both America and Africa have evolved relationships with ants, no ant-acacia systems are known from centers of *Acacia* diversity in the oriental and Australian regions. Meanwhile, ant dispersal of seeds, a novelty in England, is a major category of relationship in Australia, where an estimated 1,500 plant species rely to some degree on ants for seed transport (Berg, 1975). Likewise, mutualistic defense systems seem to be a common feature of rapidly growing plants in the tropics, but are inconspicuous in higher latitudes. Is it mere coincidence that aphid abundance and diversity follow precisely the opposite trend (Eastop, 1973)?

Obviously, one's perception of which categories of interaction are "important" are greatly biased by geographical experience. Without doubt, biogeographical studies of globally patchy insect-plant interactions will richly repay our efforts in both revealing new problems and solving old ones. I will return to this idea, with examples, in sections that follow.

FREQUENT APPROACHES TO THE STUDY OF INSECT-PLANT INTERACTION

1. Physiology and Behavior

Most research on insect-plant interaction has taken the form of detailed description of particular aspects of relationships. Such description often involves experimental elucidation of morphological (Pillemer and Tingey, 1976; Hoxie et al., 1975; Levin, 1969), chemical (Hanover, 1975; Erickson and Feeny, 1974; Self et al., 1970), or behavioral (Yeo, 1973; Dethier, 1970; Jermy, 1966; Harris and Mohyuddin, 1965) mechanisms. Thus, the literature on pollination (see previous section) and host relations of leaf feeders (Moreau and Gahukan, 1975) deals primarily with proximate questions such as how pollen gets transferred or how an insect identifies its proper host. Unfortunately, most chemists recently entering the field of insect-plant interactions have adopted this ''functional morphology'' approach to plant chemistry. The field of ''chemical ecology'' (Sodheimer and Simeone, 1970) would be more properly termed ''ecological chemistry'' because the questions attacked by most of the research are chemical or physiological in nature — not ecological or evolutionary (see Janzen, 1973b). Few studies use experimental methods to demonstrate the co-adaptive nature of entire sets of traits possessed by insects and plants (but see Janzen, 1966).

2. *Population Ecology*

Studies on interacting populations of insects and plants have been devoted primarily to protection of crops (Van Emden and Way, 1973) and forest trees (Kulman, 1971) or to the control of weeds (Wilson, 1964; Dunn, 1973; Harris, 1973). These studies do give us good examples of the impact that insects can have on plant populations. However, since data are collected to answer applied rather than evolutionary questions, such research is usually of limited use in the development and testing of general theory.

3. *Behavioral Ecology, Energetics*

The application of foraging theory to the analysis of pollination systems (Heinrich and Raven, 1972; Heinrich, 1975a, 1975b) puts the study of plant-pollinator interaction ahead of other areas of insect-plant interaction in terms of the development of testable predictions. Briefly this is a behavioral/ecological approach based on rather straightforward assumptions about how natural selection should operate on pollinator foraging tactics and plant reward structures given particular life historical and ecological constraints. It represents a special case of the application of optimality models that have received attention elsewhere in this volume (e.g., papers by Lewontin and Orians). The use of this approach in the development of predictions about plant-parasite interactions will be evaluated below.

4. *Evolutionary Systematics and Geographical Ecology*

I lump two kinds of studies under what might be called the comparative approach to analyzing insect-plant interactions. The most common of the two involves systematic studies of the literature on host relations (Brues, 1924; Ehrlich and Raven, 1965) and plant morphology or chemistry (Levin, 1971, 1973, 1976; McKey, 1974) from which evolutionary explanations for observed patterns are generated. Occasionally, such broad taxonomic surveys involve data collected by the authors (Baker and Baker, 1975; Benson, Brown, and Gilbert, 1975), but more often such a broad data base requires armchair techniques. The second method involves detailed field studies of specific interactions over time and space, holding taxonomy constant if possible. By this method, which is one facet of geographical ecology, Janzen (1972a, 1973d, 1975a) has analyzed selective pressures acting on various plants; Singer (1971) has elucidated the microevolution of host preference in *Euphydryas* butterflies; and Benson, Brown, and Gilbert (1975) have provided evidence of the dynamic nature of coevolution between *Heliconius* butterflies and *Passiflora* vines.

5. *Community and Ecosystem-Level Studies*

Finally, I would call attention to a heterogeneous collection of studies that focus on community and ecosystem-level patterns resulting from insect-plant interaction. Some areas of interest include the impact of leaf-feeding insects on energy flow (Haines, 1975; Lugo et al., 1973); the consequences of plant secondary chemicals in ecosystems (Brower et al., 1967; Janzen, 1974b); community structure of plant secondary compounds (Janzen, 1973a); the composition of insect faunas specific to unwanted weeds (Goeden and Richer, 1976; Shröder, 1969; Waloff, 1968a, 1968b; etc.); and the study of ecosystem "texture" (Gilbert, 1977; Atsatt and O'Dowd, 1976; Root, 1975).

EVOLUTIONARY AND ECOLOGICAL CONSEQUENCES OF INSECT-PLANT INTERACTION

Insects as predators, parasites, and pollinators have undoubtedly helped shape the morphological, chemical, and behavioral traits of flowering plants. Conversely, insect mouth parts, such as grasshopper mandibles (Isely, 1944) or bee tongues (Brian, 1957), and lepidopteran digestive systems (Krieger et al., 1971) seem closely adjusted to the appropriate host plants. Agricultural genetics provides us with many morphological or chemical traits of plants that confer resistance to insect attack (Gallun, 1972; Hanover, 1975), yet the study of Jones (1973) is perhaps the only one that elucidates the selective forces at work in a natural system of interacting herbivores and host plants polymorphic for a specific chemical defense.

The great interspecific variability in "secondary products" chemistry is thought to be another product of plant-insect coevolution (see review by

Levin, 1976). Since insects that specialize upon particular plants utilize some of the plant's defensive chemistry for host recognition and as cues for release of feeding behavior, there is an obvious disadvantage for other species to use similar defenses (Janzen, 1973a), and one possible selective force favoring strict reproductive isolation between congeneric plants would be the increased herbivore load that hybrid progeny, possessing the unique secondary chemistry of each parent species, would attract. Thus, coevolutionary interaction between plants and host-specific parasites would tend to generate new niches for related plant genotypes or species with disruptive chemical "aspects," provided faithful pollinator species are available, this latter contingency being more likely in tropical regions (see Janzen, 1971). The relationship between aspect diversity, species diversity, and visual predation in moths (Ricklefs, 1975) is an analogous situation. Chemical studies of closely related sympatric assemblages of plant species as suggested by Janzen (1973a) are greatly needed (e.g., Dolinger et al., 1973). Atsatt and O'Dowd (1976) have a somewhat different view of interplant relationships in communities.

The role of pollinators in the evolution of floral diversity and flowering phenology has been convincingly argued by Heinrich (1975) for certain bee-pollinated communities. Other studies of flowering synchrony (Janzen, 1967) and seasonal (Frankie, 1975; Gentry, 1974) or diurnal (Gilbert, 1975) partitioning of pollinator service suggest that the need for reproductive isolation and/or competition for insect pollinators are among the important proximate selective forces creating such phenologies.

The ecological results of insect-plant interactions are most clearly seen in the control of introduced plant populations by insects (Harris, 1973; Kok and Surles, 1975). In natural plant associations, the depression of plant density by host-specific predaceous and parasitic insects can be observed to maintain local plant diversity in ecological time (Huffaker, 1971; Burdon and Chilvers, 1974; see also Harper, 1967, Connell, 1971, and Janzen, 1970). Insects also are considered major factors in plant community succession (see Graham and Knight, 1965, Chap. 14, and discussion after Harris, 1973), though this idea is difficult to find in the recent literature of pure ecology (e.g., Horn, 1974). In the tropics, trees adapted for rapid colonization and fast growth (e.g., balsa) tend to have less-dense woods than do climax species. Therefore, the probability of damage from winds, epiphyte loads, and so on, is greater for early successional trees. Once damaged, beetles and fungi can make quick work of such light wood (see Bultman and Southwell, 1976). Thus, the short life span of early successional species is part of an evolutionary trade-off for maximum efficiency as colonizers.

Similar events occur in the temperate zone forests where, for example, defoliation of aspen by tent caterpillars leads to conifer climax (Hodson,

1941). In Texas, large patches of pine killed by bark beetles are succeeded by oak and hickory (personal observation). Thus, we can view insect-caused mortality as a kind of disturbance that helps to prevent dominance of any particular plant species. However, this is qualitatively different from the patch formation by storms in tide pools or landslides in forests that is currently receiving attention from theoreticians (Levin and Paine, 1974; 1975) because the ''disturbances'' by insects (and the pathogens they spread) are species-specific and can spread through the community until the insect is checked by its own predators and/or the decreasing density of hosts (see Birch, 1971). By differential impact on early succession trees, parasitic insects will drive a forest toward a hardwood climax, itself kept diverse largely by seed mortality, due to host-specific predaceous insects (Janzen, 1970).

THEORIES TO EXPLAIN THE EVOLUTION AND ECOLOGY OF PHYTOPHAGOUS INSECTS AND THEIR HOST PLANTS

Scientific theories simplify the analysis of complex phenomena by restricting the data-collecting to that required to test hypotheses suggested by the theory. Essentially all theories of population biology are elaborated combinations of basic natural selection theory, fundamental population growth (Lokta-Volterra equations), and probability theory. The theories developed to explain evolutionary and ecological aspects of insect-plant interactions are no exception to this generality.

As I see it, there are currently four major areas of theory contributing to our understanding of plants, plant-feeding insects, and their interactions. These are: (1) ''Insect-plant coevolution,'' (2) ''Host plants as islands,'' (3) ''Plant apparency and chemical defense,'' and (4) ''Resources predictability vs. evolutionary strategies of insects.'' In the remainder of this paper I will interpret and evaluate these four approaches in light of information that my colleagues and I have accumulated on the details of butterfly-host plant interactions.

1. Insect-Plant Coevolution

Ehrlich and Raven (1965), noting the striking correlation between higher taxa of butterflies and their host plants, proposed that such patterns are generated by a process of reciprocal adaptive radiation. (Brues, 1926, also developed this idea, but with little impact.) That is, plants evolving new and highly effective chemical defenses against herbivores have a competitive advantage over other plants. According to the theory, plant groups have undergone much of their diversification or adaptive radiation during intervals of relative escape from herbivores. Subsequently, herbivores evolving a means to circumvent the chemical defense of such plants enter a diverse adaptive zone where they increase numerically and diversify taxonomically. The fact that closely related higher taxa of butterflies are often restricted to closely related host-plant taxa is presumably due to the facts that chemical distinctiveness between plants is often correlated with their taxonomy and that host-shifts are more likely between chemically similar plants. This line of thought leads to the simplistic prediction that when two related butterflies feed on plants from different families, unknown chemical similarities must exist between the plants. There is now reason to believe that the assumptions that lead to this prediction (i.e., that shifts only occur to chemically similar plants) are violated in many insect-plant systems (see the next section).

It is of interest to ask whether the generation of broad systematic patterns of host usage is a continuing process. That is, can similar patterns also be found at lower taxonomic levels (e.g., sub-genera, sub-specific levels) that suggest reciprocal adaptive radiations are in progress? Benson, Brown, and Gilbert (1976) examined extensive host-plant data gathered for heliconiine butterflies and their host plants, primarily of the genus *Passiflora*, and attempted to answer just this question. Although there are taxonomic correlations between heliconiines and *Passiflora* of the sort Ehrlich and Raven described, there are also radiations by some heliconiines with respect to plant part that are apparently independent of the chemical and taxonomic boundaries recognized by other heliconiine groups. We find abundant evidence of coevolution between *Heliconius* and *Passiflora*, but our ability to evaluate the evolutionary historic aspects of the Ehrlich and Raven theory is highly restricted.

Just over a decade after Ehrlich and Raven coined the term coevolution, it is widely used to refer to any situation in which a pair of organisms are simultaneously or alternatively acting as selective agents for one another. Of course, in those general terms, this concept can be found in Darwin's "Origin of Species." A major contribution of the Ehrlich and Raven theory is that it relates the ecological observation of host-utilization patterns in host-specialist phytophagous insects to a process of biochemical coevolution between these insects and their food resources. As such it has important implications for the ecological analysis of community structure in insects such as butterflies (see Gilbert and Singer, 1975, pp. 377–80). Perhaps of greater significance is the fact that this paper has stimulated a general coevolutionary viewpoint in ecology.

There is no doubt that the Ehrlich and Raven paper has done more than any other single contribution to stimulate the current strong interest in insect-plant interaction. However, their coevolution theory is framed in broad, sweeping, historical terms so that the processes hypothesized are impossible to observe directly. A further weakness of the theory is that it only explains that subset of insect-plant relationships that are highly specific and taxonomically structured.

One can point to many host-specific insect groups that exhibit no particular taxonomic correlations between herbivores and plants (e.g., aphids; see Eastop, 1973). One can also point to many generalist insects that consume many chemically distinct taxa of hosts (see below).

2. *Host Plants as Islands*

Ehrlich and Raven, by emphasizing the butterfly point of view, wittingly ignore the fact that each host-plant taxon supports an entire fauna of herbivorous insects. An area of current interest and controversy involves theory to predict the diversity of a plant's fauna given information on the plant's history and distribution.

Janzen (1968, 1973c) developed the logical framework for extending the equilibrium theory of island biogeography (MacArthur and Wilson, 1967) to insect faunas on host-plant "islands." I should mention that Southwood's (1960, 1961) pioneering work in this area, like the work of Brues on host specificity, went relatively unappreciated until the theories of population biology provided both context and rationale for such a novel approach. As Stent (1972) points out, this seems to be a general phenomenon of science.

Janzen's host-plant island theory outlines the important variables that govern the colonization by an insect species of (1) new host-plant species in evolutionary time, or (2) different individuals or patches of the same host species in ecological/contemporary time. Whereas the MacArthur and Wilson theory treats islands as blank targets that vary in size and in distance from the source of colonization, host-plant islands are rather more complex. Janzen pointed out that both physical and physiological distance between different plants are important (to name just two factors of many). Moreover, host-plant islands coevolve with their colonists, whereas "true" islands do not; and Janzen (1973c) discusses the consequences of host evolutionary responses for the pattern of faunal build-up in evolutionary time.

A current controversy involves the role of evolutionary time as a factor explaining the difference in diversity between the insect faunas of different plants. Strong (1974a, 1974b,) on the basis of data on tree-feeding insects, maintains that the area occupied by host-plant species in contemporary time is sufficient to explain diversity. A major problem with this analysis is its heavy bias toward trees and other plants (sugar cane) defended by digestion reduction mechanisms or quantitative defenses (Feeny, 1975). Such plants tend to be fed upon by generalist insects that would not have to bridge severe physiological barriers to feed on a new host plant. Consequently, since trees with large geographical ranges are exposed to more other-tree species from which colonization might occur, they will have "sampled" more insect species. For example, cacao plantations in different regions over the short run will sample generalist insects from nearby forests. The more acreage of cacao, the greater interface with native insect faunas. It is thus possible that the species-area relations of cacao pests are largely due to this sampling of ecological diversity in ecological time. In fact, I would predict a "species-perimeter curve" due to the fact that many small cacao plantings would have greater interface with local vegetation, and thus more colonists, than the same acreage of cacao planted in larger blocks.

Opler's (1974) data on oak leaf miners suggest that an analogous "sampling" process may have occurred in evolutionary time. That is, widespread oaks tend to overlap many other oak species from which host shifts are possible. Once on the widespread species, a new colonist might then spread rapidly throughout the range of that host. Shifts to rare, local species would be much less likely. The pool of species of a particular feeding guild would be limited by the over-all size of the (collective) oak island, and that number sets the limits for any one oak species regardless of its area. Thus, different immigration rates of leaf miners onto oaks account for the differences between them. It is quite likely that because of the small size of leaf miners, and because of the abundance of their resource even on a single tree, extinction probability is near zero on host islands above a few acres in size. This is supported by the fact that twelve of fourteen species are still present on *Q. agrifolia* on Santa Cruz Island. On the basis of this data and a presumed separation time of Santa Cruz Island, Opler estimates a relaxation of time of 1,000,000 years. However, given the rapid evolution of host races in some insects (Bush, 1975), evolutionary diversification might counter attainment of the predicted ecological equilibrium.

Both Strong's and Opler's analyses show significant species-area relationships. However, if Strong's data are dominated, as I suspect they are, by *generalist* insects that readily shift to new hosts in ecological time, then the form of his curve does not prove or disprove the hypothesis that the numbers of *host-specific* insects on a plant will increase over evolutionary time. Opler's data likewise do not apply to testing the "time theory" since presumably all of the oaks in his study are of equivalent age and it is possible that their current relative abundances are also of long standing. Comparing the host-specific faunas of oaks of similar distributional areas but of longer evolutionary exposure time would yield a species-area curve above that of Opler's (Opler, 1974, fig. 2) if time were a factor.

Much of the current confusion in testing island biogeographic theory using host-plant islands would be eliminated if adequate attention were given to the significant diversity inherent in plants and in insect colonists. We now know that plants or plant parts may be divided into at least two groups: (1) those that are primarily defended by toxins and that require highly specialized digestive tactics by herbivores, and (2) those that are defended by digestion-reducing agents such as tannins, and that require few specialized digestive tactics (see Feeny, 1976).

This dichotomy in plant defensive chemistry is more fully explained in the following section. It suffices to say for now that the bulk of data providing generalities in the literature on host-plant islands come from group two plants — primarily trees.

Group 1 plants are primarily herbaceous, early successional, and/or annual plants that typically harbor highly specialized insect faunas. Whereas a beetle on a particular deciduous tree species in Britain might also feed on three other tree families, any insect found eating mustard leaves is likely to eat only crucifers. Group 1 also includes the butterfly host plants that best fit the Ehrlich and Raven (1965) coevolution scheme.

In *both groups*, older, more permanent parts tend toward group 2 defenses, and younger, more ephemeral parts of the same plant tend toward group 1 defenses. Likewise, a parallel dichotomy in insect digestive tactics has been demonstrated (Krieger et al., 1971). Quite obviously, the dynamics of colonization will be a function of both the category of the "island" and the counter-defensive category of the colonist. The colonization of a toxic family of vines by a particular insect may require extensive genetic alteration, and this will have to be an evolutionary event. A leaf-feeding insect that feeds on three families of tropical trees may be able to feed on cacao on first encounter with little alteration of fecundity.

These basic observations lead to the prediction that faunal build-up on an introduced herb will be much slower than that on a tree. In fact, if introduced weedy herbs accumulated herbivores as fast as cacao, we would not always be searching countries of origin for biological control agents! Area occupied quite obviously has had little to do with the accumulation of a fauna in ecological time by introduced weeds, since they often spread over vast areas and remain free of herbivory while excluding native plants. It thus seems crucial that future analyses of host-plant faunae include a factoring-out of known generalist insects from suspected specialists, and that ecological categories of plants as well as plant parts be analyzed separately. In general, plants or plant parts that require evolutionary adjustments by colonists will be colonized more slowly, and therefore would be more likely to show effects of time in addition to area in faunal accumulation.

One facet of the host plant-as-island concept is that it provides an approach to modeling insect-plant coevolution. Although the literature to date focuses primarily upon very gross lumping of taxonomic and geographical information, the actual colonization steps involve individuals and populations. Studies of insect dispersal behavior and local plant spacing patterns represent a level of data-collecting that will bear upon the microevolutionary mechanisms involved in host shifts.

For example, in order for a strongly host-specific insect to shift to a new, chemically distinct sympatric host, several simultaneous co-adapted mutations would have to appear together in a single pair of individuals. The loci involved would include (Huettel and Bush, 1972): (1) a gene controlling perference by ovipositing females, (2) a gene to allow tolerance of new host by larvae, and (3) a gene to allow larval gustatory responses to host chemistry. Several other conditions might be expected to enhance a radical host-shift: (1) occasional ovipositional "mistakes" by females, (2) male courtship around the individual host plant from which they emerge, (3) individual host plant large relative to insect so that several sibs might emerge on same plant. Items 2 and 3 above would help ensure the maintenance of rare triple mutants necessary to invade new host.

On the basis of these simple assumptions, I compared the variety of host plants used by groups of closely related butterfly species that court and mate independent of host-plant patches with those where courtship and mating are structured around individual or clumps of individual host plants. The prediction was that radical host shifts (i.e., to chemically distinct plants) should be more common in taxa containing these latter kinds of courtship behaviors.

Figure 1 shows these results. Above the line are species that mate independently of host-plant clumps. All of these species belong to higher taxa that are characterized by strong affinity for particular host families. Below the line are groups with many species for which courtship is structured around host-plant clumps. Note the variety of host families used by each genus. Individual species are strongly host-specific, yet butterfly taxonomy here does not predict host plant used by species. Such departures from the pattern emphasized by Ehrlich and Raven is rather significant, since the Lycaenidae, which accounts for about half of all butterfly diversity, includes many genera within which radical host shifts have been common.

I have concluded, based upon this study, that the physical distance and physiological (=chemical) distances between host-plant patches are not equivalent and will in fact interact in a complex way. For example, it appears that for butterflies the probability that oviposition mistakes will occur increases as time spent away from normal host plant increases. Thus the probability of oviposition on a new host is expected to go up as ovipositing females disperse away from concentrations of the preferred plant (Singer, personal communication). Conversely, the probability of actually encountering a particular new plant decreases as physical distance between it and concentrations of the normal host increase. In the event that eggs are laid and mutant larvae develop on a new host, the probability that any gene complexes enhancing the shift will be

	Selaginellaceae	Cycadaceae	Pinaceae	Cupressaceae	Liliaceae	Commelinaceae	Gramineae	Lauraceae	Anonnaceae	Monimiaceae	Cruciferae	Capparidaceae	Flacourtiaceae	Violaceae	Passifloraceae	Tiliaceae
					2	2	5	1	1	1	2	2	4	4	4	8
Pap. Parides																
Pier. Pieris											X					
Itaballia												X				
Nym. Heliconius															X	
Speyeria														X		
Danaus																
Tithorea																
Ithomia																
Euptychia	3						X									
Acraea						1								1	13	5
Charaxes							1	1	1				3			
Anaea								1		1						1
Adelpha																
Neptis																
Lyc. Lycaena																
Philotes																
Callophrys			3	4	1											
Strymoniini		1	3	4	1											
Virachola																
Aloeides																

	Euphorbiaceae	Linaceae	Celastraceae	Melianthaceae	Sapindaceae	Rhamnaceae	Urticaceae	Moraceae	Ulmaceae	Piperaceae	Polygonaceae	Ericaceae	Asclepiadaceae	Apocynaceae	Convolvulaceae	Solanaceae	Verbenaceae
	9	10	13	13	13	14	18	18	18	19	20	23	26	26	27	27	28
Pap. Parides																	
Pier. Pieris																	
Itaballia																	
Nym. Heliconius																	
Speyeria																	
Danaus													X				
Tithorea														X			
Ithomia																X	
Euptychia																	
Acraea							9	2							1		
Charaxes	1	1	1	2	2	1						1			1		
Anaea										1							
Adelpha								1	1								
Neptis	3				3				2							1	1
Lyc. Lycaena						1				1	9	1					
Philotes											7						
Callophrys											3	3					
Strymoniini	1				2				1		3	3				1	1
Virachola																	
Aloeides																	

	Connoraceae	Rosaceae	Leguminosae	Crassulaceae	Myrtaceae	Melastomaceae	Loranthaceae	Rubiaceae	Compositae	Salicaceae	Fagaceae	Aristolochiaceae	Salvadoraceae	Santalaceae
	31	31	31	31	36	36	39	44	46	47	52	35		39
Pap. Parides												X		
Pier. Pieris														
Itaballia														
Nym. Heliconius														
Speyeria														
Danaus														
Tithorea														
Ithomia														
Euptychia														
Acraea	1								1					
Charaxes			22		2								1	
Anaea			1											
Adelpha						1								
Neptis			1											
Lyc. Lycaena		2												
Philotes				1										
Callophrys		1					2							
Strymoniini		4		1			3			2	6			
Virachola			4	1			1	1	1					
Aloeides			X											

Patterns of host plant use within selected butterfly genera. X means essentially all members of the genus specialize on indicated plant family. Numbers indicate number of apparently monophagous species from a genus feeding on particular host families.

Fig. 1. Host shifts as function of male courtship behavior in butterflies. Above the line are genera that court and mate independent of host plants. Below the line are tribes or genera in which individual host plants may be used as landmarks about which mate-seeking behavior is structured. See text for further discussion (from Gilbert, in preparation).

maintained by inbreeding increases with distance from concentrations of the normal host. Therefore, at a given chemical distance (measured as number of mutations required for colonization) between old host and new host, there should be some optimal physical distance between host patches such that the probability of oviposition on the new host is high and of back-crossing to the source population is relatively low. This optimal distance *increases* as plants become more dissimilar chemically and *decreases* as butterfly courtship tactics increase the effective separation of old vs. new hosts. Because of their courtship tactics, groups below the line in figure 1 are more likely to *evolutionarily* colonize new host islands that are sympatric with old host islands.

Feeny (1975) and Janzen (1968) have indicated in broad terms the kinds of ecological changes that will influence the probability of evolutionary colonization. The decrease of the natural host of a species due to agriculture or to its extinction due to disease can bring the host-patch sizes and spacing patterns into the appropriate relationships for host shifts by particular host-specialist insects. Butterflies below the line in figure 1 should shift to radically different hosts at a much earlier stage of such host changes than those above the line. It is reassuring that two of the first host-specific colonists of *Eucalyptus* in Brazil are butterflies of the family Lycaenidae (Anonymous, 1968).

Previous analyses of host-plant islands have dealt at the very coarse level of over-all host range. The fact that host-plant islands of both the same and different "archipelagos" may or may not overlap in space has not been worked into the theory of this field. The complications that host islands can occur on real islands, that colonist species of equal dispersal abilities can vary greatly in digestive tactics, that host islands may vary in chemical quality and in the spatial texture (plant-clump size and spacing) have all been ignored in analyses of real data.

In summary, I believe that the complexities of insect plant interactions have been inadequately incorporated into attempts to test general island theory. Furthermore, I do not expect that gross level compositions are likely to reveal much about how insects colonize plants. At the very minimum we will need to consider plants other than trees before any degree of generality can be claimed. Finally, one obvious simplification of the original MacArthur-Wilson theory is the lack of any inter-trophic level interaction component to the immigration function. Heatwole and Levins (1973) have made progress in this direction. Yet, since the bulk of diversity on almost any island will ultimately be insects and host plants, general island theory will be improved immensely by an understanding of the host-plant islands within real islands.

3. *Plant Apparency and Chemical Defense*

Simple yet powerful generalities are extremely rare in science. It is therefore both reassuring and exciting that two (initially) very different lines of research and thinking have led to the same basic theory of plant defensive chemistry. The theory is elegantly stated by Feeny (1975, 1976) using terms that will probably become permanent fixtures of the field.

Starting from a very different point of view, Orians and his colleagues (Orians et al., 1975; Cates and Orians, 1975; Rhoades and Cates, 1976) have independently arrived at the same point almost simultaneously.

Feeny's ideas emerged, first, from empirical observations on the differences between chemical defenses of oak trees (Feeny, 1970) and mustard plants (Erickson and Feeny, 1974) and, second, from observed differences in the digestive strategies of specialist vs. generalist insects (Krieger et al., 1971). He developed the concepts of: (1) *quantitative defenses*, i.e., chemicals such as tannins that reduce digestability, act according to their concentration in leaves, and are costly to produce; (2) *qualitative defenses*, i.e., chemicals such as alkaloids whose activities are specific, that are effective in low concentration, and are less expensive from a plant point of view; (3) *apparent plants*, i.e., plants that because of their size and/or longevity are "bound to be found" by herbivores and use quantitative defenses to drastically reduce herbivore efficiency at removing leaves; and (4) *unapparent plants*, i.e., plants that, because they can escape in time and space from herbivores and because they typically occur in early succession where competition for light between plants is intense, are selected to devote most of their energy and matter to growth and so employ "inexpensive," qualitative toxins.

Feeny reasoned that specialists should generally feed on toxic, unapparent plants because a detoxification of a specific compound would require specialized digestive machinery. For generalists, on the other hand, the slower growth rates, which correlate with the processing of quantitatively defended leaves, are compensated by a larger, more constant resource base.

Orians began with the basic allocation theory approach that he has advocated for ecological analyses generally (Orians, 1975). This approach started with the idea that early-succession plants, since they escape in time and space, would be selected to devote most of their energy budget to rapid growth and little if any to herbivore deterrents. Conversely, late-succesion plants (trees) cannot escape herbivory and thus should be highly toxic.

Thus, the prediction was made that specialists should be associated with late-succession, and generalists with early-succession, plants. Unfortunately, a reexamination of the basic assumptions of this hypothesis was delayed because the initial tests (Cates and Orians, 1975) were carried out with banana slugs, herbivores that indeed generalize on herbaceous plants but that, like tiger moths, probably represent an atypical group of toxic plant generalists.

In retrospect (see Rhoades and Cates, 1976), the Orians/Cates/Rhoades theory becomes identical to that of Feeny when it is realized that even patchy and ephemeral plants are bound to be discovered by generalist herbivores and that the resulting selection on early-succession plants will lead to low-cost toxins that exclude generalists. Conversely, an easy-to-find, permanent plant would be ideal for all insects, and therefore, given that more resources can be allocated to defense, leaves can be toughened and packed with digestion-reducing compounds that act on all herbivores. Specialist insects often use toxic compounds as attractants, so there is obvious selection for interspecific chemical aspect diversity in early-succession plants that in turn selects for further specialization. On the other hand, there may be less diversity inherent in the kinds of defenses optimum for the quantitative defense of late-succession plants, and more plant species will be effectively identical from an insect's point of view. Associated insects are thus "generalist" with respect to plant taxa eaten.

One just beginning to read in this subject may be slightly confused by papers (Otte, 1975; Futuyma, 1976) testing the original version of Orians's hypothesis. The data of Otte (1975), for example, showed that grasshoppers find a wider range of late-succession plants acceptable than they do of early-succession plants. Given our current understanding, this is clearly the predicted result of both the Feeny (1976) theory and the updated Orians theory (Rhoades and Cates, 1976).

The plant apparency theory applies to plant parts as well as to entire plants. Thus, the unapparent new leaves of a shrub may be defended by toxin, the more apparent mature leaves by toughness and/or tannins, and so on. This may account for the observed toxic quality of young but not old leaves of some *Passiflora* vines to heliconiine larvae (Benson, Brown, and Gilbert, 1976).

I have previously suggested (Gilbert, 1975) that the increased behavioral sophistication of *Heliconius* butterflies resulting from selection for more-efficient adult foraging for pollen, a major factor in egg production (Gilbert, 1972), has important consequences for larval host plants. To express my previous argument in Feeny's terms, increased adult foraging ability effectively *increases* the apparency of *Passiflora* vines. Since *Heliconius*, like other specialists on toxic plants, are likely to be highly successful at circumventing any additional chemical defense evolved to meet the new challenge, non-chemical means of decreasing the efficiency of these important herbivores are to be expected. For example, some *Passifloras* have hooked trichomes that effectively exclude many heliconiines (Gilbert, 1971). Morphological aspect (leaf-shape) diversity and convergence on the appearances of other common plants may be other important responses (Gilbert, 1975). Other traits of *Passiflora* probably evolved in response to increased apparency to *Heliconius* are: (1) mimetic eggs that reduce oviposition rates (Benson, Brown, and Gilbert, 1976); (2) stipules that resemble tendrils and then are shed with the eggs of species that oviposit on new tendrils (stipules that bear egg mimics are not deciduous); (3) leaves that are shed in response to slight larval damage (Singer, personal communication); (4) extrafloral nectaries that attract mutualist defenders, a broad-spectrum defense similar to broad-spectrum chemical defense (Gilbert, 1971); (5) tough older leaves (*P. coriaceae*); and (6) digestion-reducing compounds in older leaves (*P. ambigua*). Thus, the elaboration of another level of diverse defense mechanisms by *Passiflora* has created additional opportunity for life-history variations among coexisting heliconiines that helps account for their local species diversity (Gilbert, 1975). More generally, the fact that insects are active year-around in wet tropical areas would tend to render unapparent plants (and plant parts) relatively more apparent. The result for early-succession plants would be a diversification of non-chemical defenses as described (though *Passiflora* may be the extreme example of this).

Futuyma (1976) hints at the possibility that rare tropical trees may be sufficiently unapparent so that diverse qualitative toxins may be the optimum defense against leaf-feeding insects. I would suggest that even the new leaves of a very rare rain forest tree would be more apparent than the new leaves of a temperate zone oak. This is an outcome of the fact that new leaves of oaks in a seasonal environment can effectively use escape in time (Feeny, 1976), whereas rain forest trees are constantly exposed to herbivorous insects for which seasonal constraints on adult reproductive longevity are practically nonexistent. I will make a wild guess and say that, when all the factors are considered, the new growth of a tropical rain forest tree will have about the same relative apparency, the same type of chemical defense, and the same predominance of coevolved specialists (see Gilbert, 1977) that characterizes temperate zone herbaceous plants.

What about the mature leaves of tropical trees? In these I would predict less commitment to quantitative defenses like tannins than in temperate trees because the competitive advantage of rapid growth should extend further into succession in high-productivity tropical environments. Likewise, in such environments, insect herbivores can be made more apparent to their predators, which are also constantly available, by the plant's investing in extrafloral nectar production (rather than in tannins). Conversely, in rain forests where productivity is severely limited by low-nutrient soils, levels of tannins and other quantitative chemcials in mature leaves would be expected to be more comparable to the concentrations in leaves of temperate zone oaks (see Janzen, 1974b).

Although the theory of apparency is likely to undergo further modification as we obtain a better picture of the kinds of plant defenses and herbivore digestive and bionomic strategies in diverse environments, I agree with Feeny (1976) that it will stand the test of time. Furthermore, the theory helps explain some interesting adaptive radiations among phytophagous insects. For example, extending the theory to another trophic level, we see that adoption of the digestive tactics necesary to be a generalist (e.g., mixed-function polysomal oxidases) may eliminate the possibility of sequestering plant compounds for the insects' defenses against birds (Alpin et al., 1975; Brower and Glazier, 1975; Brower et al., 1975; Rothschild et al., 1975). Such insects are edible and are selected to reduce their apparency (Feeny, 1976; Rothschild, 1972). For example, moths tend to be cryptic, rest during the day and fly at night. Most moths are also grass or tree feeders. Many butterflies, on the other hand, feed on early-succession plants, and all of the major unpalatable, warningly colored groups are specialists on toxin-rich herbs, shrubs, or vines that occur in early- to mid-succession vegetation (see Brower and Brower, 1964). Day-flying moths tend to be warningly colored or mimetic and feed on herbaceous vegetation. Butterflies that feed on trees or grasses (both quantitatively defended) tend to be edible and cryptic, whereas highly apparent butterflies are either mimics or fast and difficult to capture. Becoming apparent in the Lepidoptera is associated primarily with exploitation of diurnal floral resources. See Gilbert and Singer (1975) for a further discussion of this topic.

Professor Lewontin began this volume by severely questioning the use of optimization arguments as a tool for developing ecological theory. He stressed the importance of understanding the natural history of a system before such methods can safely be applied, at which point, he argues, they are no longer necessary. As an insect/plant natural historian, I certainly feel that much of the confusion in the literature on plant defensive strategies has resulted from "general" ecologists racing in to test general theory on organisms for which they have little intuition. However, Orians has proved to me that even with initial bad assumptions (about the nature of plant chemical defenses and herbivore digestive tactics) and some rabbit-chasing at the outset, his approach to generating prediction leads to empirical studies that later correctly revise the theory. I am convinced that had the young Feeny chosen another profession, Orians and colleagues would have eventually developed the apparency theory,

TABLE 2

Ecological Correlates of Plant Defensive Chemistry

Apparent	Unapparent
Predictable constant	Unpredictable ephemeral, patchy
Late succession woody perennial	Early succession herbaceous
Quantitative defense	Qualitative defense
Excludes specialists	Excludes generalists
Eaten by generalists	Eaten by specialists
Forest entomology	Agricultural entomology

as we now have it, on their own. Even so, the rate of progress toward a satisfactory theory using this approach is directly proportional to the sophistication of our natural history knowledge. Conversely, one's perception of nature is sharpened by the use of evolutionary hypotheses. In the analysis of the plant-insect interface, continued interplay between descriptive, theoretical, and experimental studies is vital.

Tables 2 and 3 summarize plant defensive strategies and insect digestive tactics discussed in this section.

RESOURCE PREDICTABILITY VS. EVOLUTIONARY STRATEGIES OF INSECTS

A fourth major area of theory focuses upon the way in which environmental predictability (expressed in terms of host plant availability) should influence the evolution of insect feeding and reproductive tactics.

Host Specialization

Using host plant family records for Lepidoptera, Futuyma (1976) has evaluated the "conventional wisdom" that unpredictable resource spectra select for generalist feeding habits. Measuring generality of feeding as a function of number of plant families utilized, he shows that, contrary to expectation, groups using predictable plants tended to be generalists and those using unpredictable plants tended to be specialists. The take-home message of this paper is that specialization defined at the level of host-plant families is governed by interactions with host-plant chemistry (Brues, 1926; Ehrlich and Raven, 1965; Feeny, 1975) rather than by competitive interactions or relative abundances of host families.

However, *within* a host family are numerous relatively similar taxa from a chemical point of view. Thus, it might be expected that "conventional wisdom" might apply to patterns at this lower level. For example, consider the "x" in figure 1, which indicates that all heliconiines are specialists on Passifloraceae. Now realize that this "x" stands for at least 383 known interactions between some 66 heliconiines and 123 Passifloraceae species

TABLE 3

FEEDING TACTICS OF PHYTOPHAGOUS INSECTS

Generalist	Specialist
Eat plants of diverse taxa, chemistries	Eat plants of one or few taxa of similar chemistry
Eat any plant lacking repellents	Respond to plants with particular attractants
Mixed-function polysomal oxidases	Specific excretory or sequestering mechanisms
Usually unapparent to predators	Often apparent to predators

(Benson, Brown, and Gilbert, 1976: Appendix table, pp. 678–79) over a vast region from southern South America to Texas.

If you next examine this latter table, you will see, for example, that *Heliconius melpomene* is recorded from 31 *Passiflora* species from about fifteen areas. *Heliconius cydne* is recorded from 8 *Passiflora* species from 4 areas. What can we conclude from these much more detailed data? Nothing! for one thing, we have abundant evidence for interpopulation variation in host use for *Heliconius* species as well as for other butterflies (Gilbert and Singer, 1975). Thus, species with many hosts recorded over their ranges may either be generalists or extreme specialists with ecogenotypic variation over their range. The extreme specialization by some *H. melpomene* populations (see below) would not be obvious even from our detailed breakdown. Furthermore, within geographical areas (e.g., "Costa Rica") used in Benson, Brown, and Gilbert (1976), there is much local variation in *Passiflora* community composition and in heliconiine species present.

Consequently, to test ideas about host specialization *within* the set of chemically suitable host species available, we must deal on the actual scale at which hosts are encountered by ovipositing female *Heliconius*. We know this to be extremely local (Ehrlich and Gilbert, 1973); thus, it would seem that for these insects we must examine the details of within-habitat interactions in order for host records to make any sense in evaluating current ecological theory. As Futuyma (1976) laments, host-plant records are not taken in a way that makes them of much use for testing ecological theory.

Two lines of evidence from *Heliconius* community studies confirm that if Lepidoptera-host plant relationships are examined at a level of organization analagous to a gleaning guild of birds, then the predictions of general theory seem to apply to the problem of specialization on food resources. First, if one examines in detail a local guild that feeds on the new growth of *Passiflora*, then, indeed, *Heliconius* species that encounter an unpredictable, low-density set of *Passiflora* species tend to be generalists, and those that encounter a more-abundant and predictable supply of *Passiflora* species tend to specialize. For example, John Smiley has found in our Costa Rican study site that *H. cydno*, primarily a forest species, encounters a wide variety of adult and seedling *Passiflora* but at a low rate with no species highly predictable. Smiley has elegantly shown that *H. cydno* oviposit on over 90 percent of available *Passiflora* species and that the larvae do equally well on all host species (fig. 2). But in the same study, he has shown that *H. melpomene*, which chooses to fly with its model *H. erato* along edges and in secondary vegetation, oviposits only on *P. oerstedii* even though most other *Passiflora* species are commonly encountered in its microhabitat. Since *H. melpomene* larvae grow on other available *Passiflora* with equal speed and efficiency (fig. 2), host restriction is due to ecological factors rather than chemical differences between hosts. It

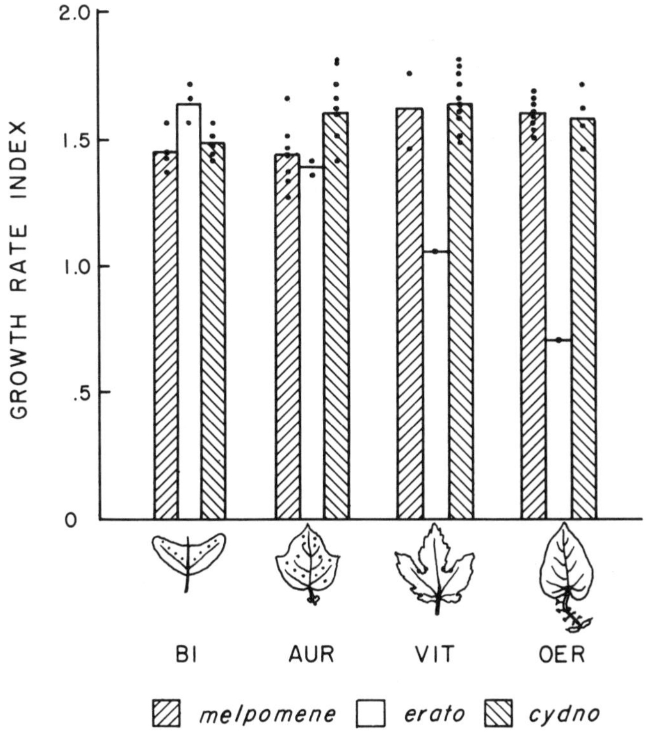

Fig. 2. Growth-rate indexes for larvae of three *Heliconius* species on each of four sympatric *Passiflora* species (*biflora, auriculata, vitafolia* and *oerstedii*). J. Smiley's (unpublished) data shows that specialization by *H. melpomene* on *P. oerstedii* is ultimately due to ecological factors, and *H. erato*'s avoidance of certain *Passiflora* is likely due to chemical differences between species (see text for further discussion). Smiley's growth rate index (G) averages relative size increase per instar over the 5 instars of larval development, and it corrects for species differences in egg weight and final pupal weight. Thus,

$$G = \frac{I}{D}$$

where I is the within instar relative size increase, approximated by the fifth root of the ratio of pupal weight to egg weight minus 1, and D is the average days per instar, i.e., the number of days from egg hatching to pupation divided by the number of instars (5). G allows direct comparison of the relative success of different butterfly species in minimizing the time to pupation (thus reducing exposure to predation) while maximizing final pupal weight (thus increasing reproductive potential) on the various available host species. This index is therefore a more ecologically relevant measure of host-plant suitability than are simple measures of conversion efficiency.

should be stressed however that this "realized niche" is enforced by an evolved chemical preference for *P. oerstedii* by female *H. melpomene*. Smiley's study further indicates a possible reason for the high desirability of *P. oerstedii*. It ranks lowest of all sympatric nectary-bearing *Passiflora* in the activity levels of predaceous ants (this must be a long-standing condition given *melpomene's* evolved preference).

Benson (1977) in an important study has examined local *Heliconius* communities from a broader geographical perspective. He compares average host-breadths of heliconiines between climatically predictable (Trinidad/Costa Rica) and climatically less predictable (Rio de Janeiro) habitats. As theory would predict if resource predictability is governed by climate, the number of hosts per species increases by a factor of 2 in the less predictable habitat. Benson also details the partitioning of *Passiflora* by the various competing heliconiines, and it is clear that, from the standpoint of ecological analysis, plant part utilized can be a major niche dimension to consider (see also Gilbert and Singer, 1975).

At a very coarse level of comparison, Scriber (1973) gave evidence that species of papilionid butterflies are more specialized in tropical than in temperate regions. This result probably reflects the increasing toxicity and greater inter-taxa chemical diversity of trees that are expected from the apparency theory (see above), combined with the high diversity of *Aristolochia* specialists in the tropics. Since Scriber lumps several different feeding "guilds" of Papilionidae, these results cannot in my opinion be safely applied to evaluating ecological theory.

I propose that the term "ecological monophagy" be used to refer to an insect's restriction to one particular host when other potential hosts are available but not used due to ecological rather than chemical/nutritional differences among them. Important examples of this category of specialists are *Euphydryas editha* (Singer, 1971) and, of course, Smiley's *H. melpomene* example. Other probable cases are discussed by Gilbert and Singer (1975) and Wicklund (1974, 1975).

Likewise, the general restriction of *heliconius* to *Passiflora* or of any species of insect to a species of plant because of larval digestive specializations to host chemistry should be termed "coevolved monophagy." Obviously, to accurately analyze community interactions among phytophagous insects, it will be important to understand what aspects of feeding specialization are due to coevolution with host and what aspects are due to behavioral and microevolutionary responses to ecological differences among resources (see Gilbert and Singer, 1975).

Incidently, the *melpomene-cydno* dichotomy elucidates the assumptions about insect-plant interaction upon which Levins and MacArthur (1969) based their theory of monophagy. They were implicitly concerned with monophagy at a plant family level, yet carried out a strategic analysis more appropriate to the type of "ecological monophagy" developed by *melpomene* in Costa Rica. Tiger moths may be one of the few lepidopteran groups that, because they (as species) are capable of eating a wide range of toxic plants (see Cook, 1961), might be approximately analyzed by the Levins-MacArthur theory.

Bionomic Strategies

Recent attempts to extend modern ecological generalities concerning life history strategies to phytophagous insects (see Southwood et al., 1974) have resulted in an equally confused literature. The difficulties result from:

1. an inadequate knowledge of the resource ecology of phytophagous insects;
2. inadequate consideration of the impact of complex life cycles, and, in particular, adult vs. larval requirements;
3. a failure to recognize that particular reproductive tactics (e.g., egg size, clutch size) may be evolved in response to host traits (defenses, density, etc.) rather than to some gross measure of habitat predictibility.

Feeny (1976) near the end of his excellent paper makes the generality that ephemeral plants impose ephemeral, ''r-selected'' life histories on many associated insects, whereas more permanent ''apparent'' plants impose slow growth, higher predation, and thus more ''K-selected'' life histories on their insect herbivores. A consideration of heliconiines indicates precisely the opposite conclusion. Species that utilize mature, ''apparent'' foliage tend to lay many tiny eggs individually (*Eueides* spp.) or to oviposit in large clutches on older foliage (*Dione juno*, *Heliconius doris*). Such species are relatively ''r-selected'' when compared with *Heliconius* species that utilize the more ephemeral, unapparent meristematic regions of *Passiflora*, and are constantly reproductive at a lower rate over a longer adult life (Gilbert, 1972).

The original application of the ''r- and K-selection'' concept (see Pianka, 1970) was primarily aimed at drawing broad comparisons between vastly different organisms. Moreover, entire suites of life history traits are considered in concert in these original comparisons. It is impossible to pronounce one *Papilio* species more ''K-selected'' than another on the basis of just one of a few larval traits, as Feeny does. Considering the entire life histories of tree-feeding vs. herb-feeding swallowtails, I see no clear-cut r-K dichotomy. In fact, my own observations of adult Paplionidae had led me to the subjective impression that, if anything, herb and vine feeders are more ''K-selected'' than tree-feeding species. Because adult butterflies are the stage of dispersal and reproduction, any simple categorization that in essence summarizes the life history strategy of a species must stress the traits of the adult stage (I thus see things quite differently from Southwood, 1976). The larval stage is involved only to the extent that its feeding tactics may constrain the adult's total energy and matter budget. Depending upon long-term ecological circumstances, adults may evolutionarily "choose" to a variable degree to put energy and matter derived from the larval stage into the reproductive (vs. the somatic maintenance) part of its budget. Within a given type of larval strategy, the over-all life history may vary from "r" to "K" depending on how the adult uses its resources. In particular, if high quality, predictable, adult resources are available, the adult has the evolutionary option of extending its reproduction over a more-extended life-span and expending more time and energy on "parental care" (in butterflies, this translates to "careful egg placement"). In *Heliconius*, the increased foraging capabilities and the extended adult life-span, which result ultimately from pollen feeding (Gilbert, 1975), increase the effective availability of even the most ephemeral and rare *Passiflora* sp. or parts of *Passiflora* sp. Consequently, life histories of various *Heliconius* may or may not follow predictions based on subjective assessment of plant predictibility.

Another possible pitfall is the assumption that resource availability may be more predictable in forest (permanent) as opposed to successional (temporary) habitats. Thus, if we predicted that the forest-restricted *Heliconius cydno* would lay larger eggs than would the more "r-selected" early-succession species *H. melpomene*, we would be correct. However, we would be correct not because large eggs are part of a "K-selected" strategy evolved in the predictable forest habitat but because, to the contrary, *Passiflora* vines in the forest are rare and more difficult to find relative to those in more successional habitats nearby (Smiley, unpublished). Smiley concludes that since plants may be discovered at a low rate relative to oviposition capabilities, and since the larval development period is a decreasing function of egg size, natural selection will favor larger eggs in species that use these less-predictable resources. Singer (personal communication) has found similar egg weight and fecundity gradients in tropical grass-feeding satyrine butterflies: i.e., forest species lay larger and fewer eggs than do early-succession species. However, the "K" tactics of the forest species are due to the unpredictable nature of those particular resources important to satyrines, not to the over-all predictability of the forest habitat, as theory would predict. It just happens that for satyrines, heliconiines, and perhaps other forest groups that do not feed on trees, the gradient of resource predictability runs opposite to that of the forest environment viewed as a whole.

Detailed studies on tree-feeding butterflies indicate equally complex relationships between resource ecology and life history strategies. C. Jordan (unpublished data) has studied pierid butterfly species whose larvae feed on the Mexican capparidaceous tree *Capparis frondosa*. In Tamaulipas, this small understory tree is utilized by five different pierid species, no two of which share identical dispersal, host range, or reproductive tactics. Jordan's study helps reemphasize the need for higher resolution in data for testing life-history theory in phytophagous insects. For example, consider two of the *Morisonia* "guild," *Itaballia viardi* and *Ascia josephina*. If we compare the various traits of these two species, it is very difficult to designate one as the more "r-selected" over-all. Jordan has found that *I. viardi* oviposits only on buds, and larvae can only utilize young leaves. Its resources are thus more unpredictable than are the mature leaves utilized by *A. josephina*. *I. viardi* is smaller as an adult and produces a smaller egg (.25 mg) than does *A. josephina*, which is roughly 2.5 times larger and produces an egg which is six times the size of an *I. viardi* egg (1.41 mg). Although *I. viardi* would thus be rated the more "r-selected" species on body-size and egg-size criteria, it is highly sedentary, more host restricted, probably more host-plant limited, probably longer lived, has more constant population size, and its females mimic *Heliconius charitonia*. Thus, on these criteria (dispersal, host specialization, longevity, predator avoidance, etc.), one would be inclined to rate *I. viardi* as the more "K-selected" of the two species. In the absence of fecundity data, I favor this latter conclusion, since egg size is almost certainly related to host toughness rather than to host predictability. Jordan has shown that the leaves eaten by *A. josephina* (large eggs) require more than seven times the force to shear than do the tender leaves used by *I. viardi* (small eggs). These observations are consistent with the fact that large eggs in butterflies often endow first instar larvae with larger, more powerful heads and mandibles. Therefore, *A. josephina* may well be producing the smallest egg possible given its host quality. Conversely, eggs one-half the size of those produced by *I. viardi* would undoubtedly survive on the new leaves. *I. viardi* is apparently selected to produce larger, *not smaller*, eggs (and thus more rapid attainment of maturity) by the temporary nature of the young leaves!

It is not my intention to attack the procedure of testing ecological generalities on phytophagous insects. I merely wish to emphasize that the few phytophagous insect-host plant systems analyzed at the population level reveal important detail and complexity that must be considered in interpretation of tests of theory based upon relatively crude estimates of resource availability and insect life-history tactics. Both with respect to feeding specialization and life-history strategies of herbivorous insects, viewing resource patterns through the eyes of the insect seems the best way to ensure that we are testing hypotheses at the appropriate level in ecological or evolutionary analysis. From the limited cases available, it seems important to make tests of general theory within guilds defined by host-plant preferences, between local habitats within which actual resource patterns are known, and with the over-all life cycle of the insect fully considered.

CONCLUDING REMARKS

I have tried to show that the theories that seem most promising in organizing the ecological analysis of insect-plant interactions are modifications of the basic theories of population biology. These theories are not as bound to particular idealized life histories as are the standard approaches to testing these theories. I have used examples from butterflies to show that appropriate tests of hypotheses about phytophagous insects require careful consideration of the taxonomic and ecological levels of organization from which data is gathered. Unfortunately, most of the descriptive data on host relations were not collected with such hypotheses in mind and are of marginal value at best, misleading at worst.

At the beginning of this paper, I briefly described the overwhelming species and life history diversity represented among phytophagous insects. Unfortunately, my own perspective, and that of the literature as a whole, is biased by the disproportionately well-developed taxonomic and host-plant information available for Lepidoptera. To test general theories about phytophagous insects, we need much additional life history and ecological work on other major herbivore groups in natural rather than agricultural settings. Studies such as those of Isely (1946), Williams (1954), and Ueckert and Hansen (1971) on Orthoptera; Le Quesne (1972), Janzen (1972b), and Root and Chaplin (1976) on Hemiptera; Knerer and Atwood (1973) on diprionid sawflies; and Rockwood (1974), Rathcke (1976), and Strong (1976) on Coleoptera, provide progress in that direction.

Further advance in the understanding of insect-plant interactions will also depend upon the ability of mathematicians to incorporate the complexities of insect life histories into general models, the ability of physiologists to elucidate the digestive mechanisms of phytophagous insects (see Beck and Reese, 1976), and a shift in interest by phytochemists from systematic and phylogenetic questions to ecological questions requiring focus on intra-community and intra-species patterns of plant defensive chemistry.

This paper has only superficially sketched the boundaries of a broad field that still undoubtedly will be an exciting frontier of ecology when our great grandchildren begin to ask about butterflies.

ACKNOWLEDGMENTS

I thank P. Atsatt, W. Benson, R. Cates, P. Feeny, D. Futuyma, B. Heinrich, C. Jordan, D. Levin, R. Root, J. Smiley, and D. Strong for giving me access to their unpublished data and/or manuscripts. Discussions over the years with W. Benson, G. Bush, P. Ehrlich, D. Janzen, B. Heinrich, P. Morrow, G. Orians, M. Singer, J. Smiley, and many others have helped to shape my thinking about insects and plants. Hopefully I have done justice to their contributions. Several persons kindly read parts of this manuscript and made constructive comments.

These included C. Boggs, D. Feener, A. Joern, C. Jordan, P. Levin, and J. Smiley. The work described has been supported by N.S.F. grants GB4074X-P and GB39270, University of Texas U.R.I. funds, and most recently from my own pocket. I would like to thank my wife, Christine, for typing the manuscript and for tolerating me during its production.

1. In this paper I include as phytophagous any insect that feeds on plant parts, propagules, or secretions.

REFERENCES

Alpin, R. T., R. d'Aray Ward, and M. Rothschild. 1975. Examination of the large white and small white butterflies (*Pieris* spp.) for the presence of mustard oils and mustard oil glycosides. J. Ent. 50:73–78.

Anonymous. 1968. Report of Servicio Forestal da Cia Siderugica Belgo—Mineira, July 1968 (Brazil).

Atsatt, P. R., and D. J. O'Dowd. 1976. Plant defense guilds. Science 193:24–29.

Baker, H. G., and I. Baker. 1975. Studies of nectar-constitution and pollinator-plant coevolution. *In* L. E. Gilbert and P. H. Raven (eds.), Coevolution of animals and plants. University of Texas Press, Austin and London.

Barras, S. J. 1975. Release of fungi from mycangia of southern pine beetles observed under a scanning electron microscope. A. Ang. Ent. 79:173–76.

Beattie, A. J. 1971. Pollination mechanisms in *Viola*. New Phytologist 70:343–60.

Beattie, A. J., and N. Lyons. 1975. Seed dispersal in *Viola* (Violaceae): adaptations and strategies. Amer. J. Bot. 62:714–22.

Beck, S. D., and J. C. Reese. 1976. Insect-plant interactions: nutrition and metabolism. *In* J. Wallace and R. Mansell (eds.), "Biochemical interactions between plants and insects." Recent Advances in Phytochemistry 10:1–40.

Benson, W. W. 1976. Resource partitioning by passion vine butterflies. Unpublished.

Benson, W. W., K. S. Brown, Jr., and L. E. Gilbert. 1976. Coevolution of plants and herbivores: passion flower butterflies. Evolution 29:659–80.

Bentley, B. L. 1976. Plants bearing extra floral nectaries and the associated ant community: inter habitat differences in the reduction of herbivore damage. Ecology 57:815–20.

Berg, R. Y. 1975. Myrmecochorous plants in Australia and their dispersal by ants. Aus. J. Bot. 23:475–508.

Birch, L. C. 1971. The role of environmental heterogenity and genetical heterogeneity in determining distribution and abundance. *In* P. J. den Boer and G. R. Gradwell (eds.), Dynamics of populations. Centre for Agricultural Publishing and Documentation Wageningen. Pp. 109–28.

Breedlove, D. E., and P. R. Ehrlich. 1971. Coevolution: patterns of legume predation by a lycaenid butterfly. Oecologia 10:99–104.

Brian, A. D. 1957. Differences in the flowers visited by four species of bumblebees and their causes. J. Anim. Ecol. 26:71–98.

Brower, L. P., and J. V. Z. Brower. 1964. Birds, butterflies, and plant poisons: a study in ecological chemistry. Zoologica 49:137–59.

Brower, L. P., J. V. Z. Brower, and J. M. Corvino. 1967. Plant poisons in a terrestrial food chain. Proc. Nat. Acad. Sci. USA 57:893.

Brower, L. P., and S. C. Glazier. 1975. Localization of heart poisons in the monarch butterfly. Science 188:19–25.

Brower, L. P., W. N. Ryerson, L. L. Coppinger, and S. C. Glazier. 1968. Ecological chemistry and the palatability spectrum. Science 161:1349–51.

Brower, L. P., M. Edmunds, and C. M. Moffitt. 1975. Cardenolide content and palatability of a population of *Danaus chrysippus* butterflies from West Africa. J. Ent. (A) 49:183–96.

Brues, C. T. 1924. The specificity of food-plants in the evolution of phytophagous insects. Amer. Natur. 58:127–44.

Brues, C. T. 1972. Insects, food, and ecology. Dover Publ. New York, 466 pp.

Bultman, J. D., and C. R. Southwell. 1976. Natural resistance of tropical American woods to terrestrial wood-destroying organisms. Biotropica 8:71–95.

Burdon, J. J., and G. A. Chilvers. 1974. Fungal and insect parasites contributing to niche differentiation in mixed stands of eucalypt saplings. Aus. J. Bot. 22:103–14.

Bush, C. L. 1975. Sympatric speciation in phytophagous parasitic insects. *In* P. W. Price (ed.), Evolutionary strategies of parasitic insects and mites. Pp. 187–206. Plenum Press, New York.

Cates, R. G., and G. H. Orians. 1975. Successional status and the palatability of plants to generalized herbivores. Ecol. 56:410–18.

Cavers, P. B. 1973. The effects on reproduction of removal of plant parts by natural or artificial means. *In* P. H. Dunn (ed.), Proc. second int'l. symp. on biol. control of weeds. Pp. 140–44. Misc. Publ. No. 6 Commonwealth Inst. Biol. Control.

Center, T. D., and C. D. Johnson. 1974. Coevolution of some seed beetles (Coleoptera: Bruchidae) and their hosts. Ecol. 55:1096–1103.

Cherrett, J. M. 1968. The foraging behavior of *Atta cephalotes* L. (Hymenoptera, Formicidae): foraging pattern and plant species attacked in a tropical rain forest. J. Animal Ecol. 37:387–403.

Connell, J. H. 1971. On the role of natural enemies in preventing competitive exclusion in some marine animals and in rain forest trees. *In* P. J. den Boer and G. R. Gradwell (eds.), Dynamics of populations. Centre for Agricultural Publishing and Documentation, Wageningen.

Cook, L. M. 1961. Food-plant specialization in the moth *Panaxia dominula* L. Evolution 15:478–85.

Cromartie, W. J., Jr. 1975. The effect of stand size and vegetational background on the colonization of cruciferous plants by herbivorous insects. J. Appl. Ecol. 12:517–33.

DeLong, D. M. 1965. Ecological aspects of North American leafhoppers and their role in agriculture. Bull. Entomol. Soc. Am. 11:9–26.

DeLong, D. M. 1971. The bionomics of leafhoppers. Ann. Rev. Entomol. 16:179–210.

Dethier, V. G. 1970. Chemical interactions between plants and insects. *In* E. Sondheimer and J. B. Simeone (eds.), Chemical ecology. Academic Press, New York.

Dolinger, P. M., P. R. Ehrlich, W. L. Fitch, and D. E. Breedlove. 1973. Alkaloid and predation patterns in Colorado lupine populations. Oecologia 13:191–204.

Dunn, P. H. (ed.). 1973. Proceedings of the second international symposium on biological control of weeds. Misc. Publ. No. 6 Commonwealth Institute of Biological Control (C. A. B.), 225 pp.

Eastop, V. F. 1973. Deductions from the present day host plants of aphids and related insects. *In* H. F. Van Emden (ed.), Insect/plant relationships. John Wiley & Sons, New York.

Ehrlich, P. R., and P. H. Raven. 1965. Butterflies and plants: a study in coevolution. Evolution 18:586–608.

Ehrlich, P. R., and L. E. Gilbert. 1973. Population structure and dynamics of the tropical butterfly *Heliconius ethilla*. Biotropica 5:69–82.

Erickson, J. M., and P. O. Feeny. 1974. Sinigrin: a chemical barrier to the black swallowtail butterfly, *Papilio polyxenes*. Ecology 55:103–11.

Feeny, P. O. 1970. Seasonal changes in oakleaf tannins and nutrients as a cause of spring feeding by winter moth caterpillars. Ecol. 51:656–81.

Feeny, P. O. 1976. Plant apparency and chemical defense. *In* J. Wallace and R. Mansell (eds.), Biochemical interactions between plants and insects. Recent Advances in Phytochemistry 10:1–40.

Feeny, P.O. 1975. Biochemical coevolution between plants and their insect herbivores. *In* L. E. Gilbert and P. R. Raven, eds. Coevolution of animals and plants. University of Texas Press, Austin and London.

Free, J. B. 1974. Insect pollination of crops. Academic Press, London, New York, 544 pp.

Freeland, W. J., and D. H. Janzen. 1974. Strategies in herbivory by mammals: the role of plant secondary compounds. Amer. Natur. 108:269–89.

Futuyma, D. J. 1976. Food plant specialization and environmental predictability in Lepidoptera. Amer. Natur. 110:285–92.

Gallun, R. L. 1972. Genetic interrelationships between host plants and insects. J. Environ. Qual. 1:259–65.

Gentry, A. H. 1974. Flowering phenology and diversity in tropical Bigoniaceae. Biotropica 6:64–68.

Gilbert, L. E. 1971. Butterfly-plant coevolution: has *Passiflora adenopoda* won the selectional race with heliconiine butterflies? Science 172:585–86.

Gilbert, L. E. 1972. Pollen feeding and the reproductive biology of *Heliconius* butterflies. Proc. Nat. Acad. Sci. USA 69:1403–7.

Gilbert, L. E. 1975. Ecological consequences of a coevolved mutualism between butterflies and plants. *In* L. E. Gilbert and P. H. Raven (eds.), Coevolution of animals and plants. University of Texas Press, Austin and London.

Gilbert, L. E. 1977. The role of insect-plant coevolution in the organization of ecosystems. *In* V. Labeyrie (ed.), Le comportement des insectes et les signaux issus du milieu trophique. C. N. R. S., Paris.

Gilbert, L. E., and M. C. Singer. 1973. Dispersal and gene flow in a butterfly species. Amer. Natur. 107:58–72.

Gilbert, L. E., and M. C. Singer. 1975. Butterfly ecology. Ann. Rev. Ecol. and Syst. 6:365–97.

Goeden, R. D., and D. W. Richer. 1976. The phytophagous insect fauna of the ragweed, *Ambrosia dumosa,* in Southern California. Environ. Entomol. 5:45–50.

Graham, S. A., and F. B. Knight. 1965. Principles of forest entomology. McGraw-Hill, New York.

Haines, B. 1975. Impact of leaf-cutting ants on vegetation development at Barro, Colorado Island. *In* F. B. Golley and E. Medina (eds.), Tropical ecological systems. Pp. 99–111. Springer-Verlag, New York.

Hanover, J. W. 1975. Physiology of tree resistance to insects. Ann. Rev. Ent. 20:75–95.

Harper, J. L. 1969. The role of predation in vegetational diversity. Brookhaven Symp. Biol. 22:48–62.

Harris, P. 1973. Insects in the population dynamics of plants. *In* H. F. Van Emden (ed.), Insect/plant relationships. John Wiley & Sons, New York.

Harris, P., and A. I. Mohyuddin. 1965. The bioassay of insect feeding tokens. Can. Entomol. 97:830–33.

Heatwole, H., and R. Levins. 1973. Biogeography of the Puerto Rico bank: species turn-over on a small sandy cay. Ecology 54:1042–55.

Heinrich, B. 1975a. The role of energetics in bumblebee-flower interrelationships. In L. E. Gilbert and P. H. Raven (eds.), Coevolution of animals and plants. University of Texas Press, Austin and London.

Heinrich, B. 1975b. Energetics of pollination. Ann. Rev. Ecol. and Syst. 6:139–70.

Heinrich, B. 1975c. Bee flowers: a hypothesis on flower variety and blooming times. Evolution 29:325–34.

Heinrich, B., and P. H. Raven. 1972. Energetics and pollination ecology. Science 176:597–602.

Hering, E. M. 1951. Biology of the leaf mines. W. Junk, The Hague, 420 pp.

Hocking, B. 1970. Insect associations with swollen thorn acacias. Trans. R. Ent. Soc. Lond. 122:211–55.

Hocking, B. 1975. Ant-plant mutualism: evolution and energy. *In* L. E. Gilbert and P. H. Raven (eds.), Coevolution of animals and plants. University of Texas Press, Austin and London.

Horn, H. S. 1974. The ecology of secondary succession. Ann Rev. Ecol. and Syst. 5:25–37.

Hoxie, R. P., S. G. Wellso, and J. A. Webster. 1975. Cereal leaf beetle response to wheat trichome length and density. Environ. Entomol. 4:365–70.

Huettel, M. D., and G. L. Bush. 1972. The genetics of host selection and its bearing on sympatric speciation in *Procecidochares* (Diptera, Tephritidae). Entomol. Exp. Appl. 15:465–80.

Huffaker, C. B. 1971. The phenomenon of predation and its role in nature. *In* P. J. den Boer and G. R. Gradwell (eds.), Dynamics of populations. Pp. 327–43. Centre for Agricultural Publishing and Documentation, Wageningen.

Isely, F. B. 1944. Correlation between mandibular morphology and food specificity in grasshoppers. Ann. Ent. Soc. Amer. 37:47–66.

Isely, F. B. 1946. Differential feeding in relation to local distribution of grasshoppers. Ecol. 27:128–38.

Janzen, D. H. 1966. Coevolution of mutualism between ants and acacias in Central America. Evolution 20:249–75.

Janzen, D. H. 1967. Synchronization of sexual reproduction of trees with the dry season in Central America. Evolution 21:620–37.

Janzen, D. H. 1968. Host plants as islands in evolutionary and contemporary time. Amer. Natur. 102:592–95.

Janzen, D. H. 1969. Allelopathy by myrmecophytes: the ant *Azteca* as an allelopathic agent of *Cecropia*. Ecology 50:147–53.

Janzen, D. H. 1970. Herbivores and the number of tree species in tropical forests. Amer. Natur. 104:501–28.

Janzen, D. H. 1971a. Euglossine bees as long-distance pollinators of tropical plants. Science 171:203–5.

Janzen, D. H. 1971b. Seed predation by animals. Ann. Rev. Ecol. and Syst. 2:465–92.

Janzen, D. H. 1972a. Association of a rainforest palm and seed-eating beetles in Puerto Rico. Ecology 53:258–61.

Janzen, D. H. 1972b. Escape in space by *Sterculia apetala* seeds from the bug *Dysercus fasciatus* in a Costa Rican deciduous forest. Ecology 53:350–61.

Janzen, D. H. 1973a. Community structure of secondary compounds in plants. Pure Appl. Chem. 34:529–38.

Janzen, D. H. 1973b. Embryonic field of ecology. Science 182:1125–26.

Janzen, D. H. 1973c. Host plants as islands II: competition in evolutionary and contemporary time. Amer. Natur. 107:786–90.

Janzen, D. H. 1973d. Dissolution of mutualism between *Cecropia* and its *Azteca* ants. Biotropica 5:15–28.

Janzen, D. H. 1974a. Epiphytic myrmecophytes in Sarawak: mutualism through the feeding of plants by ants. Biotropica 6:237–59.

Janzen, D. H. 1974b. Tropical blackwater rivers, animals, and mast fruiting by the Dipterocarpaceae. Biotropica 6:69–103.

Janzen, D. H. 1975a. Behavior of *Hymenaea courbaril* when its predispersal seed predator is absent. Science 189:145–47.

Janzen, D. H. 1975b. Interactions of seeds and their insect predators/parasitoids in a tropical deciduous forest. *In* P. W. Price (ed.), Evolutionary strategies of parasitic insects and mites. Pp. 154–86. Plenum Press, New York.

Jermy, T. 1966. Feeding inhibitors and food preferences in chewing phytophagous insects. Entomol. Exp. Appl. 9:1–12.

Jones, D. A. 1973. Coevolution and cyanogenesis. *In* V. H. Heywood (ed.), Taxonomy and ecology. Pp. 213–42. Academic Press, London.

Knerer, G., and C. E. Atwood. 1973. Diprionid sawflies: polymorphism and speciation. Science 179:1090–99.

Kok, L. T., and W. W. Surles. 1975. Successful biocontrol of musk thistle by an introduced weevil, *Rhinocyllus conicus*. Environ. Entomol. 4:1025.

Krieger, R. I., P. O. Feeny, and C. F. Wilkinson. 1971. Detoxification enzymes in the guts of caterpillars: an evolutionary answer to plant defenses? Science 172:579–81.

Kulman, H. M. 1971. Effects of insect defoliation on growth and mortality of trees. Ann. Rev. Ent. 16:289–324.

Le Quesne, W. J. 1972. Studies on the coexistence of three species of *Eupteryx* (Hemiptera: Ciccadellidae) on nettle. J. Entomol. (A.) 47:37–44.

Levin, D. A. 1969. The effect of corolla color and outline on interspecific pollen flow in *Phlox*. Evolution 23:444–55.

Levin, D. A. 1971. Plant phenolics: an ecological perspective. Amer. Nat. 105:157–81.

Levin, D. A. 1973. The role of trichomes in plant defense. Q. Rev. Biol. 48:3–15.

Levin, D. A. 1976a. Alkaloid-bearing plants: an ecogeographic perspective. Amer. Natur. 110:261–84.

Levin, D. A. 1976b. The chemical defenses of plants of pathogens and herbivores. Ann. Rev. Ecol. and Syst. 7:121–59.

Levin, D. A., and D. E. Berube. 1972. *Phlox* and *Colias:* the efficiency of a pollination system. Evolution 26:242–50.

Levin, S. A., and R. T. Paine. 1974. Disturbance, patch formation and community structure. Proc. Nat. Acad. Sci. USA 71:2744–47.

Levin, S. A., and R. T. Paine. 1975. The role of disturbance in models of community structure. *In* S. A. Levin (ed.), Ecosystem analysis and prediction. Pp. 56–67. S. I. A. M., Philadelphia.

Levins, R., and R. MacArthur. 1969. An hypothesis to explain the incidence of monophagy. Ecol. 50:910–11.

Lugo, A. E., et al. 1973. The impact of the leaf-cutter ant *Atta colombica* on the energy flow of a tropical wet forest. Ecol. 54:1292–1301.

MacArthur, R. H. 1972. Geographical ecology. Harper & Row, New York, 269 pp.

Mani, M. S. 1964. Ecology of plant galls. W. Junk, The Hague.

Martin, M. M. 1974. Biochemical ecology of the attine ants. Accounts of Chemical Research 7:1–5.

McKey, D. 1974. Adaptive patterns in alkaloid physiology. Amer. Nat. 108:305–20.

McKey, D. 1975. The ecology of coevolved seed dispersal systems. *In* L. E. Gilbert and P. H. Raven (eds.), Coevolution of animals and plants. University of Texas Press, Austin and London.

Meeuse, A. D. J. 1973. Anthecology, floral morphology, and angiosperm evolution. *In* V. H. Heywood (ed.), Taxonomy and ecology. Academic Press, London and New York.

Moreau, J. P., and R. T. Gahukar. 1975. Importance des méthodes expérimentales dans les études des relations plantes-insectes. Ann. Zool. Ecol. Anim. 7:119–36.

Munro, J. 1967. The exploitation and conservation of resources by populations of insects. J. Anim. Ecol. 36:531–48.

Murdoch, W. W., F. C. Evans, and C. H. Peterson. 1972. Diversity and pattern in plants and insects. Ecology 53:819–28.

Opler, P. A. 1973. Fossil lepidopterus leaf mines demonstrate the age of some insect-plant relationships. Science 179:1321–23.

Opler, P. A. 1974. Oaks as evolutionary islands for leaf-mining insects. Amer. Sci. 62:67–73.

Orians, G. H. 1974. An evolutionary approach to the study of ecosystems. *In* W. H. van Dobben and R. H. Lowe-McConnel (eds.), Pp. 198–200. Proc. Intl. Congr. Ecol., 1st Wagenigen: Cent. Agric. Publ. Doc., W. Junk, The Hague (pp. 198–200).

Orians, G. H., R. G. Cates, D. F. Rhoades, and J. C. Schultz. 1974. Producer consumer interactions. *In* W. H. van Dobben and R. H. Lowe-McConnel (eds.), Unifying concepts in ecology. Pp. 213–17. Proc. Intl. Congr. Ecol., 1st Wagenigen: Cent. Agric. Publ. Doc., W. Junk, The Hague.

Otte, D. 1975. Plant preference and plant succession: a consideration of evolution of plant preference in *Schistocerca*. Oecologia 18:129–44.

Pianka, E. R. 1970. On r- and k-selection. Amer. Natur. 104:592–97.

Pillemer, E. A., and W. M. Tingey. 1976. Hooked trichomes: a physical barrier to a major agricultural pest. Science 193:482–84.

Platt, W. J., G. R. Hill, and S. Clark. 1974. Seed production in a prairie legume (*Astragalus canadensis* L.). Oecologia 17:55–63.

Pliske, T. 1975. Pollination of pyrrolizidine alkaloid-containing plants by male Lepidoptera. Env. Ent. 4:474–79.

Rathcke, B. J. 1976. Competition and coexistence within a guild of herbivorous insects. Ecol. 57:76–87.

Rhodes, D. F., and R. G. Cates. 1976. Toward a general theory of plant antiherbivore chemistry. *In* J. Wallace and R. Mansell (eds.), Biochemical interactions between plants and insects. Recent Advances in Phytochemistry 10:168–213.

Ricklefs, R., and K. O'Rourke. 1975. Aspect diversity in moths: a temperate-tropical comparison. Evolution 29:313–24.

Rockwood, L. 1973. The effects of defoliation on seed production of six Costa Rican tree species. Ecology 54:1363–69.

Rockwood, L. 1974. Seasonal changes in the susceptibility of *Crescentia alata* leaves to the flea beetle, *Oedionychus* sp. Ecology 55:142–48.

Rockwood, L. 1976. Plant selection and foraging patterns in two species of leaf cutting ants (*Atta*). Ecology 57:48–61.

Root, R. B. 1975. Some consequences of ecosystem texture. In S. A. Levin (ed.), Ecosystem analysis and prediction. Pp. 83–97. S. I. A. M., Philadelphia.

Root, R. B., and S. J. Chaplin. 1976. The life-styles of tropical milkweed bugs, *Oncopeltus* (Hemiptera: Lygaeidae) utilizing the same hosts. Ecology 57:132–40.

Rothschild, M. 1972. Some observations on the relationship between plants, toxic insects, and birds. *In* J. B. Harborne (ed.) Phytochemical ecology. Proceedings of the Phytochemical Society, No. 8, Academic Press, New York and London.

Rothschild M., J. Von Euw, T. Reichstein, D. A. Smith, and J. Pierre. 1975. Cardenolide storage in *Danaus chrysippus* (L.) with additional notes on *D. plexippus* (L.). Proc. R. Soc. Lond. B. 190:1–31.

Saxena, K. N. 1954. Feeding habits and physiology of digestion of certain leafhoppers Homoptera: Jassidae. Experientia 10:383–84.

Scriber, J. M. 1973. Latitudinal gradients in larval feeding specialization of the world Papilionidae (Lepidoptera). Psyche 80:355–73.

Schröder, D. 1969. Studies on phytophagous insects of *Euphorbia* spp. *Chamaesphecia empiforius* (Esp.) Progr. Rep. Commonwealth Inst. Biol. Control 23, 13 pp.

Self, L. S., F. E. Guthrie, and E. Hodgson. 1964. Metabolism of nicotine by tobacco feeding insects. Nature (London) 204:300.

Singer, M. C. 1971. Evolution of food-plant preference in the butterfly *Euphydryas editha*. Evolution 25:383–89.

Smart, H., and N. F. Hughes. 1973. The insect and plant: progressive palaeoecological integration. *In* H. F. Van Emden (ed.), Insect/plant relationships. John Wiley & Sons, New York.

Smith, K. M. 1965. Plant virus-vector relationships. Adv. Virus Research 11:61–96.

Sondheimer, E., and J. B. Simeone. 1970. Chemical Ecology. Academic Press, New York.

Southwood, T. R. E. 1960. The evolution of the insect/host tree relationship—a new approach. XI Int. Cong. Ent., Vienna 1:651–55.

Southwood, T. R. E. 1961. The number of species of insects associated with various trees. J. Anim. Ecol. 30:1–8.

Southwood, T. R. E. 1973. The insect/plant relationship—an evolutionary perspective. *In* H. F. Van Emden (ed.), Insect/plant relationships. John Wiley & Sons, New York.

Southwood, T. R. E. 1976. Bionomic strategies and population parameters. In R. M. May (ed.), Theoretical ecology. Blackwell Scientific Publications, Oxford.

Southwood, T. R. E., R. M. May, R. M. Hussell, and G. R. Conway. 1974. Ecological strategies and population parameters. Amer. Natur. 108:791–804.

Stent, G. S. 1972. Prematurity and uniqueness in scientific discovery. Scientific American 227:84–93.

Strong, D. R. 1974a. Nonasymptotic species richness models and the insects of British trees. Proc. Nat. Acad. Sci. USA 71:2766–69.

Strong, D. R. 1974b. Rapid asymptotic species accumulation in phytophagous insect communities: the pests of cacao. Science 185:1064–66.

Strong, D. R. 1977. Rolled leaf hispine beetles (Coleoptera, Chrysomelidae) and their Zingiberales (Monocotyledonae) plants in Middle America. Biotropica 9:156–69.

Tevis, L., Jr. 1958. Interrelations between the harvester ant *Veromessor pergandei* (Mayr) and some desert ephemerals. Ecology 39:695–704.

Ueckert, D. N., and R. M. Hansen. 1971. Dietary overlap of grasshoppers on sandhill rangeland in northeastern Colorado. Oecologia 8:276–95.

Van Emden, H. F., and M. J. Way. 1973. Host plants in the population dynamics of insects. *In* H. F. Van Emden (ed.), Insect/plant relationships. John Wiley & Sons, New York.

Waloff, N. 1968a. A comparison of factors affecting different insect species on the same host plant. *In* T. R. E. Southwood (ed.), Insect abundance. Symp. Royal Entomol. Soc. London 4:76–87.

Waloff, N. 1968b. Studies on the insect fauna on scotch broom *Sarothamnus scoparis* (L.) Wimmer. Adv. Ecol. Res. 5:87–208.

Wettstein, R. R. 1889. Pflanzen und Ameisen. Naturwissenschaft 29:309–27.

Wheeler, W. M. 1973. The fungus growing ants of North America. Dover Publ., New York, 136 pp.

Wicklund, C. 1974. Oviposition preferences in *Papilio machaon* in relation to the host plants of the larvae. Ent. Ex. Et. Appl. 17:189–98.

Wiklund, C. 1975. The evolutionary relationship between adult oviposition preference and larval host plant range in *Papilio machaon* L. Oecologia 18:185–97.

Williams, L. H. 1954. The feeding habits and food preferences of Acrididae and the factors which determine them. Trans. Ent. Soc. Lond. 105:23–454.

Yeo, P. F. 1973. Floral allurements for pollinating insects. *In* H. F. Van Emden (ed.), Insect/plant relationships. John Wiley & Sons, New York.

Zwölfer, H., and P. Harris. 1971. Host specificity determination of insects for biological control of weeds. Ann. Rev. Entomol. 16:159–78.

GORDON H. ORIANS AND NOLAN E. PEARSON

On the Theory of Central Place Foraging

6

A rapidly growing body of literature deals with the ways in which animals select prey items and foraging patches (Emlen, 1966; Charnov, 1976; MacArthur, 1972; MacArthur and Pianka, 1966; Pulliam, 1974; Pearson, 1974, 1976; Schoener, 1971). The rationale behind most of these theories is that individual fitness is increased if an organism maximizes its net rate of energy intake while foraging. This appears to be an appropriate maximization provided that either (a) there are competing uses of time, (b) there are risks associated with foraging, or (c) there are uses of additional energy that contribute positively to fitness. A full discussion of this rationale is provided by Schoener (1971).

As discovered independently by several investigators, this maximization is achieved by ranking prey by their expected net energy content (C) divided by expected handling time (t_H), where handling time is taken to include pursuit, capture, eating, and any mandatory post-consumption pauses. Prey included in the optimal set are those whose C/t_H is greater than the expected rate of energy intake if the prey item were ignored and the predator sought only prey with higher C/t_H. A corollary is that whether or not a prey item should be taken is independent of its own abundance but depends only on abundances of higher-ranked prey types. The predator expands its optimal diet to include lower-ranked items when higher-ranked items become scarcer, and it contracts its diet when higher-ranked items become more abundant. Proof for

Gordon H. Orians, Department of Zoology, University of Washington, Seattle, Washington 98195.

Nolan E. Pearson, Center for Quantitative Science, University of Washington, Seattle, Washington 98195.

these assertions can be found in Charnov (1976), Schoener (1971), MacArthur (1972), Pearson (1974, 1976), and Pulliam (1974).

Many foragers, however, do not consume their prey where they are captured but return with them to some fixed central place where they are eaten, stored, or fed to dependent offspring (Central Place Foraging). Most organisms with parental care return with food to a central place at least part of the time. Also some mammals (Brown & Lieberman, 1973; Smith, 1968) and birds (Durango, 1951; Michener, 1945; Schuster, 1950; Swanberg, 1951) cache food that is eaten during periods of reduced prey availability.

The problems posed by Central Place Foraging (CPF) can be divided into two major categories. The first considers that an organism has a central place to which it returns with prey and asks how it should select patches, prey, and its load size. The second assumes a set of potential foraging patches and asks where an organism should locate its central place. All these decisions should be influenced by patch qualities, effects of the predator on prey availability within patches, effects of competitors on prey availability, and risks to the predator while foraging, traveling, and resting at the central place. In addition, these decisions require the predator to assess patch and prey qualities either directly by statistical sampling or indirectly by use of cues such as weather, time of day, signs of prey availability, success of other individuals, and so forth.

Central Place Foraging is governed by many of the same factors as non-central place foraging. In fact, non-central place foraging is a special case of CPF in which traveling time (T_T) is zero. In this paper we develop some aspects of CPF relating to prey selection, patch selection, optimal load size, and selection of foraging itineraries. Additional problems, such as selection of a central place and methods of environmental assessment, will receive cursory attention. As in most general foraging theory, we assume that a forager increases its fitness by maximizing rate of delivery of energy to the central place.

THEORY

The basic unit in CPF is the *round trip*, composed of (a) an outbound trip, (b) a foraging period, and (c) a return trip. Energy is expended in all three phases of the trip but is acquired only during the second. Because of load cost, a return trip is more expensive per unit distance traveled than an outbound trip. Additionally, if a central place forager consumes some food during a trip, its internal weight is likely to be greater when returning than when outbound.

Selection of Prey during a Trip

Determination of rules by which a central place forager should select prey requires some simplifying assumptions. We assume that the predator can search simultaneously for prey items of many different types though only a single prey can be pursued and handled at a time. We also assume that the predator is at equal risk per unit time while at the patch, in transit, or at the central place. Finally, we assume that prey types available in a patch are encountered randomly in accord with a Poisson process, so that the expected time to the next encounter with a prey item of type *i* is not conditioned on the length of time the predator has been searching unsuccessfully in a patch. This should be nearly true if patches are relatively large and if the predator lacks precise knowledge of the number of prey items in a patch when it starts foraging there. It would not hold true if, for example, a predator knew that a specific number of prey items were present in a small patch, in which case the probability of an encounter would increase as the predator unsuccessfully searches the patch.

Given the unavoidable cost of a round trip, it would normally be advantageous for a predator to return with more than one prey. However, return with a single prey item may be favored if capture of the first item seriously encumbers a predator's prey-capturing apparatus, thus making additional captures so difficult that a higher rate of delivery of energy to a central place is achieved by returning with the first prey. The best examples are predators that capture relatively large prey and that carry prey in the appendages used for capture, such as hawks, owls and carnivorous mammals.

If holding previously captured prey items adversely affects a predator's ability to capture additional items, the actual C/t_H of a prey item changes as a function of the number of prey already held by the predator. If some prey types affect subsequent captures more than other types, the over-all rate of energy capture during a trip may be increased by selecting those prey with minimal adverse effects on subsequent captures at the beginning of a foraging sequence and only taking other types toward the end of a sequence when foraging is about to be terminated for other reasons. In the most extreme cases, the predator gains by returning with the first prey item and becomes a single-prey loader.

If its mouth is also its capture apparatus, a predator must either devote some trips to feeding only itself or must meet its energy needs by consuming prey at the beginning of some foraging bouts. Since there is no advantage to returning to the central place without any food, a predator should feed itself at the beginning of enough foraging bouts to meet its energy needs and then complete the bout by loading food for the central place. Greatest efficiency of digestion and traveling might be achieved by taking a modest amount of food at the beginning of most foraging bouts rather than filling the gut completely during a smaller number of trips; but if prey are large, a predator may have no option but to fill its gut whenever it eats. In eastern Washington, foraging Yellow-headed (*Xanthocephalus xanthocephalus*) and Brewer's (*Euphagus

TABLE 1

PROPORTION OF TRIPS IN WHICH ADULT BLACKBIRDS FED THEMSELVES PRIOR TO LOADING PREY FOR NESTLINGS (COLUMBIA NATIONAL WILDLIFE REFUGE, WASHINGTON)

Species	Sex	Location	Year	Fed First No.	Fed First %	Loaded Prey No.	Loaded Prey %
Yellowhead	♀	Coot Lake	1968	53	77	16	23
Yellowhead	♂	Coot Lake	1968	7	50	7	50
Brewers	♀	Pit Lake	1976	41	55	34	45

cyanocephalus) blackbirds feed themselves first on more than half of trips in which they gather food for nestlings or fledglings (table 1).

Prey Selection by Single-prey Loaders (Handling Time Neglected)

A solution to the problem of prey selection by single-prey loaders is graphed in figure 1, in which traveling time is plotted to the left and time in the patch to the right of the origin. For the present, time spent pursuing and capturing prey is presumed negligible. The lower of the two curves plots the energy (C) of a given prey type versus expected time (t'_s) to encounter any prey possessing energy greater than or equal to C. The exact shape of the curve depends on prey densities in the patch, but regardless of actual prey densities the curve will inevitably be monotone increasing (or monotone

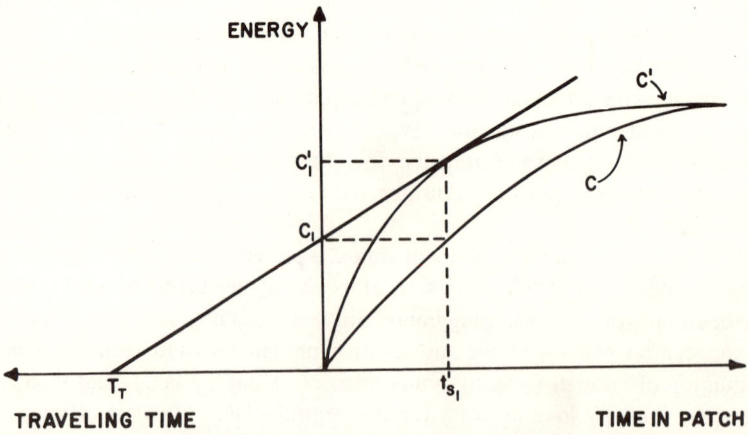

Fig. 1. Prey selection by single-prey loaders. T_T= traveling time. Predator should take any prey with energy $\geq C_1$. After many trips, predator can expect an average energy $= C_1'$, resulting in the maximum expected energy delivery rate =

$$\frac{C_1'}{T_T + t_{S_1}}$$

non-decreasing, if a finite number of discrete prey types is being considered).

The upper of the two curves represents expected energy (C') versus expected time to encounter any prey of energy greater than or equal to C. The expected energy, C', resulting from a policy of taking any prey with energy $\geq C$ takes on a value between C and the maximum prey energy in the patch, as explicitly derived in the Appendix. The two curves, C and C', are intimately related by the property that a tangent through any point on the upper curve, say point (t_{s_1}', C') intersects the oridinate at C_1, the corresponding value of the lower curve at t_{s_1}' (see Appendix). This guarantees that C' will increase at a decreasing rate.

Thus, it can be seen from figure 1 that a predator that must spend T_T units of time traveling to and from a patch should pursue only prey of energy C_1 or greater to maximize its expected delivery rate $= C_1'/(T_T + t_{s_1}')$ to the central place.

The expected delivery rate is, of course, the average energy per unit time delivered over many trips. It might be argued that this rate could be improved upon if the predator varied its prey selection in response to length of time already spent in the patch unsuccessfully searching for a prey of energy $\geq C_1$. In figure 2, the predator has just encountered a prey of energy C_1 at an elapsed search time T_s, and must decide whether to take C_1 or adopt a

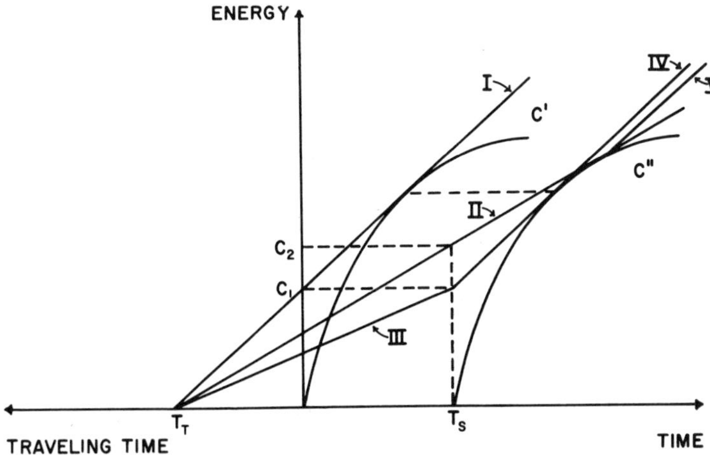

Fig. 2. Altering diet in response to unsuccessful foraging can increase energy delivery rate for that particular trip, but reduces energy rate averaged over future trips. T_T = traveling time. T_S = actual time in patch when prey of energy C_1 is encountered. Slope of I = expected energy rate at beginning of trip. Slope of II = expected energy rate if diet is altered to maximize delivery rate for the trip. Slope of III = energy rate if C_1 is taken (< slope of II). Line IV is expected total energy vs. elapsed time if C_1 is taken. Line V = expected total energy vs. total elapsed time for altered diet.

new policy of taking only those prey with energy $\geq C_2$. The curve C' is as previously defined for figure 1, expected energy versus expected search time to encounter prey with energy $\geq C$. The curve C'' is the horizontal translation of C' through T_s units of time, and therefore represents expected energy versus expected search time beyond T_s to encounter prey with energy $\geq C$. The slope of line I gives the maximum delivery rate expected *before* the predator enters the patch and is fulfilled by a dietary tactic of taking any prey with energy $\geq C_1$. The slope of II gives a revised maximum delivery rate for that particular trip in which no appropriate prey were encountered for T_s units of time in the patch. The revised maximum rate for that trip requires restricting the diet to prey of energy $\geq C_2$. The slope of line III gives the actual delivery rate for that trip if energy C_1 is taken when encountered at time T_s, a rate that is clearly less than that given by line II.

However, a comparison of lines IV and V, which for the two dietary strategies give future projections of cumulative energy versus total elapsed time, distinctly favors line IV corresponding to the policy of taking C_1 (or prey of greater energy) whenever encountered. It seems anomalous that maximization of energy per unit time per trip need not result in maximizing total energy delivered per total time foraging. However, since Energy/Time = (Energy/Time)/Trip × (# of trips), not only should energy per time per trip be maximized but more trips should be taken in the total time available particularly to swamp out the effects of a bad trip. Since the two quantities cannot be independently maximized, only a fuller analysis, such as the graphical one presented in figure 2, can provide the correct tactic.

As distance, and hence traveling time, from patch to central place is increased, the greater must be the prey energy selected by the predator. Graphical demonstration of this is given in figure 3, in which the patch quality is held constant and distance from patch to central place is varied. A corollary is that if traveling time is reduced to zero, a predator should take any prey encountered, which is the result predicted for non-central-place foragers when handling time of prey is nil.

Prey Selection by Single-Prey Loaders (Handling Time Included)

If time to pursue and capture prey is included, some conditions yield qualitatively different solutions to the problem of prey selection by single-prey loaders. The case for two prey types is graphed in figure 4. Prey type 1 has energy C_1, handling time t_{H_1}, and expected search time t_{S_1}, while prey type 2 has energy C_2, handling time t_{H_2}, and expected search time t_{S_2}. A predator that takes either prey type 1 or 2 upon encounter will have an expected energy gain of $\overline{C}$ and an expected time in patch (searching plus

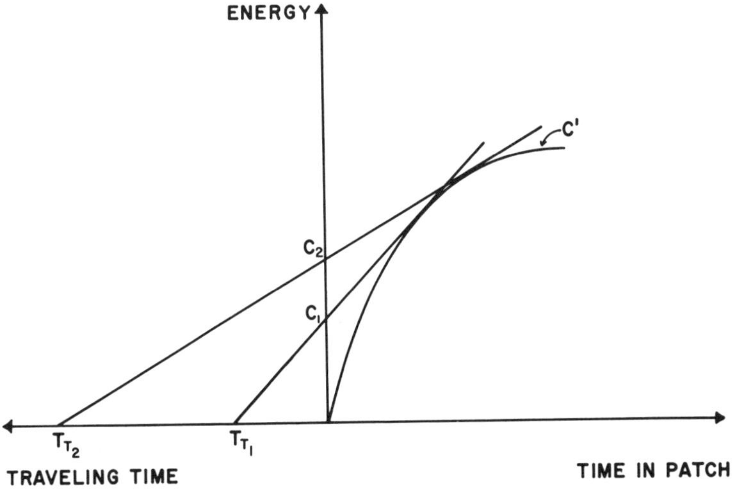

Fig. 3. For single-prey loaders optimal prey selection narrows with increasing distance between central place and patch. For a central place requiring a round-trip time of T_{T_1}, prey with energy $\geq C_1$ should be taken. A round trip T_{T_2} dictages selection of energy $\geq C_2$.

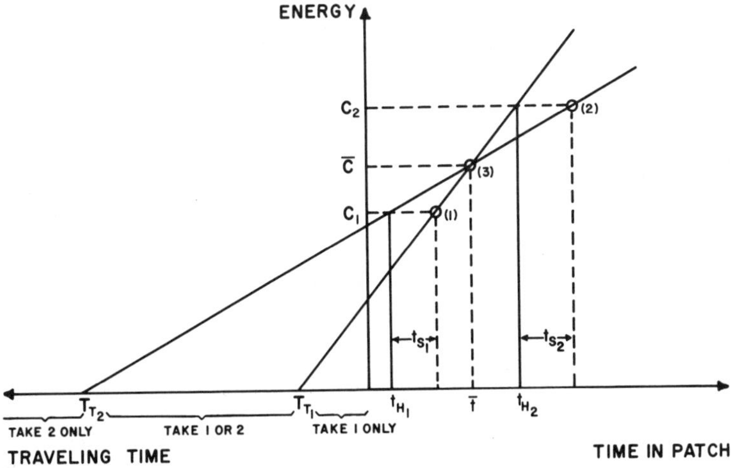

Fig. 4. Prey selection by single-prey loaders where handling time (pursuit and capture) is not negligible. C_i=energy in prey i. t_{S_i}=handling time for prey i. $\bar{C}$=expected energy if predator takes prey 1 or 2. t=expected time in patch (searching and handling) if predator takes 1 or 2. If traveling time to patch is short, predator takes prey 1 only, since

$$\frac{C_1}{t_{H_1}+t_{S_1}} > \frac{C_2}{t_{H_2}}.$$

If traveling time is long, predator takes prey 2 only, since $C_2 > C_1$. For intermediate traveling times predator takes either prey 1 or 2 when encountered.

handling) of $\bar{t}$. The three circled points in figure 4 represent expected energy versus expected time in patch for the three foraging tactics, (1) take prey 1 only, (2) take prey 2 only, and (3) take either prey 1 or 2 whenever encountered. If round-trip traveling time is short ($< T_{T_1}$), the predator should take only prey 1, which has lower energy and handling time than prey 2, but a larger ratio of energy to handling time. If traveling time is long ($> T_{T_2}$), the predator should take only prey 2 because of its higher energy, since, for long traveling times, handling time is comparatively unimportant. Traveling times intermediate to T_{T_2} favor a policy of pursuing either prey type 1 or 2.

To generalize the theory of optimal single-prey loading to more than two prey types we note that each prey type i has associated with it a fixed energy C_i and a fixed time expenditure $T_T + t_{H_i}$, in addition to a search time expenditure t_{S_i} which varies with prey abundance. We may thus apply already-developed non-central-place foraging theory by ranking prey in descending order of $C_i/(T_T+t_{H_i})$, rather than the standard ranking based on C_i/t_{H_i}. Thus the same prey type may be ranked differently in patches at different distances from the central place. For short traveling times, superiority of prey hinges on energy-per-unit handling time. For long traveling times, superior prey are those of higher energy, regardless of their handling times. Actual choice of diet in a given patch depends on availability of higher-ranked prey, insufficient numbers of highly ranked prey leading to broadening of the diet to include lesser-ranked prey.

Optimal Load for Multiple-Prey Loaders

A solution to the problem of optimal load for multiple-prey loaders is given in figure 5. The curve (C') represents the expected energy obtained during a foraging bout versus time spent in the patch. It is assumed that prey are small relative to the prey-retaining apparatus, and, though there is some loss in efficiency in searching for and handling additional prey items, it is not necessary to terminate the foraging bout after the first prey is captured. The fact that there is loss of efficiency in handling subsequent prey, as well as some depletion of the resource, requires that C' increase at a decreasing rate. As in previous graphs, optimal solutions are given by the point of tangency of straight lines originating to the left of the origin by one round-trip time from the central place. If patch quality remains constant, optimal load increases with increasing distance of the patch from the central place, that is, a predator should continue to load even though its rate of loading is dropping. As a corollary, time in a patch increases with increasing distance of the patch from the central place. As seen in figure 6, for constant distance of the patch from the central place, load increases slightly with patch quality, whereas time spent in the patch decreases significantly. Figure 7 illustrates that including

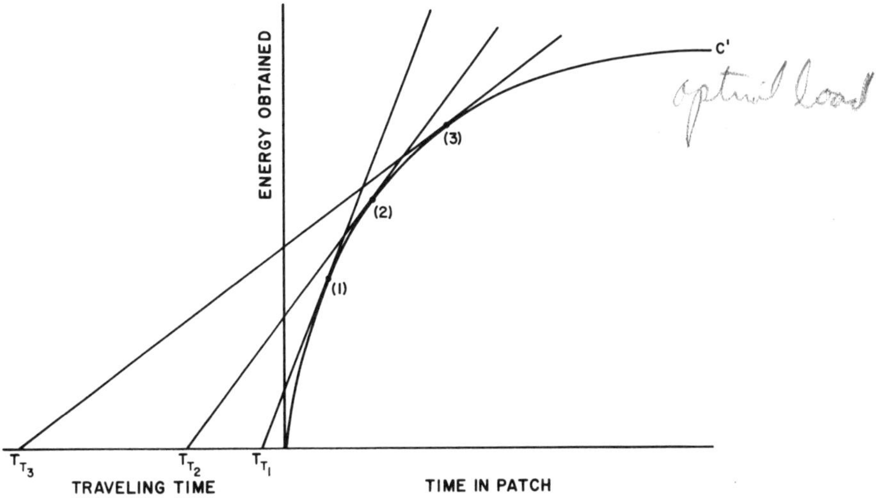

Fig. 5. Optimal load size related to distance from central place for multiple prey loaders. Optimal load increases from (1) to (2) to (3) as round trip distance to central place increases from T_{T_1} to T_{T_2} to T_{T_3}.

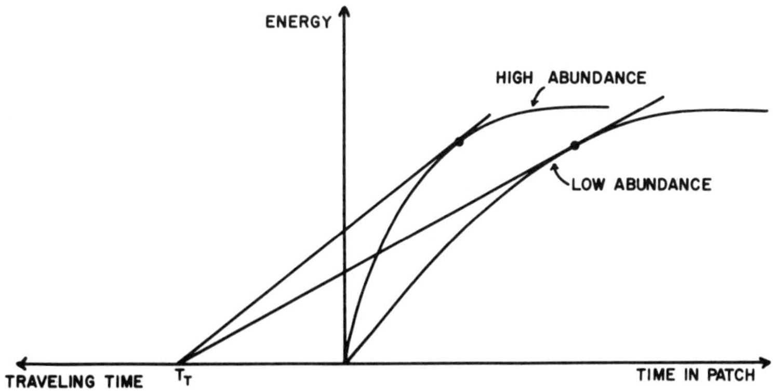

Fig. 6. Optimal load as a function of patch quality for multiple-prey loaders. Optimal load increases and optimal time in patch decreases with patch quality.

costs of traveling and searching lowers net energy gain per unit time spent foraging, which is equivalent to lowering patch quality, thereby favoring longer time in the patch.

Single-prey vs. Multiple-prey Loading

Some animals are morphologically capable of loading either a single large prey item or several items of smaller prey. One or the other tactic, or a

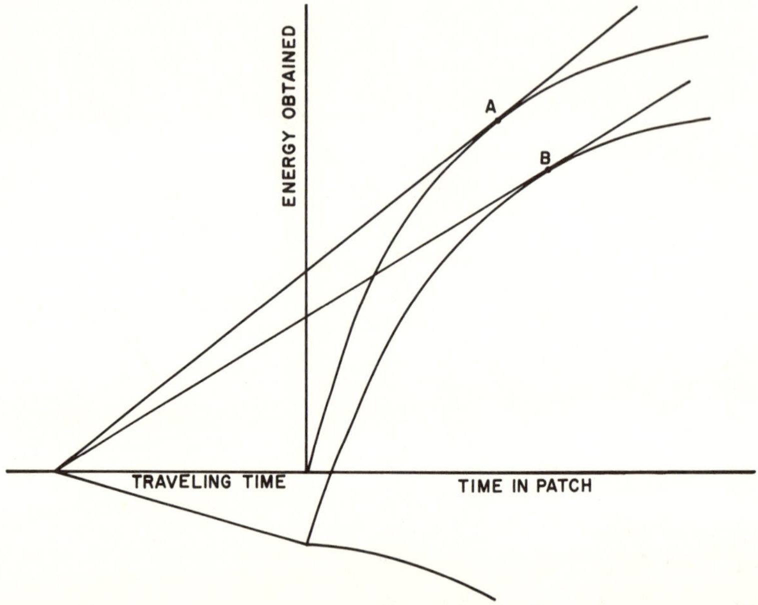

Fig. 7. Optimal load for a multiple-prey loader with and without consideration of overhead costs. Overhead increases optimal time in a patch from A to B.

mixture of both, may be favored depending on prey energies and abundances, traveling time, and the degree to which previously captured small prey hamper pursuit and capture of additional prey. Although a full analysis considers all these factors, we will only treat the following special case:

1. There are two prey types, a small one that may be loaded multiply and a large one that must be loaded singly.
2. The large prey item is energetically greater than a full load of small items.
3. Handling time is negligible while collecting small prey items.
4. The large prey item can be captured only at the expense of giving up any previously captured smaller items.

Under these assumptions, figure 8 demonstrates the relative merits of single- and multiple-prey loading for three different traveling times. Curve C_M is expected energy intake for multiple-prey loading, and C_S is expected energy for the large, singly-loaded prey. For short traveling times (T_{T_1}), maximum delivery rate is achieved by multiple-prey loading of the small prey to capacity. If there is a chance encounter with the large prey, already loaded prey should be dropped and the large prey taken for its high energy. For

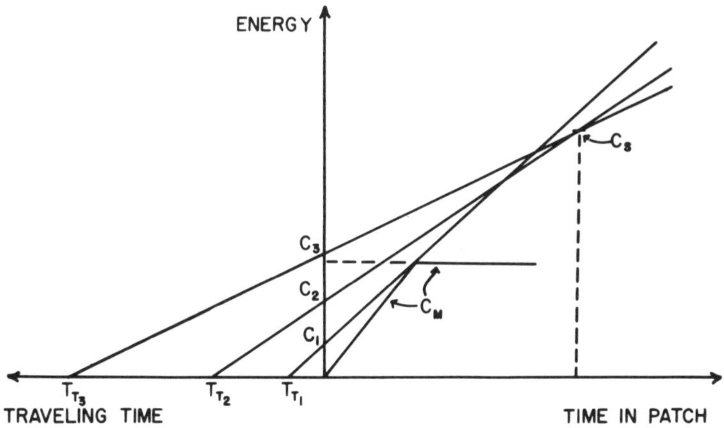

Fig. 8. Contrast between single-prey and multiple-prey loading. C_M=expected energy for multiple-prey loading. C_S=expected energy for single-prey loading. For traveling time $>T_{T_3}$, single prey only should be taken. For traveling time $<T_{T_3}$, either prey should be taken when encountered.

intermediate traveling times (T_{T_2}), expected delivery rate appears to be maximized by rejecting small prey and taking only the large prey. This is not true, however, since the present value (C_2) of an unencountered large prey is less than that of a capacity load of small prey. Consequently, the small prey should be taken when encountered and delivered to the central place when fully loaded. As usual, the large prey must be taken whenever encountered. For long traveling times (T_{T_3}), the energy of a full load of small prey is less than the present value (C_3) of an unencountered large prey, thus favoring a tactic of pursuing large prey only.

It is possible that, though the large prey may effectively prevent capture of any additional prey items, a partial or even full load of small items will not seriously hamper pursuit and capture of large prey, so the predator need not drop previously captured small prey to take the larger one. If so, small items should always be taken when encountered, provided that handling such items does not seriously detract from search time. If the large item is rare, it should be taken whenever encountered; but if it is abundant, capture should be postponed until a full load of small prey is obtained, especially when travel time is great.

Optimal Patch Choice

A lower rate of energy delivery to a central place is inevitable with increasing distance of the foraging patch from the central place for both single- and multiple-prey loaders unless patch quality increases. This can be seen more

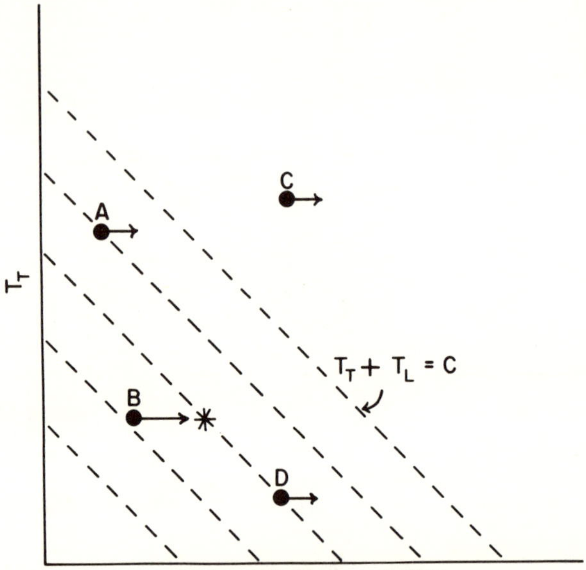

Fig. 9. Plot of traveling time against time to obtain a fixed optimal load in a patch (adapted from Charnov, pers. comm.). Patches of equal quality are connected by straight lines with slopes of minus one. Patch D is not utilized until the time to acquire the optimal load in patch B increases to *.

clearly by plotting time to obtain a fixed optimal load against traveling time (figure 9). $T_T + T_L = C$, where T_L is the time to obtain some fixed maximum load and C is a constant. On such a graph, patches yielding a constant rate of energy delivery to a central place are connected by straight lines with slopes of minus one. More distant patches are not utilized until T_L increases sufficiently to compensate for the greater T_T. The predator adds or subtracts more distant sites depending on relative levels of prey availability.

Optimal patch choice for a single-prey loader is determined graphically in figure 10, where expected energy versus expected searching time is plotted for each of several patches, then translated horizontally through an amount equal to the patch's round-trip traveling time. The maximum expected delivery rate is given by the slope of the line tangent to C_{II}'. In figure 10, maximum delivery rate occurs when patch II is searched for prey of energy $\geq C$.

If the predator happens to perceive a prey of energy $\geq C$ while traveling to the optimal patch, it will clearly save time by taking the prey. However, such a prey need not have energy as great as C, and should be taken if its energy per round-trip time from the central place to its point of discovery exceeds the maximum delivery rate from the optimal patch, as is represented by the shaded region of figure 10.

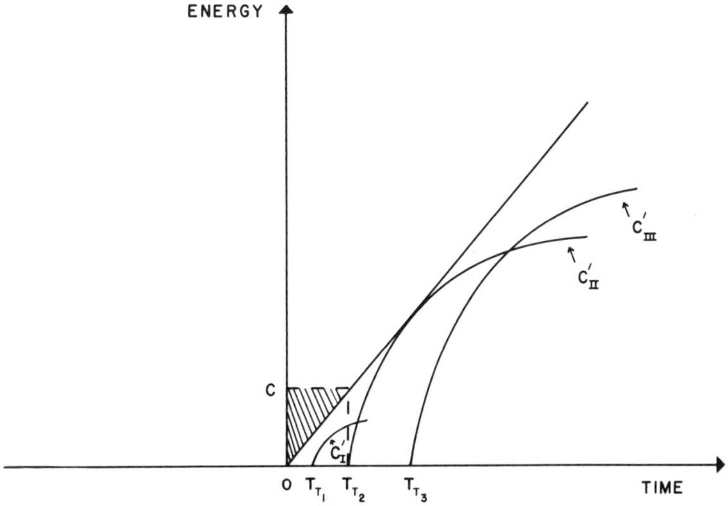

Fig. 10. Patch selection by single-prey loaders. Curves C_I', C_{II}', and C_{III}' are expected energy vs. expected search time, each displaced to the right by an amount equal to the round-trip traveling time to patch. The optimal strategy in this case requires that prey of energy $\geq C$ be sought in Patch II. If a predator encounters prey while traveling to Patch II, it should be taken if the energy-traveling time relationship places it in or above the shaded region.

Optimal Itineraries

If a single patch is visited during a trip, the optimal itinerary is determined by optimal patch choice plus optimal load. However, a predator may depress prey availability within the first patch before it has acquired a full load. For example, the quantity of resource within a single flower or even all flowers on a plant may not fill the honey stomach or pollen baskets of a foraging bee. If so, the bee should visit more than one flower or plant before returning to its nest.

The rules for selecting subsequent patches are complex, but there is one obvious general prediction. Since the predator weighs more on its return trip than on its outbound trip, it can increase its rate of net energy delivery to the central place by shortening the return trip relative to the outbound one. Therefore, foragers should, other things being equal, travel initially to the farthest site to be exploited during a trip and then forage while moving in the general direction of the central place. To our knowledge this prediction has never been tested rigorously.

However, other things may not be equal. Lighting conditions for perception of prey and wind may influence the preferred foraging direction. For example, birds that hunt by searching for prey while flying have a slower ground speed when flying into the wind. Also foraging into the wind may be preferred because strong tail winds ruffle feathers or fur and often cause equilibrium problems. These types of factors may exert more influence on direction of foraging relative to the central place than pure energy considerations, especially if the distance covered during the searching phase is small relative to the distance travelled to and from the central place.

TESTS OF THEORY

If capture of a single prey item terminates a foraging bout, a predator should be more selective in the prey it pursues than a multiple-prey loader foraging under the same circumstances. We know of no tests of the specific rules we have suggested for determining minimum sizes of prey selected by single-prey loaders, but comparative data are available for Red-winged Blackbirds (*Agelaius phoeniceus*) in different regions. In Costa Rica, a single prey item is brought to the nest in most trips. Prey availability in marsh vegetation near nests was sampled with a standard insect sweep net, and prey taken were sampled with pipe-cleaners placed around the necks of nestlings (Orians, 1972). It is uncertain whether foraging blackbirds encounter prey of different sizes in the same proportions as a net, but it is clear that few small grasshoppers were delivered to nestlings compared with their relative abundances in the foraging areas.

In contrast, Red-wings feeding nestlings in eastern Washington normally bring an average of nearly twenty prey items per trip. These prey range in size from dragonflies and cicadas as large as 800 calories per individual to small flies of 10 to 15 calories. Due to a marked diurnal cycle of emergence of aquatic insects (Orians and Horn, 1969), prey of damselfly size or larger ($\geq$50 calories) are much more abundant during mid- to late morning than earlier and later in the day. Therefore, if rejection of small prey were to occur, it is expected during peak emergence periods of damselflies. Although the proportion of large prey in food samples taken from nestling Redwings and Yellowheads increases during emergence periods (table 2), there is no evidence that small prey are ever rejected. In fact, number of prey $\geq$50 calories delivered per hour per nest is higher in samples containing both large and small prey than in samples containing only large prey (Orians, in prep.), the opposite of the pattern predicted if small prey were being dropped from the diet because of increased capture rates of larger prey.

Emerging aquatic insects do not have to be pursued by blackbirds because they are not fully hardened and can fly only weakly or not at all. For such prey, multiple captures are easy, and optimal loads should be large. In con-

TABLE 2

PERCENTAGE OF PREY DELIVERED TO NESTLING BLACKBIRDS ≧ 50 CALORIES (COLUMBIA NATIONAL WILDLIFE REFUGE, WASHINGTON)

Species	Year	No. of Samples	Time of Day: Dawn-0700	Time of Day: 0700-1300	Time of Day: 1300-Dusk
Redwing	1964-65	86*	36.4	82.1	40.8
Yellowhead	1963	74	81.6	99.3	91.6
Yellowhead	1964-65	57	50.8	80.7	71.4
Yellowhead	1968	50	27.7	92.7	87.0

*Prey sampled by placing collars of segments of pipe cleaners around necks of nestlings. A sample consists of all prey delivered to nestlings in a nest over a one-hour period.

trast, prey captured by Redwings in Costa Rica are either nocturnal and hidden during the day in the sheathing leaf bases of stalks of emergent vegetation (Lepidoptera) or among detritus on the surface of the water of the marshes (spiders) where they must be exposed by gaping movements of the bill, or they are active (grasshoppers) and require pursuit. In both cases, capture of additional items once one prey is already held in the bill is presumably much more difficult. This probably explains the great differences in loads for birds in the two areas.

Few data are apparently available on load sizes as a function of distance in central place foragers, but number of prey delivered per trip by foraging Brewer's Blackbirds in eastern Washington increases with distance from the nest (table 3). Cody (1973) pointed out a correlation between distance from nest at which adult murres, guillemots, puffins, and auklets (Alcidae) forage and number of prey brought to the nest per trip. In this case, there are additional constraints on foraging because adults of the smaller species are at such high risk of gull predation when they return to their nests that feedings

TABLE 3

NUMBER OF PREY PER TRIP DELIVERED TO NESTLING BREWER'S BLACKBIRDS BY ADULTS FORAGING AT DIFFERENT DISTANCES FROM THE NEST (COLUMBIA NATIONAL WILDLIFE REFUGE, WASHINGTON)

Date	Sex	Close to Nest (< 50 Meters): No. of Trips	Close to Nest (< 50 Meters): Prey/Trip	Medium Distance from Nest (50-100 Meters): No. of Trips	Medium Distance from Nest (50-100 Meters): Prey/Trip	Far from Nest (> 100 Meters): No. of Trips	Far from Nest (> 100 Meters): Prey/Trip
May, 1975	♂	6	3.4	21	(16.3)*	12	(15.1)
May, 1975	♀	25	2.8	. . .		. . .	
May, 1976	♀	15	3.9	2	6.5	46	12.3
June, 1970	♂	. . .	. . .	. . .	. . . . .	14	20.0
June, 1970	♀	. . .	. . .	. . .	. . . . .	17	18.6

*Figures in parentheses represent averages of estimates based on size of load when bird arrived at the nest from foraging trips where captures could not be monitored.

may be restricted to a single nocturnal visit per night per adult. This stringent restriction gives strong selective advantage to morphological features that increase effectiveness of capture of the first prey item. Cody (op. cit.) in fact suggested that the bizarre alcid bill shapes evolved primarily because of selection on central place foraging efficiency, though alternative theories exist to explain the same phenomenon (Bedard, 1976).

Holding prey in other than the capturing apparatus should be strongly favored among central place foragers. The stomach and upper portions of the digestive tract are obvious candidates for suitable storage sites, and many birds and carnivorous mammals swallow prey and later regurgitate them to their young. Why this habit is virtually lacking among insectivorous birds is puzzling.

Selection of the Central Place

If maximization of rate of energy delivery to a central place is the sole criterion influencing selection of a central place, then the optimal central place is one that minimizes traveling time, that is, the site that lies at the "center of gravity" of the food distribution (Horn, 1969). If foraging is possible in all directions around the central place, the exploitable area in a thin ring of width Δr at distance r from the central place is $2\pi r \Delta r$. However, birds in a colony on a coastal cliff or an island very close to shore have only a half-circle available for exploitation. In extreme cases, such as linear colonies on a coastal cliff, the available foraging area in a given increment, Δd, from the central places may be constant.

Such arrangements of foraging areas prevent selection of central places close to the potential center of gravity of food distributions. In addition, the central place is usually either a cache, in which case the quality of the storage area and its defendability are of great importance, or it is a nest with dependent offspring, in which case site safety is important. In fact, the extent to which risks influence choice of central place can be estimated by the nature of deviation of the central place from the location predicted on purely energetic criteria.

For many predators, a central place is a much safer place to be than any of its potential foraging sites, and food is taken back to a central place to be eaten only because of the danger of remaining outside to eat it. The effect of risk on foraging itineraries has been explored by Covich (1976). Rate of arrival of prey to a given circle of radius r around a central place should be strongly positively correlated with risk to a forager of spending a unit of time outside the central place because both density of prey for a central place forager and density of its own predators should increase with environmental productivity.

Therefore, if a central place forager is at higher risk when foraging than when in the central place, it should be willing to travel farther from the central place to pursue a prey item in a poorer than in a more productive environment. Support for this prediction is provided by foraging patterns of Australian trap-door spiders. Species in richer environments do not venture out of their burrows to capture prey, whereas members of two genera living in arid regions construct twig lines radiating from the burrow entrance and run out on these lines to capture prey detected by their vibrations (Main, 1957).

When dependent young are being provisioned and guarded at a central place, their vulnerability in the absence of the parents may strongly influence optimal lengths of foraging trips. For example, adults may forage so as not to be absent from the offspring longer than a certain maximum time. Such a pattern might be favored if the probability that a predator could successfully find and eat the offspring were less if adults made frequent short visits to the nest rather than fewer though longer ones. Alternatively, the conspicuousness of returning adults may increase the probability that predators locate the offspring, favoring reduced visitation rates (Skutch, 1949).

DISCUSSION

Our analysis of the special problems posed by CPF has demonstrated that both CPF and non-CPF are governed by the same rules if rate of delivery of energy to the central place is to be maximized. In fact, non-CPF is a special case of CPF where traveling time is negligible. Similarly, single-prey loading is a special case in which actual C_i/t_{H_i} ratios are sufficiently reduced by capture of the first prey that rate of energy delivery is maximized by returning to the central place with the first prey. Our graphs provide a precise solution to the problem of when it pays to be a single-prey loader and when to load more than one prey and relates this to distance from central place and relative abundances of prey of different caloric values. These results provide a broader generality to optimal foraging theory than previously available.

Given the premium on increasing load size for central-place foragers, it is perhaps surprising that so many species of birds appear to have bills adapted for efficiency of capture of the first prey item. A possible reason for this is that even a slight loss in efficiency of capture of the first prey item may be more important than enhancement of the ability to capture additional items. Capture efficiencies of first prey items should be especially important for individuals during their first months of independence when mortality rates are usually higher than later and when individuals must learn how to hunt. For example, it may take up to several years for young birds to learn to capture prey as efficiently as adults (Ashmole and Tovar, 1968; Orians, 1969; Recher and

Recher, 1969). Very strong selective pressure when breeding would be required to override the effects of loss in efficiency of prey capturing during those first critical learning months. However, selection is powerful enough on breeding alcids, which are unusually vulnerable to predators when returning to their nests, that bill shapes which presumably decrease capture efficiency of the first prey but enhance abilities to capture additional prey are favored.

In fact, very few animals are central-place foragers throughout their life cycles (social Hymenoptera are apparent exceptions), and conflicts between selection during CPF and non-CPF periods are probably general. The nature and strength of these conflicting selective pressures are likely to influence both optimal morphology and patterns of community structure. For example, consider birds (finches) with deep-heavy bills adapted for crushing hard seeds. Most of these species feed their young on insects and therefore act as central-place foragers exploiting prey for which their bills would seem to be poorly adapted. The success of such finches may be due to strong selection during the winter, a period of greatly reduced arthropod abundances (Fretwell, 1972). Similarly, possession of a seed-eating bill may enable individuals to return to high latitude breeding areas and establish territories early in the season before arthropod foods are available. They may also be able to gain weight and produce eggs early in the season by utilizing seeds persisting in the soil from previous years. In this context it is interesting that finches that feed their young insects are dominant members of breeding bird communities in grasslands and desert scrub especially at high latitudes (Cody, 1968, 1974; Wiens, 1973). These habitats have good supplies of seeds but few arthropods early in spring. In a Saskatchewan grassland Cody (1968) censused a plot with four species of breeding finches, one pipit, and one lark. The latter two species also feed extensively on seeds early in the season. In Minnesota a plot held two sparrows and a bobolink, an icterid with a heavy finch-like bill. Further south in Kansas he found one sparrow, Horned Lark (*Eremophila alpestris*), and Western Meadowlark (*Sturnella neglecta*), the latter another icterid that eats seeds in winter. Breeding bird communities at IBP grassland sites contain 3-6 species of gleaning insectivorous birds of which 1-3 are finches and the others are usually species, such as Horned Larks and Western Meadowlarks, that are primarily granivorous in the winter (Wiens, 1973). In scrub habitats the importance of finches that gather insects while foraging at a central place is greater at high- than at low-latitude sites (table 4).

Alternatively, a finch-like bill might be considered better adapted for multiple prey captures than the longer, thinner bills of strictly insectivorous species. This possibility is difficult to evaluate on present evidence, though blackbirds, which do feed extensively on seeds during colder months of the year, are capable of holding large numbers of insects with little apparent

TABLE 4

COMPOSITION OF GLEANING INSECTIVOROUS BIRD COMMUNITY DURING THE BREEDING SEASON IN VARIOUS SCRUB HABITATS (DATA FROM CODY, 1974) (THE CHILEAN SITES ARE ALL AT RELATIVELY LOW LATITUDES.)

	NUMBER OF SPECIES OF BREEDING						
Site	Flycatchers*	Finches	Blackbirds	Ovenbirds, Wrens, and Gnatcatchers	Titmice and Wrentits	Warblers and Vireos	Thrushes, Thrashers, Tapaculos
Wyoming, willows	...	5	...	...	...	3	1
Wyoming, sagebrush	...	3	1	...	...	...	...
Colorado, saltbush	...	2	1	...	...	...	1
Mohave Desert	...	1	...	1	2	...	1
Calif., chaparral	...	3†	...	1	3	2	...
Sonoran Desert	...	1	...	2	1	1	1
Arizona mesquite	...	3	...	2	...	...	2
Coastal scrub, Chile	1	2	...	...	...	...	1
Chile, matorral	2	2	1	3	...	...	3
Chile, savannah	2	2	2	3	...	...	1

*Gleaning species only.
†All terrestrial.

diminution in ability to caputre additional prey items. However, Brewer's Blackbirds, which have relatively thin bills, appear to be as effective at this as Redwings, which have thicker bills.

Although most of the theory presented here is yet to be tested, opportunities to do so are excellent. Among social Hymenoptera there are both single-prey loaders (ants) and multiple-prey loaders (bees), and many species lend themselves readily to manipulative studies of their foraging behavior. Birds are less easy to manipulate in this regard, but there is a richness of single- and multiple-prey loaders that are conspicuous when foraging and returning to their nests. Seed-caching rodents are also excellent research subjects whose food supplies can be altered experimentally.

There is an interesting implication of the theory for problems of competition among organisms. For example, since prey selection may change as a function of distance from the central place, the nature and extent of depression of prey by potential competitors may also change as a function of distance from their respective central places. This could permit, for example, the existence, in a highly competitive situation, of colonies of a small central place forager in a region dominated by a larger one provided the smaller one's colonies were located some minimal distance from those of the larger.

ACKNOWLEDGMENTS

Many of our ideas had their genesis in a discussion group on foraging theory that included E. L. Charnov and C. W. Fowler. During subsequent development of our ideas we have profited greatly from suggestions by J. L.

Gould, H. S. Horn, and J. F. Quinn. T. W. Schoener and E. L. Charnov provided valuable suggestions during the final stages of manuscript preparation. Field work on blackbirds has been generously supported by grants to Orians from the National Science Foundation. K. J. Orians, D. R. Paulson, and S. Syrjala helped with the field work.

APPENDIX

Proof of statement in main text that a tangent though a point C_1' *on the upper curve in figure 1 intersects the ordinate at* C_1, *the corresponding value of the lower curve at* t_{s_1}'.

The proof will be performed for an array of discrete prey types, assumed to be very closely spaced in available energies, thus closely approximating a continuum of energy values. Let

C_k=energy obtainable from prey type k
$<C_{k+1}$ for every k.

Also let

$C'_k = E$ [Energy obtainable | Prey encountered has energy $\geq C_k$]
$t'_{s_k} = E$ [Search time before encountering prey of energy $\geq C_k$]
$\lambda_k = E$ [Number of encounters with prey type k per unit time searching].

We wish to show that for any pair of adjacently ranked prey types, k and $k+1$, the line passing through C'_k and C'_{k+1} intersects the ordinate axis at an energy C_k (figure 11).

For a Poisson encounter process, we have

$$C_k' = \frac{\sum_{i \geqslant k} C_i \lambda_i}{\sum_{i \geqslant k} \lambda_i} \tag{1}$$

and

$$t_{s_k}' = \frac{1}{\sum_{i \geqslant k} \lambda_i} \tag{2}$$

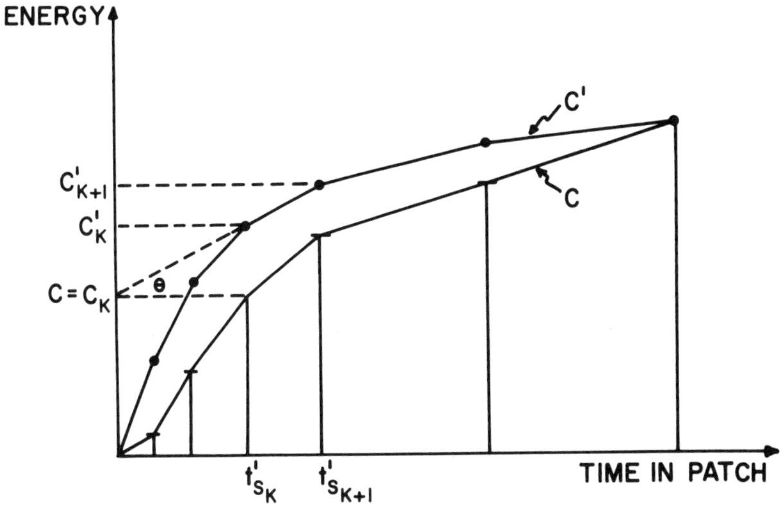

Fig. 11. The discrete analog to figure 1. C and C' are envelopes of C_k and C_k' for every prey type k. For any choice of k the line connecting C_k' and C'_{k+1} intersects the ordinate at C_k.

From figure 11 it is clear that

$$\tan\theta = \frac{C'_{k+1} - C}{t'_{s_k+1}} = \frac{C_k' - C}{t'_{s_k}}. \tag{3}$$

Substitution of equations (1) and (2) into equation (3) leads to:

$$C = C_k,$$

which was to be proved.

As more and more prey types are added, figure 11 will approach more and more closely the continuous analog presented in figure 1, with the line tangent to C' intersecting the ordinate at the corresponding value of C.

It is this property that requires the *slope* of C' to be monotone decreasing, that is, that C' has no inflection points. The reader may verify this for himself by assuming a curve C' with inflection points and noting that in those regions where the slope of C' is increasing, the intersection of tangent lines with the ordinate axis is decreasing. This implies that the corresponding values of C are decreasing, which violates the condition that C must be monotone increasing.

LITERATURE CITED

Ashmole, N. P., and H. Tovar. 1968. Prolonged parental care in royal terns and other birds. Auk 85:90–100.

Bedard, J. 1976. Coexistence, coevolution, and convergent evolution in seabird communities: a comment. Ecology 57:177–84.

Brown, J. H., and G. A. Lieberman. 1973. Resource utilization and coexistence of seed-eating desert rodents in sand dune habitats. Ecology 54:788–97.

Charnov, E. L. 1976. Optimal foraging theory: attack strategy of a mantid. Amer. Natur. 110:141–51.

Cody, M. L. 1968. On the methods of resource division in grassland bird communities. Amer. Natur. 102:107–47.

Cody, M. L. 1973. Coexistence, coevolution, and convergent evolution in seabird communities. Ecology 54:31–44.

Cody, M. L. 1974. Competition and the structure of bird communities. Princeton University Press, Princeton, N.J. 318 pp.

Covich, A. P. 1976. Analyzing shapes of foraging areas: some ecological and economic theories. Ann. Rev. Ecol. and System. 7:235–57.

Durango, S. 1951. Om törnskatans (*Lanius collurio* L.) spetsning av bytesdjur. Fågelvarld 10:49–65.

Fretwell, S. D. 1972. Populations in a seasonal environment. Princeton University Press, Princeton, N.J. 217 pp.

Horn, H. S. 1968. The adaptive significance of colonial nesting in the Brewer's Blackbird (*Euphagus cyanocephalus*). Ecology 49:682–94.

MacArthur, R. H. 1972. The economics of consumer choice. In Geographical Ecology. Pp. 59–69. Harper and Row, New York.

Main, B. Y. 1957. Biology of Aganippine trapdoor spiders (Mygalomorphae: Ctenizidae). Austr. J. Zoology, 5:402–73.

Michener, H., and J. R. Michener. 1945. California jays, their storage and recovery of food, and observations at one nest. Condor 47:206–10.

Orians, G. H. 1969. Age and hunting success in the Brown Pelican (*Pelecanus occidentalis*). Anim. Behav., 17:316–19.

Orians, G. H. 1972. The Red-winged Blackbird in tropical marshes. Condor 75:28–42.

Orians, G. H., and H. S. Horn. 1969. Overlap in foods and foraging of four species of blackbirds in the Potholes of central Washington. Ecology 50:930–38.

Pearson, N. E. 1974. Optimal foraging theory. Quantitative Science Paper No. 39, University of Washington, Seattle. 48 pp.

Pearson, N. E. 1976. Optimal foraging: some theoretical consequences of different feeding strategies. PhD thesis, University of Washington.

Pulliam, H. R. 1974. On the theory of optimal diets. Amer. Natur. 108:59–74.

Recher, H. F., and J. A. Recher. 1969. Comparative foraging efficiency of adult and immature Little Blue Herons (*Florida caerulea*). Anim. Behav. 17:320–22.

Schoener, T. W. 1971. Theory of feeding strategies. Ann. Rev. Ecol. and Systematics 2:369–404.

Schuster, L. 1950. Über den Sammeltrieb des Eichelhähers (*Garrulus glandarius*). Vogelwelt 71:9–17.

Skutch, A. F. 1949. Do tropical birds rear as many young as they can nourish? Ibis 91:430–55.

Smith, C. C. 1968. The adaptive nature of social organization in the genus of tree squirrels *Tamiasciurus*. Ecol. Monogr. 38:31–63.

Swanberg, P. O. 1951. Food storage, territory, and song in the Thick-billed Nutcracker. Proc. X. Int. Orn. Congr, 1950:545–54.

Wiens, J. A. 1973. Pattern and process in grassland bird communities. Ecol. Monogr. 43:237–70.

PHILIP C. MILLER

Quantitative Plant Ecology

7

A HISTORICAL PERSPECTIVE ON GENERAL AND PLANT ECOLOGY

Ecology has developed several traditions or schools of thought, including: population-community, ecosystem, plant community ecology, and physiological plant ecology (fig. 1). The population-community tradition is concerned with organisms, i.e., how organisms act in populations, how populations fit

TRADITIONS IN ECOLOGY

POPULATION/ COMMUNITY	ECOSYSTEM	PRODUCTION PROCESSES	PLANT ECOLOGY
control of population numbers, evolutionary questions	flows of water and energy	control of primary production (of crops)	succession, community description

PHYSIOLOGICAL ECOLOGY

evolutionary questions

Fig. 1. Some intellectual traditions in ecology.

together, evolutionary questions, and generalizable organizational principles concerning the coexistence of populations. The ecosystem tradition is concerned with the flow of matter and energy between biotic and abiotic parts of the ecosystem and how these flows are controlled, and with interactions between species in similar trophic levels or functional groups.

The population-community and ecosystem traditions are based on quantitative theories. Most early theories and some contemporary theories in each

Philip C. Miller, Department of Biology, San Diego State University, San Diego, California 92182.

tradition were simple enough so that one person could grasp the totality of the model system and understand its strengths and weaknesses as an explanation of real world events. Later theories incorporated more mechanisms and were less descriptive, but became more complex. As the ideas and data of more people are incorporated in a complex model, they become less intuitive and are often less satisfying to any one person. The simple models help clarify central critical issues, but complex models more often provide a means of directly testing these broad ideas.

The plant community tradition began with concerns about the functional relations between individual plants, the factors determining the distribution of species, and the temporal sequence of species, and developed into a concern with the organization of landscapes into meaningful, and real, units. The initial subjective evaluations and groupings were later evaluated with objective, and quantitative procedures (Williams, 1964; Grieg-Smith, 1964; Goodall, 1952). The physiological plant ecology tradition has been concerned with adaptations of plants to their environment, and takes an approach similar to the evolutionary studies of population ecologists.

All four ecological traditions search for general organizational principles that are independent of the actual species considered. These organizational principles provide short cuts to a complete and exhaustive understanding of the physiology of the individuals. One such principle or hypothesis of population-community ecology is that as the utilization of resources becomes maximal, species enter or leave a community (MacArthur, 1970), a hypothesis pertinent to the mechanism of plant succession. Within the ecosystem tradition we find a second hypothesis: the biota of the ecosystem evolves so as to maximize energy turnover (energy out of a trophic level divided by biomass of the trophic level), even at the expense of reduced efficiency (energy out of a trophic level divided by energy into the trophic level) (Odum and Pinkerton, 1955). In succession, population replacement is by species with higher rates of energy turnover. Maximizing energy turnover in plants involves maximizing the efficiency with which energy from the sun is captured (energy converted to photosynthate divided by the incoming flux).

While these traditions were developing within the field of ecology, mathematical models of primary production processes developed. This theory began with simple concepts for light penetration through monospecific canopies, and for photosynthesis and respiration of individual leaves. Field and laboratory experiments were used to develop more inclusive and complex models from these early efforts, in order to elaborate upon the physical and physiological controls on primary production. Much of this theoretical framework was developed in agricultural research. Physiological ecologists neglected models of canopy production processes. The isolation of these two groups ended with the impress of the IBP in which the modeling tradition of primary production process research worked with the experimentation talents of physiological plant ecologists to make the physiological plant ecologists aware of the production modeling concepts.

Production process research developed in much the same way as did other areas of ecology. Production process modeling began with Monsi and Saeki (1953) who used a mathematical model of light penetration in vegetation canopies and of leaf photosynthesis-light relation to explore the interrelations between leaf inclination, leaf area index, incoming light intensity, and primary production. They showed that canopy characteristics, which were often found under different light intensities, would maximize primary production in these conditions. A debate followed, in which diverse counter examples of the relation of canopy structure to maximum production were presented (Donald 1961). This debate was resolved, more or less, as a clearer understanding of respiration acclimation developed (McCree and Troughton, 1966). Davidson and Phillip (1958) developed independently a production model for pasture growth and used it to simulate strategies for maximizing primary production with different grazing regimes. These analyses of canopy architecture and the grazing strategy for maximum production are similar to current ideas in ecology, which are often proposed without the more solid mathematical, physical, and physiological bases of the earlier work.

The parallel concerns in ecological and production process research, and the capability of using production process models to account for a diverse set of physiological and environmental data indicates a potential contribution of the production process models to ecological questions. If the utilization of resources becomes maximal as species enter and leave a community, there must be a stable canopy configuration that utilizes some resources fully. For plants, these resources should usually include light energy, water, nitrogen, and phosphorus. Light energy and water, together with temperatures, were recognized as factors controlling plant succession and plant distribution from the beginning (Schimper, 1898; Warming, 1895; Clements, 1916; Braun-Blanquet, 1951). The utilization of these resources can be studied within the framework of production process models, which will, in addition to pointing out unused resources, provide a basis for understanding the mechanism of resource utilization by coexisting plant species, plant succession, plant distribution, and plant adaptation. The framework permits quantification of many interactions discussed subjectively in early ecology, and provides an objective, quantifiable framework for concepts of "empty niches" or "open" communities (Elton, 1958). The two general principles proposed by the population-community and the ecosystem traditions can be tested if species in a community can be characterized independently of the outcome of the competitive encounter or ecosystem development. An independent physiological characterization of a species represents its fundamental niche. If the physiological characteristics are encoded in mathematical models, the behavior of the species in simulation experiments that include various competing species represents the realized niche of the species in the simulated situation (Botkin 1975). The simulation models perform the role of integrating basic physiological relations and bringing them to bear on questions of resource utilization by populations or in ecosystems. These models can be developed partially from the tradition of production process research.

Research on production processes can be conveniently grouped by component processes (fig. 2): canopy processes—radiation exchange, turbulent exchange of heat, water vapor and carbon dioxide; soil processes—the diffusion of heat and water through the soil; individual plant processes—water balance, energy and carbon balance, nitrogen and phosphorus balance, growth, and reproduction; single species population dynamics—the initiation and mortality of plant parts, which may be stems, leaves, inflorescences, seeds, tillers, or new autonomous individuals; and multispecies population dynamics, which has usually involved studies of yields of pasture species or annual weeds and crops.

The different groups of processes have different spatial and temporal scales. Canopy processes are conceived in terms of land surface area, because the light and water are received on this basis, and in terms of minute changes on daily time sequence. Annual summaries are not very interesting. Soil processes are conceived in terms of soil volumes for decomposition, root growth, and the release of nitrogen and phosphorus and in terms of land surface area for heat and water diffusion, on time scales of minutes or days. Individual processes of water balance, photosynthesis and respiration are considered on an individual plant basis, usually on a minute-to-minute basis but also on a daily basis. Growth may be considered on a minute-to-minute

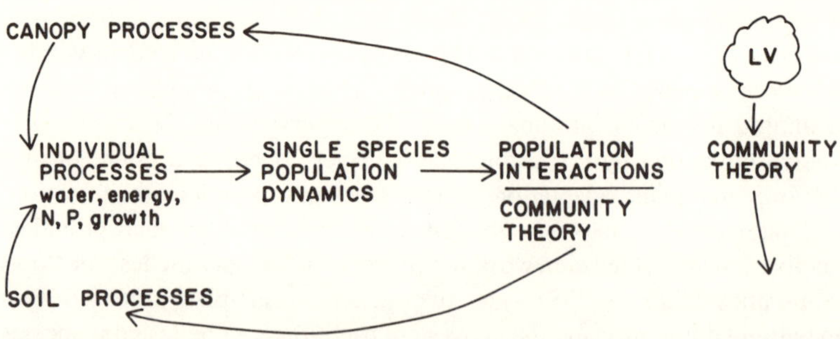

Fig. 2. Major categories of processes in production process research in relation to community theory of the Lotka-Volterra (LV) heritage.

basis, but it is usually considered on a daily basis and sometimes on an annual basis. Population processes are treated as number of plant parts per individual or numbers of individuals per land surface area, on time scales of days, weeks, or years. Population processes bridge a gap between individual physiology and long term processes on the land surface area. Answers to questions of current resource utilization can be approached by considering canopy, soil, and individual processes, but answers to questions of the mechanisms by which the current resource utilization comes about, or of predictions of species composition as resources and constraints are changed, require inclusion of population processes.

The objective of this paper is to illustrate the application of a methodology developed from production process research to problems of how resources are utilized in plant communities. The principal example will be the resource utilization of mediterranean scrub vegetation, with brief references to the tundra vegetation. The concepts presented here have been developed from research in both these vegetation types as well as in mangrove scrub forests. The relationship of evergreen and deciduous-leaved plant forms to patterns of water availability and temperature will be examined at the regional level, and the utilization of light, water, nitrogen, and phosphorus will be considered at the local level. The total utilization and the division of resources among species will involve two times scales: (1) the seasonal pattern of resource availability during a year, and (2) the year-to-year variability in resource availability. Both involve consideration of the synchrony of uptake of the resource from the environment and utilization of the resource in growth or reproduction.

The approach has been to develop models from field measurements and use these models to synthesize and generalize the data. The primary production model was first developed with sponsorship of the AEC to study heat and gas transport processes of the red mangrove canopies in areas receiving thermal effluents from a power plant (Miller, 1972 a,b). This early research increased the basic understanding of the primary production system as well as suggesting less-environmentally-damaging ways of treating heated water in mangrove areas (Miller, 1975). In addition to the AEC research, the models of primary production processes and of interactions between the vegetation and the physical environment have been developed and used in research with the Tundra Biome (Miller and Tieszen, 1972; Miller et al., 1976a) and the U.S. Origin and Structure of Ecosystem programs of the IBP (Miller and Mooney, 1975). These team research programs have provided major advances in model development.

The models generally have been developed *a posteriori,* and form a complex set of hypotheses about the mechanisms and behavior of the plant production system. The modeling efforts have provided a basis both for developing fundamental concepts and for applying research results in a diversity of vegetation types. The models include: (1) the effects of the vegetative structure on profiles of solar and infrared radiation, wind, air and soil temperatures, and vapor density (Miller, 1972; Miller, 1975; Ng and Miller, 1975, 1977; Roberts, 1974); (2) the effects of these microenvironmental variables and plant surface properties on the exchange of heat, water vapor, and carbon dioxide by leaves and stems in different positions in the canopy (Miller, 1972, 1975; Miller and Tieszen, 1972; Miller et al., 1976); (3) the effects of internal morphological and physiological plant properties on rates of photosynthesis and respiration, and on the uptake and loss of water and minerals (Miller et al., 1976); (4) the effects of daily and seasonal patterns of allocating carbohydrates and minerals in plant growth on primary production, including some of the physiological mechanisms and microenvironmental variables affecting these patterns (Stoner et al. 1976; Stoner et al., 1978; Miller et al., 1976); (5) the effects of different reproductive patterns of plant survival and dispersal in different environments (Lawrence et al., 1970); and (6) the accumulative effects of 2, 3, 4, and 5 on community composition, canopy structures, and soil thermal and chemical properties and the competition of plant species within an ecosystem for light, water, nutrients, and heat.

THE MEDITERRANEAN SCRUB ECOSYSTEM

World View

Mediterranean scrub ecosystems occur in five disjunct regions of the world: around the Mediterranean Sea, on the southwest coast of Africa, in southwestern and southern Australia, in central Chile, and in southern California. They have attracted the interest of plant geographers and ecologists because of the great similarity in the form of the vegetation, which originated from quite different phylogenetic lines. The regions have cool, wet winters with low solar irradiance, and hot, dry summers with high irradiance (fig. 3). Total annual precipitation is 300–700 mm received in monthly increments of up to 100 mm in midwinter. Monthly mean temperatures are 5° to 20°C. This climatic regime evidently constrains plant and ecosystem processes so specific plant forms are at a great selective advantage.

The patterns of vegetation and climate in southern California and Chile have been compared in some detail (Mooney, 1977; Thrower and Bradbury, 1977). Both regions are located poleward of the subtropical high-pressure zone with a cold ocean current offshore. In both regions the coastal plain extends inland about 20 km. Foothills rise to mountains with altitudes reaching about 2,000 m. On the inland side of these coastal mountains, the land surface descends to low elevations and inland desert in southern Califor-

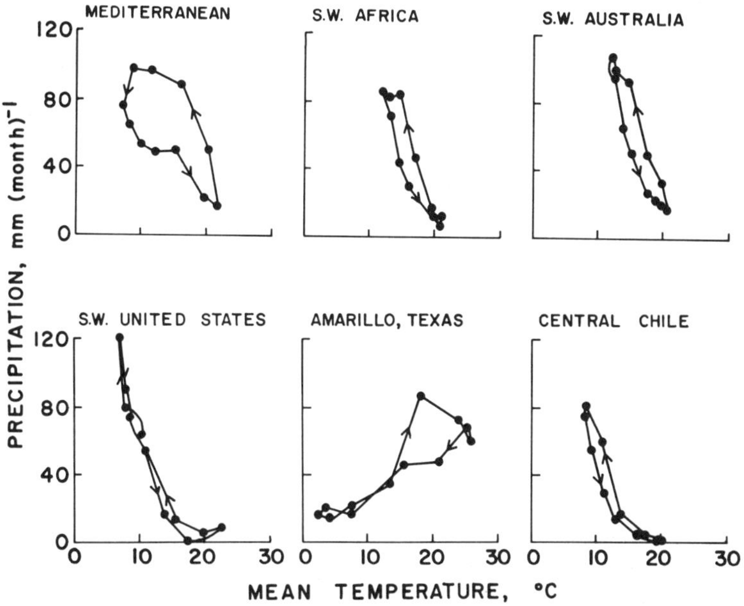

Fig. 3. Seasonal progression of monthly mean temperatures and precipitation in five mediterranean scrub regions of the world and Amarillo, Texas, a grassland region.

nia and to a central valley and then the Andes in central Chile. In both regions evergreen, sclerophyllous shrubs predominate inland from the coasts at about 1,000 m elevation. Along the coasts at lower elevations deciduous shrubs occur. In the mountains, at about 2,000 m, deciduous trees and herbs appear in the vegetation (Parsons and Moldenke, 1974; Parsons, 1973).

Climatic patterns shift from the coast inland to the mountains. Annual precipitation increases with elevation. Temperatures along the coasts are more moderate and inland more variable, due to the decreasing influence of the ocean. A temperature inversion occurs below about 700 m in California and 1,000 m in Chile, maintaining cool temperatures along the coast in summer. The most extreme temperatures occur in the foothills, just above the inversion. Above the inversion temperatures decrease adiabatically with altitude.

Bioclimatic Patterns and Leaf Form in Southern California

The geographic pattern of evergreen and deciduous-leaved plant forms can be related to patterns of water availability and temperature through the influences of the latter on the carbon balance of these plant forms. Miller and Mooney (1974) suggested that the segregation of these forms was due to the lengths of the drought along the coast and inland, and the carbon costs of maintaining or re-creating leaves. The evergreen form predominates where the carbon costs of maintaining leaves through the unfavorable period is less than the carbon cost of re-creating leaves after the drought. The deciduous form predominates where the carbon cost of maintaining leaves through the unfavorable period was greater than the cost of re-creating the leaves.

Their analysis is extended here by elaborating on the interactions between temperature, precipitation, soil moisture, photosynthesis, and growth. The elaboration was suggested in part by studies in which argo-climatic zones were developed from precipitation, potential evapo-transpiration, and potential soil moisture availability in mediterranean, sub-sahelian, and Australian regions (Wallen, 1968; Wallen and Brichambaut, 1962; Cocheme, 1968; Cocheme and Franquin, 1967; McAlpine, 1970). Although the analysis is based on a generalized view of physical and physiological processes, it is supported by measurements. An analysis such as this may be used (McAlpine, 1970) for situations in which data are not adequate for a detailed analysis.

Precipitation and altitude. Precipitation records were collected from 150 stations in San Diego County in southern California. Records from stations located in a transect from the coast inland were plotted against altitude, considering the location of the station in relation to surrounding topography and possible rain shadow effects. Temperature records were collected for 30 stations. Annual and monthly mean maximum and minimum temperatures were plotted in relation to altitude. Mean daily temperatures were calculated as the mean of the maximum and minimum temperatures. Mean daylight temperatures were calculated as 0.75 times the maximum plus 0.25 times the minimum temperatures (van Keulen, 1975).

On the coastal slope of the coastal mountains, precipitation increases with altitude at a rate of about 313 mm per 1 km elevation (fig. 4). Along the coast annual precipitation is about 240 mm. Stations on the lee side of the mountains receive 75–100 mm less precipitation than coastal-facing stations at similar elevations. Most of the precipitation in southern California occurs from November to April.

Potential evaporation. Potential evaporation was calculated by the Penman equation (1948). This approach has been documented for arid regions by Stanhill (1961) and Omar (1968). The potential evaporation depends on net radiation and the drying power of the air. Solar radiation was measured at the coast and inland at 1,000 m and 2,000 m with pyrheliographs from April 1971 to December 1975. Along the coast the commonly observed pattern is for fog or clouds to move inland from the ocean to the foothills at night. These clouds are evaporated during the morning, from inland toward the coast. The marine air and clouds are trapped within the inversion. The albedo of the vegetation and ground was assumed to be 0.25, in order to make the calculations consis-

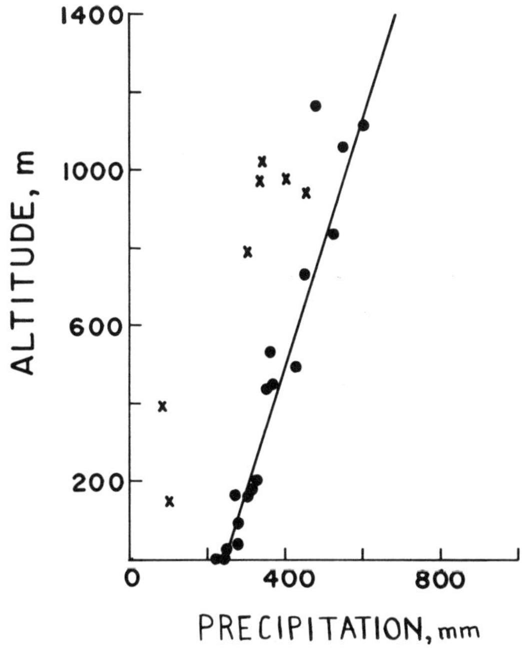

Fig. 4. Annual precipitation and altitude in southern California, based on 20 years' records.

tent with earlier studies (Cocheme, 1968; Cocheme and Franquin, 1967; Wallen, 1988; Wallen and Brichambaut, 1962), even though measured values for the albedo of the vegetation were 0.10–0.20 (Roberts, unpublished data). Solar radiation was assumed to increase from low elevations at the coast to higher elevations at the top of the inversion in the foothills (fig. 5). Above the inversion solar radiation was assumed constant until 1,000 m, above which it decreased with increasing elevation. Measurements of solar radiation at the coast, 1,000 m, and 2,000 m gave absolute values of solar irradiance at different sites along the altitudinal gradient. Net infrared radiation was calculated by the Brunt equation (1932), which expresses a relation between infrared radiation from the sky and vegetation, air temperature, vapor pressure, and cloud cover. Net infrared radiation decreased (became more negative) from the coast inland, corresponding with decreasing cloud cover and decreasing air temperatures. Along the coasts I assumed cloudy nights and clear days and in the mountains clear nights and cloudy days. The contribution to total potential evaporation of net radiation decreases from the top of the inversion to higher elevations and decreases slightly with lower elevations.

The drying power of the air depends on the vapor pressure deficit and wind. The mean daily vapor pressure of the air was calculated as the saturation vapor

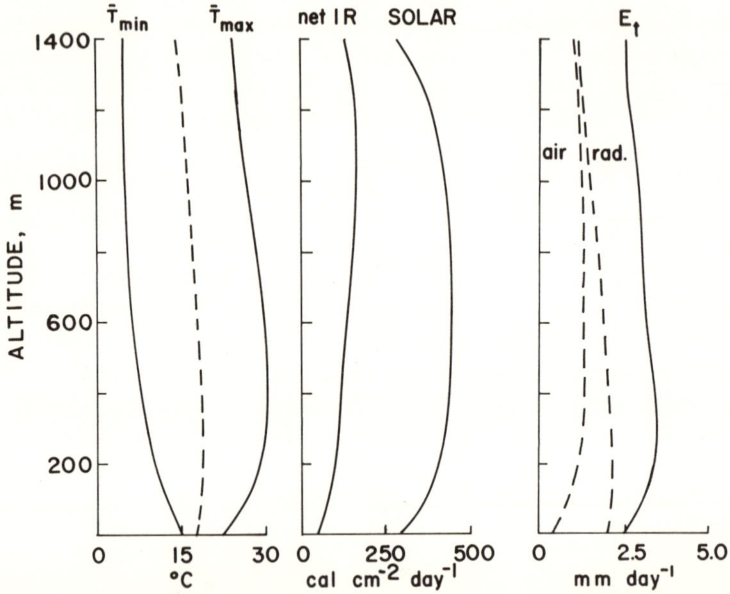

Fig. 5. Mean annual maximum ($\bar{T}_{max}$) and minimum ($\bar{T}_{min}$) temperatures, solar radiation (SOLAR), net infrared radiation (net IR), and potential evapotranspiration (E_t), and the radiation term (rad.) and air term (air) in the Penman equation.

pressure at the minimum air temperature. Measured relative humidities were usually 100 percent at night, beginning at about 2200 hr. Temperatures after humidities were 100 percent decreased slightly, about 2°C, until morning. Through the day vapor pressures were nearly constant, increasing 1–2 mb through the course of the day. Wind was assumed to be equal at all altitudes in lieu of measurements along the gradient, since wind was only measured at the coast and at 1,000 m. The drying power of the air follows the daily maximum air temperature. It is relatively constant above the top of the inversion, and decreases within the inversion.

With these assumptions mean annual potential evaporation is greatest, about 3.5 mm day^{-1}, at about 400 m and decreases to about 2.5 mm day^{-1} at the coast and in the mountains.

The drought period. The potential duration of drought was calculated by assuming that the total annual precipitation occurred in a 6-month period, during which time evaporation proceeded at ½ the mean potential annual rate. The moisture surplus, which was left in the soil in the spring when precipitation stopped, was assumed to evaporate at 1.5 times the mean annual rate. In this way the period of the year with soil moisture, and then the period without rain or soil moisture, i.e., the period of drought, were calculated.

The drought period was 5.5 months at the coast and decreased to 1 month in the mountains. At 1,000 m the calculated length of drought was about 3.5 months, similar to the drought measured by soil moisture and plant water potentials (Poole and Miller, 1975). The calculated length of drought at the coast is 1 month longer than that described by Poole and Miller (1975). During the drought, photosynthesis should have stopped because of stomatal closure (Poole and Miller, 1975; Harrison, 1971), and growth ceased.

Air temperature. Annual mean maximum air temperatures increased with altitude through the inversion, then above about 400 m elevation, they decreased at a rate of 5.4°C per 1 km elevation (fig. 5). Mean minimum air temperatures decreased with elevation, more rapidly within the inversion and near the coast and less rapidly above the inversion. Mean temperatures during the daylight period are above 15°C throughout the year along the coast (fig. 6). Temperatures above 17.5°C occur only during the drought. Inland, and with increasing elevation, mean daytime temperatures during the winter decrease, such that at about 1,100 m, temperatures are below 10°C during the winter. However, with higher elevation, temperatures become warmer before

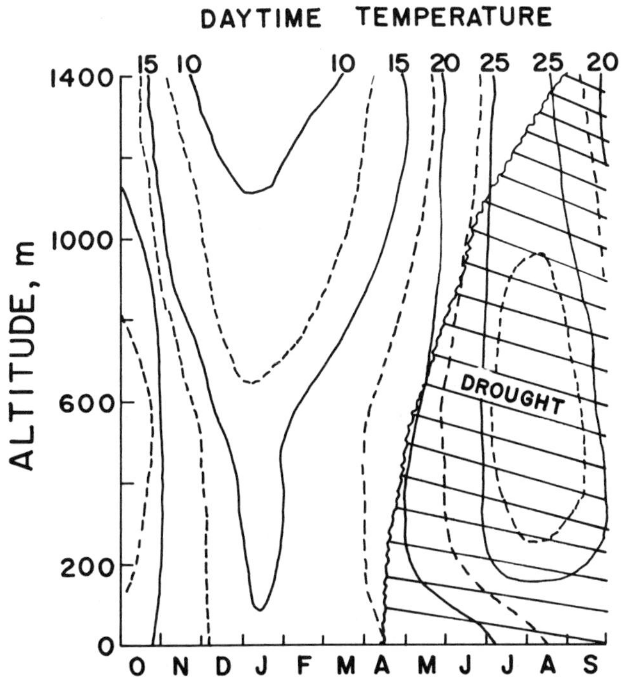

Fig. 6. Monthly mean temperatures (in °C) through the daylight period with altitude and drought period.

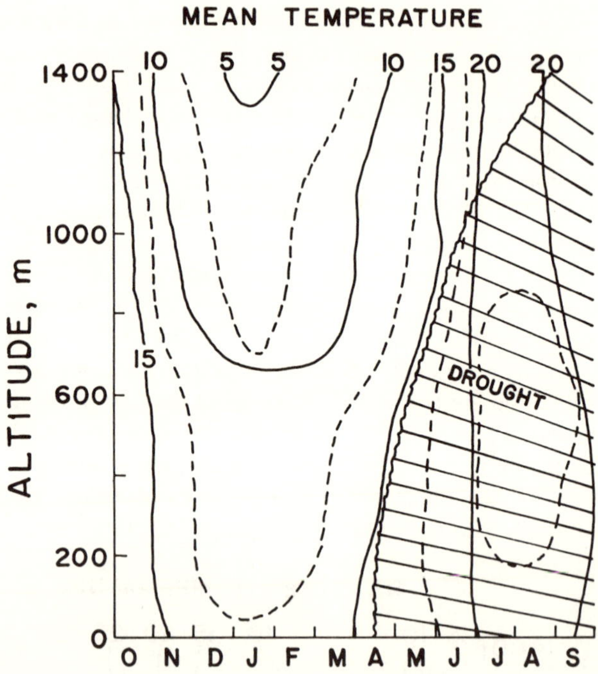

Fig. 7. Monthly mean temperatures with altitude and drought.

the drought period begins. At 700 m temperatures reach 20°C before the drought and at 1,100 m temperatures reach 25°C before the drought. Mean daily temperatures follow a similar pattern but are 2–5°C lower (fig. 7). Thus, the season of moderate temperatures and available water is winter along the coast, spring at midaltitudes, and late spring to summer at the higher altitudes. The daytime temperatures of this favorable period tend to increase with altitude.

Carbon balance and leaf form. Data for citrus (Reuther, 1973) were available and used to represent the photosynthesis-temperature relations of evergreen chaparral shrubs. Photosynthesis rates depend on the daytime temperatures. The photosynthesis-temperature relation was similar to that measured by Harrison (1971) on *Heteromeles.* In winter photosynthesis decreases with altitude (fig. 8), and all temperatures are below optimum. As temperatures rise in the spring, photosynthesis increases at all altitudes, but the increase is truncated at the coast by the drought before high rates can occur. At higher elevations relatively high rates occur in the spring and summer, before the drought. These rates of photosynthesis were assumed to include respiratory

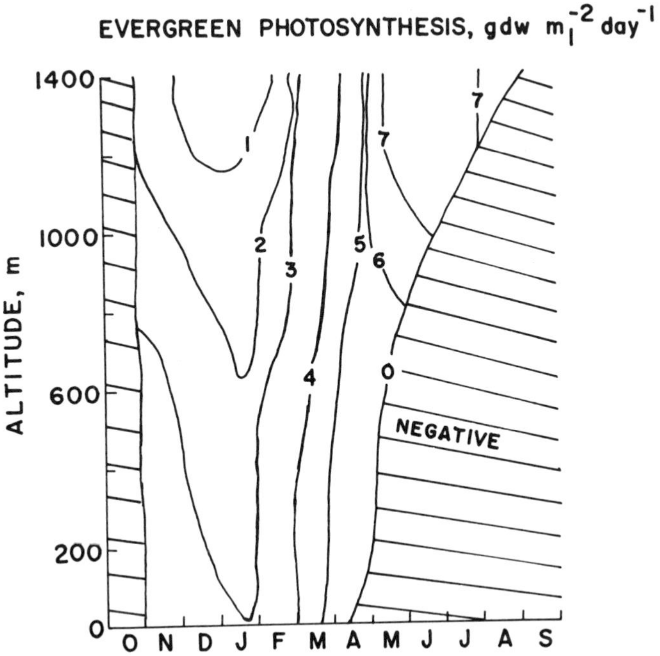

Fig. 8. Potential photosynthesis of evergreen shrubs with altitude and time of year.

costs for maintaining leaf tissue during the daytime period. Additional respiratory costs for maintaining the tissue during the night were subtracted, varying through the season with night length. Dry weight gain varied between 0.00 g dry wt. m^{-2} leaf day^{-1} in winter to 0.07 g dry wt. m^{-2} leaf day^{-1} in spring and summer at high elevations, and was 0.03–0.05 g dry wt. m^{-2} leaf day^{-1} along the coast when the drought was absent. During the drought photosynthesis was not assumed to occur and maintenance respiration was subtracted for 24 h/day. Annual dry weight gain increased with elevation from 252 g dry wt. m^{-2} leaf yr^{-1} at the coast to 1,083 g dry wt. m^{-2} leaf yr^{-1} at 1,400 m.

The photosynthesis of a deciduous shrub depends on the leaf biomass present and this depends on the growth of new leaves after the beginning of the precipitation period. Growth was assumed to occur through the 24 h period and to be proportional to the daily mean temperature when soil moisture was available. Daily mean temperatures were slightly above 12.5°C along the coast through the winter, but not above 10°C in winter at elevations above 700 m (fig. 8). Temperatures were below 5°C at the higher elevations. However, below 900 m temperatures hardly reach 15°C in the spring before the drought period begins. At higher elevations they reach 20°C before the drought. Precipitation begins in the fall when temperatures are about 15°C at the coast but below 12.5°C at 700 m and above. Thus, temperatures adequate for growth can be expected throughout the precipitation period along the coast, briefly in the fall and again in the spring at midaltitudes and only in the spring and summer at the higher altitudes.

The relation of leaf growth to temperature for citrus (Reuther, 1973) was used to represent the relation for chaparral shrubs, because data for the chaparral were not available. The maximum growth rate in grams dry weight per plant per day was calculated from data presented in Mooney et al. (1976). The maximum rate for chaparral shrubs was about 10 g dry wt. per plant per day at 25°C and assumed to decrease rather linearly to zero growth at 11.5°C. There is a gradual increase in leaf biomass during each month of the precipitation period along the coast (fig. 9). At midaltitudes leaves grow in the fall, but not through the winter. At high altitudes, growth does not occur until spring and summer. The temperatures along the coast during the growth period are moderate, and a total of about 360 g dry wt. per plant per year is produced. At

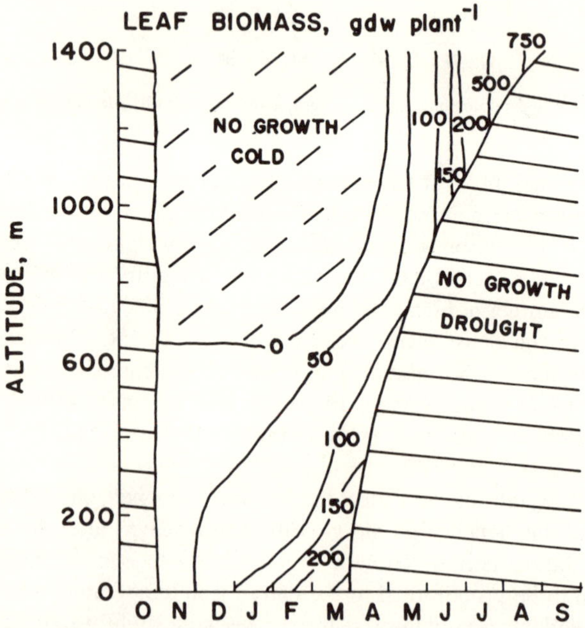

Fig. 9. Potential leaf biomass for a deciduous shrub with altitude and time of year, affected by cold temperatures and drought.

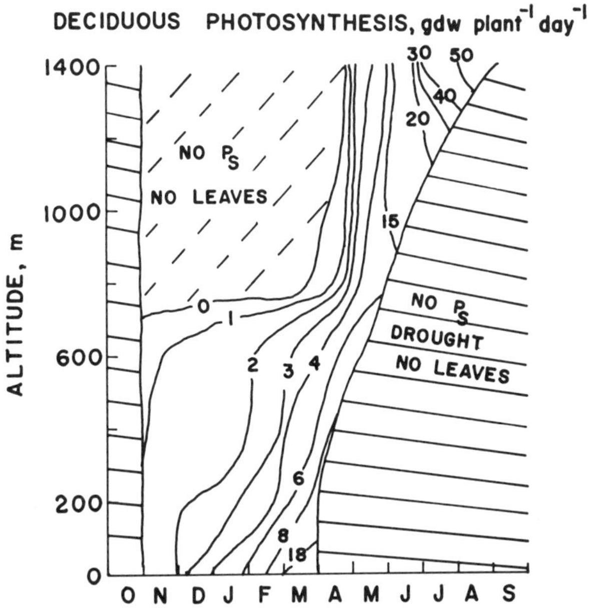

Fig. 10. Potential photosynthesis of a deciduous shrub with altitude and time of year.

midaltitude, 800 m, annual growth is depressed to 90 g dry wt. per plant per year. Temperatures are most favorable for growth at the higher elevations, and even though growth does not begin until spring, the total growth is about 790 g dry wt. per plant per year.

For a positive carbon balance to occur, annual photosynthetic gain should be at least equal to the dry weight of the leaves plus 30 percent for respiratory costs of synthesis (Penning de Vries, 1972). The rate of photosynthesis of the deciduous shrubs was assumed to be twice the rate of evergreen shrubs (Miller and Mooney, 1974; Mooney, 1972), or twice the rate for citrus, and be proportional to the standing crop of leaf biomass. Photosynthesis varied from 17.7 g dry wt. per plant per year at the coast to 2.6 at midaltitude and 57.8 at the highest altitudes (fig. 10). Subtracting the growth and respiratory costs from the production to arrive at the dry weight balance for the deciduous shrub and assuming that the specific leaf density is 0.212 g dry wt. per m^2 leaf, gives 0.53 kg dry wt. per plant per year at the coast, 0.08 kg dry wt. per plant per year inland and 1.73 kg dry wt. per plant per year at the highest elevations (fig. 11). Subtracting 0.212 g dry wt. per m^2 leaf created and converting to a meter square basis, the dry weight balance varies from 0.85 kg dry wt. per m^2

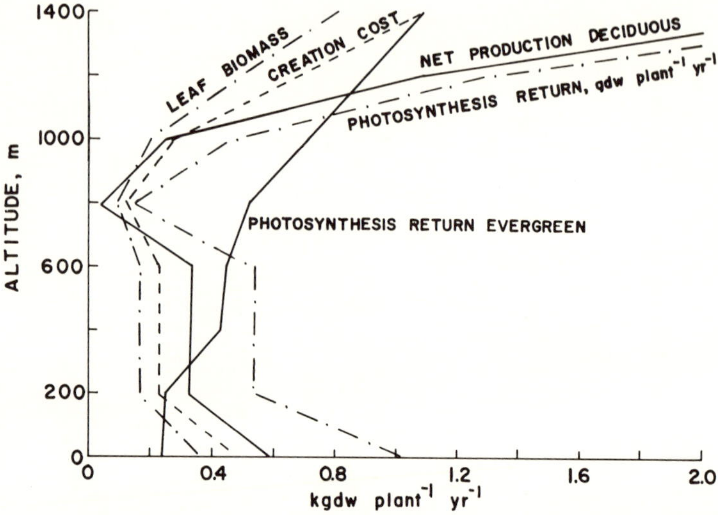

Fig. 11. Leaf biomass, cost of creating new leaves, photosynthesis, and net production for a deciduous shrub and photosynthetic return for an evergreen shrub (kg dw m^2 leaf yr^{-1}) with altitude.

per year at the coast to -0.06 at 800 m, to 3.31 kg dry wt. m^2 per year at the highest elevations.

Clearly, the deciduous shrub is at a disadvantage at the midaltitudes and favored along the coast and at the higher elevations. At the coast the growing season occurs throughout the winter. At midaltitudes the growing season is only in the spring and is too short for much photosynthesis to occur after the leaves are formed. At higher elevations the growing season is in the spring and summer, and temperatures are warm enough for reasonable growth and photosynthesis to occur. If the annual dry weight balance relates to the abundance or competitive advantage of one form over another, along the coast the deciduous form is favored, at midaltitudes the evergreen form is favored, and at the high altitudes the deciduous form is favored. Transitional areas occur along the gradient. This is the pattern that is observed (fig. 12).

PATTERNS OF LIGHT UTILIZATION

For heuristic purposes, photosynthesis will be regarded as a central focus in the following discussion, and regarded as due to a relation to an available resource and a series of efficiencies which are determined by plant and environmental properties (Monteith, 1973).

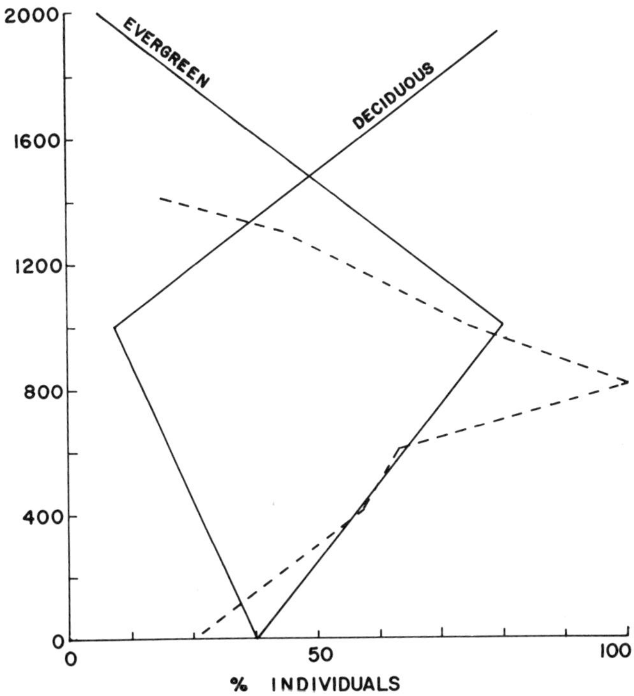

Fig. 12. Percentage of individual plants that were deciduous or evergreen with altitude (from Parsons and Moldenke, 1974). The dashed line shows the percentage of the total potential net annual production of deciduous and evergreen shrubs due to evergreen shrubs, assuming one deciduous shrub per m^2 and a leaf area index of one for evergreen shrubs.

Background on Mechanistic Models of Photosynthesis and Light

The canopy photosynthesis model was developed first for the mangroves (Miller, 1972, 1973) and used in the Arctic tundra (Miller and Tieszen, 1972; Miller et al., 1976) and chaparral (Miller and Mooney, 1974). The model is based on the energy exchange processes of a single leaf located at various heights in the canopy. Previous canopy photosynthesis models have also been based on the photosynthesis relations of a single leaf within the canopy. Monsi and Saeki (1953) developed a model to calculate stand photosynthesis, in which they assumed uniformly overcast skies and a constant rate of light penetration into the canopy regardless of the time of day or depth in the canopy. They analyzed the interrelations between leaf area index, leaf inclination, incoming light intensity, and primary production. Davidson and Phillip

(1958) developed a model, based on similar assumptions, with which they analyzed the interrelations between leaf area index, grazing schedules and primary production. de Wit (1965) proposed a more realistic model including direct and diffuse light, light scattering in the canopy, variable solar altitudes, variable leaf inclinations, and variable resistances of the air above the canopy to the transport of carbon dioxide into the canopy, and then used the model to calculate the annual course of potential production for different regions. Monteith (1965) pointed out the error in calculating photosynthesis for the total canopy using a single exponential equation for average solar intensity, as used in the previous models, and proposed a binomial expansion to express the decrease in direct and diffuse solar radiation within the canopy. Duncan et al. (1967) proposed a photosynthesis model that included direct and diffuse light, variable leaf angles, and variable solar altitudes. Lemon (1967), Stewart and Lemon (1972), and Murphy and Knoerr (1972) have been concerned with the aerial transfer of carbon dioxide and the sources and sinks of carbon dioxide in the canopy, which shows the relative contribution of different levels in the canopy to the production of the crop and the significance of the production of carbon dioxide by the soil.

The canopy radiation model was developed in the red mangrove ecosystem in south Florida (Miller, 1975) and applied to the canopies of the Arctic and alpine tundra (Tieszen et al., 1975; Miller et al., 1976a; Ng and Miller, 1975) and to the canopy of the chaparral (Lawrence, 1975; Miller and Mooney, 1974). A model of radiation in nonhomogeneous canopies and individual shrubs was developed for the chaparral (Roberts, 1974). The radiation models were developed from concepts of light penetration in continuous vegetation canopies that have a broad base of usage in temperate zone crops (Monsi and Saeki, 1953; Davidson and Phillip, 1958; de Wit, 1965; Anderson, 1966; Duncan et al., 1967; Anderson and Denmead, 1969; Yim et al., 1969; Acock et al., 1970; Idso and de Wit, 1970). Our models included infrared radiation in the canopy, which was not included previously, and the effects of stems, which has rarely been included. The model was tested and developed with data from the mangrove canopies and modified slightly for the chaparral canopy. The exceedingly small leaf size and the clustered nature of leaves are a major problem in applying the model to the chaparral.

Photosynthesis models for a leaf have been proposed by several persons (Gaastra, 1963; Lommen et al., 1971; Slatyer, 1970) and are usually based on this diffusion of carbon dioxide from free air to the chloroplast through several morphological structures that act as resistance to the flow of this gas.

Seasonal Patterns of Light Utilization

Photosynthesis is related to the fraction of the incoming solar radiation that is intercepted by the total vegetation canopy (interception efficiency), the fraction of the total intercepted radiation that is intercepted by the leaves of a particular species, the fraction of that intercepted by a species that is absorbed by the leaves of the species, the ratio of the energy converted to chemical form to the radiation energy absorbed, and the grams of photosynthesis produced per unit of energy converted. Explicitly stated:

$$\text{Photosynthesis} = (\text{Incoming solar})\left(\frac{\text{Total intercepted}}{\text{Incoming solar}}\right)\left(\frac{\text{Intercepted leaf}}{\text{Total Intercepted}}\right)\left(\frac{\text{Absorbed leaf}}{\text{Intercepted leaf}}\right)\left(\frac{\text{Energy converted}}{\text{Energy absorbed}}\right)\left(\frac{\text{Glucose}}{\text{Energy converted}}\right) \quad (1)$$

Photosynthesis expressed as grams dry weight per square meter per year and approximations to serve as baseline data for a discussion of seasonal dynamics are: photosynthesis about 2.5 kg dry wt. m^{-2} yr^{-1}; incoming solar, 147×10^3 cal m^{-2} yr^{-1}; interception efficiency, 0.5; leaf interception, 0.6; absorption efficiency, 0.5; conversion efficiency, 0.00045; and photosynthate to energy converted, 0.25 cal kg^{-1}.

In addition to the photosynthetic efficiencies as defined above, a production efficiency, referring to the gram dry weight of new biomass produced, could be considered as the above terms times an additional one.:

$$\text{Production} = \text{Photosynthesis} \times \frac{\text{Production}}{\text{Photosynthesis}}$$

The production efficiency is related to the amount of photosynthate used to maintain existing plant structure and to the biochemical costs of creating new tissue. As will be seen later the production efficiency, on an annual basis, is approximately:

$$\frac{\text{Production}}{P_s} = (1/1.5)\ [1 - (\Sigma R_m/P_s)]$$

where ΣR_m is the sum of the maintenance respiration costs for all plant structures and the value 1.5 includes the biochemical growth costs and the maintenance of new tissue until it is classified "old" and maintained in ΣR_m. Production will be approximately 600 g dry wt. m^{-2} yr^{-1} and the ΣR_m, about 3.65 g dry wt. (g dry wt.)$^{-1}$ yr^{-1} with a biomass of about 2,000 g m^{-2} in mature chaparral.

Incoming solar. The incoming solar radiation varies through the season from 263 to 693 cal cm^{-2} day^{-1} in the winter and summer, respectively (fig. 13). Maximum intensities are about 70 cal cm^{-2} day^{-1} higher. Midday intensities vary between 0.9 cal cm^{-2} min^{-1} in winter to 1.4 cal cm^{-2} min^{-1} in summer. Day lengths vary from 10 h to 14 h through the year. Winters are

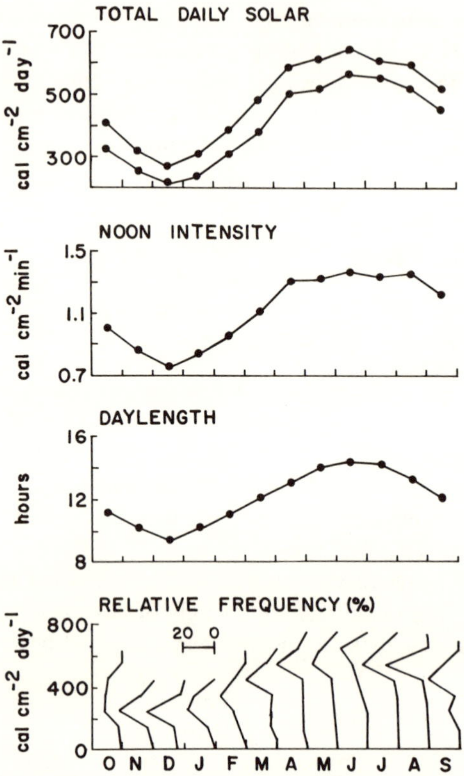

Fig. 13. Seasonal progression of total daily solar irradiance (absolute maximum for the month is given by top line and mean daily irradiance by the lower line), intensity of solar radiation at noon, daylength, and relative frequency of different daily irradiances.

cloudy as fronts pass across the region. Late summers are characteristically clear. Thus, the winter daily radiation varies more than the late summer radiation.

Interception efficiency. The fraction of the incoming solar radiation intercepted by each species depends on the solar altitude, foliage area index, and foliage inclination. At latitude 32.5°, the altitude of the sun at solar noon is about 45° in winter and 72° in summer. The mean solar altitude for the day varies between 20° and 42° in winter and summer, respectively (fig. 14). Interception is highest at low solar altitude because the path length of the solar beam through the canopy is longest. The mean daily interception efficiency is highest in winter and lowest through the summer. Through the year the variation is about 20 percent. The interception efficiency is about 40 percent

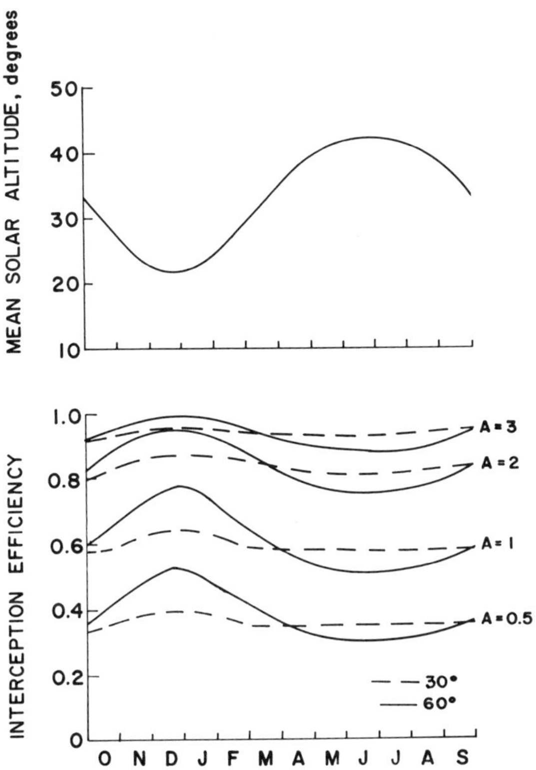

Fig. 14. Mean solar altitude and interruption efficiency with different leaf areas indices (A) and leaf inclination (30° and 60°).

with a foliage area index (one side) of 0.5, 60 percent with a foliage area index of 1.0, and 95 percent with a foliage area index of 3.0. More horizontally inclined leaves reduce the interception efficiency in summer and increase the interception efficiency in winter.

Leaf interception/Total interception. The ratio of radiation intercepted by leaves to the radiation intercepted by the total canopy is proportional to the ratio of leaf and stem areas, modified by the position of the leaves with respect to the stem. Leaf area was about 65 percent of total leaf and stem area (range 47–75 percent, with most measurements above 60 percent) (Mooney, 1977). Leaf area densities of chaparral shrubs are about 190 g m^{-2} for hard-leaved shrubs and 86 g m^{-2} for soft-leaved shrubs (Miller and Mooney, 1974).

Energy absorbed/Leaf interception. A portion of the intercepted light energy is reflected, and a portion is transmitted. The rest is absorbed. The absorptance of leaves is about 0.5, varying between 0.45 and 0.6 (Birkebak and Birkebak, 1964). The absorption in the visible is about 0.8 and in the near infrared, about 0.20. Visible energy is absorbed by the leaf pigments, and some is available for photosynthesis. The near infrared is absorbed by the water in the leaf and warms the leaf (Gates, 1962).

Energy converted/Energy absorbed. The conversion of energy absorbed by chloroplasts to chemical energy depends on the carbon dioxide available through diffusion from mitochondria, or from the outside air. For a positive carbon balance to occur, CO_2 must diffuse from the air, through the canopy, across the leaf boundary, through the stomates and cuticle, across the cell wall, membranes, and cytoplasm, to the chloroplasts. The diffusion is a physical process depending on a gradient in CO_2, and impeded by resistances at each of these locations. Depending on the formulation the internal resistances are subdivided into one (Gaastra 1963) to many (Chartier, 1970; Lommen et al., 1971; Slatyer, 1970) resistances. The internal resistances may be wholly related to internal leaf morphology (Nobel et al., 1975) or may be biochemical.

Photosynthate/Energy converted. The ratio of grams photosynthate produced to energy converted is 0.25 mg cal^{-1}, regarding glucose as the product of photosynthesis.

Dry Matter Production from Photosynthate

The energy fixed through photosynthesis can be converted to new biomass, but the energetic costs of producing an increment of biomass (growth respiration) and of maintaining that biomass (maintenance respiration) must be considered. This section explores some of these concepts in light of data previously presented on growth and carbohydrate allocation.

Respiration can be divided into at least three components: (1) maintenance respiration, (2) growth or construction respiration, and (3) translocation respiration (Penning de Vries, 1974; Penning de Vries et al., 1974a; Penning de Vries, 1975a,b). Maintenance respiration sustains the existing live biomass at a specified level of metabolic activity. It includes the energy requirements of basic metabolic processes, particularly the replacement or turnover of structural and functional proteins. In leaves maintenance respiration is included in net photosynthesis, while for nonphotosynthetic structures it is difficult to distinguish from the other respiratory components. However, from studies of temperate crop plants (McCree, 1970; Penning de Vries, 1973) the proportionality constant for maintenance respiration has been estimated at 0.015 g g^{-1} day^{-1} for tissues with 20–25 percent protein. Maintenance respiration for stems and woody roots is about 0.003 g g^{-1} day^{-1} (Penning de Vries et al., 1974b). The proportionality constant may be modified by environmental fac-

TABLE 1

BIOSYNTHETIC COSTS FOR VARIOUS PLANT CONSTITUENTS (BASED ON PENNING DE VRIES [1973, 1974]) AND BIOCHEMICAL COMPOSITION OF PLANT PARTS OF TWO CHAPARRAL SPECIES EACH FROM NORTH- AND SOUTH-FACING SLOPES

	Lignin	Cellulose	Hemi-cellulose	Lipid	Protein	TNC	Tannin & Cyanogenic Glycosides	GLUCOSE REQUIRED FOR 1 g BIOMASS	
								Synthesis ($g\ g^{-1}$)	Maintenance ($g\ g^{-1}\ day^{-1}$)
					Biochemical costs				
Glucose required (g)	2.15	1.17	1.17	3.03	1.62	1.13	3.03		
CO_2 evolved (g)	0.63	0.08	0.08	1.61	0.4	0.07	1.61		
O_2 required (g)	0.25	0.06	0.06	0.35	. . . .	0.05	0.35		
					Biochemical composition (%)				
					Adenostoma (N slope)				
Leaves, new	8	21	4	9	6	(37)	(15)	1.71	0.004
Stems	12	47	12	4	4	(21)	(0)	1.37	0.002
Roots	12	54	14	0	2	(18)	(9)	1.29	0.001
					Adenostoma (S slope)				
Leaves	8	21	3	9	5	(39)	(15)	1.79	0.003
Stems	14	44	8	5	3	(26)	(0)	1.40	0.002
					Arctostaphylos (N slope)				
Inflorescence	11	19	2	9	4				0.002
Leaves, new	11	18	9	9	5	(34)	(15)	1.72	0.003
Leaves, old	12	15	4	11	4	(39)	(15)	1.78	0.002
Stems	12	40	12	3	3	(30)	(0)	1.34	0.002
Roots	14	53	12	0	1	(20)	(0)	1.30	0.001
					Arctostaphylos (S slope)				
Inflorescence	7	17	0	9	4				0.002
Leaves, new	5	17	5	7	5	(46)	(15)	1.64	0.003
Leaves, old	11	17	9	10	3	(44)	(15)	1.73	0.002
Stems	12	38	15	2	2	(31)	(0)	1.32	0.002

NOTE. Plant samples collected 14 November 1975. Tannins and cyanogenic glucosides were estimated from Dement and Moonet (1974). TNC is estimated by difference.

tors such as temperature and O_2 level, although many plants appear to be able to adjust or acclimate within a given range so as to restore a value approaching the norm. This constant will also depend on the level of metabolic activity of the tissues or relative amounts of structural material. Fully differentiated, less-active tissues such as wood have lower values than meristematic and other metabolically active ones. Because of the low protein contents in the chaparral leaves, leaf maintenance is about 0.003 g g^{-1} day^{-1}. Penning de Vries (1972b, 1974, 1975a,b) and Penning de Vries et al. (1974a) have formulated the energy costs, as glucose equivalents, of synthesizing various classes of biochemical compounds (table 1). In synthesizing 1 g of lipid, 3.03 g of glucose are used, involving a cost of 2.03 g of glucose and a carbon dioxide evolution of 1.16 g of CO_2. Protein synthesis is regarded as two steps: (1) the formation of transportable organic nitrogenous compounds (AA_t) from glucose and NH_3 or NO_3, and (2) the interconversion of these compounds and the formation of the proteins. An average composition of chaparral leaves in November is 8–10 percent lignin, 15–20 percent cellulose, 3–9 percent hemicellulose, 7–11 percent lipid, 4–6 percent protein (P. C. Miller, unpublished data), and 5–12 percent tannins and 2–7 percent cyanogenic glycosides (Dement and Mooney, 1974). Thus, the synthesis of 1 g of leaves requires about 1.7 g glucose; 1 g of stems and roots is synthesized from about 1.3 g glucose.

Biosynthetic processes under anaerobic conditions are much less efficient because of the incomplete oxidations that can be performed. Processes performed under anaerobic conditions can take 9–17 times more glucose equivalent than under aerobic conditions (Penning de Vries, personal communication) and additional energy expenditures are needed to excrete toxic end products.

Translocation of minerals and organic compounds also requires energy, particularly for the loading and unloading of phloem elements. Translocation respiration is proportional to the translocation rate; available estimates suggest a value of 0.05 for the proportionality constant (Penning de Vries, 1973). Because loading and unloading costs are involved, it will require 1 g of glucose to translocate 10 g from source to sink. There are also energy requirements associated with nutrient uptake and translocation. For example, in the case of nitrogen, the energy cost of uptake has been established as 0.05 g glucose per gram of amino acid and the cost of translocation as 0.04 g glucose per gram amino acid (Penning de Vries, 1973, 1974).

An annual production efficiency can be estimated by assuming that photosynthate is first used to maintain the existing plant tissue, and the excess after maintenance, if any, is used for growth of new tissue. Allocation to new tissues is about 0.47 to leaves, 0.20 to stems and woody roots, 0.23 to absorbing roots, and 0.1 to inflorescences. The aboveground distribution is based on Mooney (1977). The root portion is obtained from root:shoot ratios of 0.40 (Miller and Ng, 1976; Kummerow et al., 1976), of which about 0.2 is estimated to turnover each year (Caldwell and Fernandez, 1975). Maintenance cost is assumed to apply to mature tissue, until death 1 year later. In addition, new tissue must be maintained during the period of growth. If the growth period is 3 months, maintenance of new tissue is 0.25 [1.5(0.47) + 0.25(0.20) + 0.30(0.23) + 0.30(0.1)] times the new biomass produced. This means that dry matter production can be (1/1.5) $(P_s - \Sigma R_m)$ or the production efficiency (production $/P_s$) is (1/1.5) $(1-(\Sigma R_m/P_s)]$.

As the biomass of stems and woody roots increases, the proportion of photosynthate available for new production decreases because of the increased maintenance of old stems and roots. Even though the maintenance costs for these tissues is low, the biomass becomes so large that new growth cannot be sustained; and with the death of old leaves, the stand senesces. Senescence in chaparral has been observed in 50-year-old stands without fire (Hanes, 1971; Muller, 1966; Christensen and Muller, 1975). Although the causes of senescence have been attributed to the accumulation of allelopathic chemicals and low nitrogen in the soils, the senescence can be explained almost wholly on the basis of simple carbon balance calculations.

Increasing the Photosynthetic Efficiency

The annual carbon dioxide uptake and light energy utilization can be increased by: (1) Increasing interception by increasing leaf biomass and placing the leaf biomass at the top of the canopy. This requires increased stems and branches which include lignin, a compound with a relatively high synthesis cost. (2) Increasing the ratio of leaf interception to total interception by increasing leaf biomass. Leaf biomass is expensive in terms of nitrogen and phosphorus, as will be seen, and, therefore, the total leaf biomass may be limited by the supply of these nutrients. (3) Increasing the ratio of energy converted to energy absorbed by increasing the capacity for CO_2 diffusion. This capacity is increased by decreasing the leaf resistance to CO_2 and water diffusion, which increases transpiration losses, and by decreasing the internal resistances to CO_2 diffusion. This latter is effected by altering the internal leaf structures to increase the cell wall diffusing surface relative to the leaf diffusing surface, and by increasing the chloroplast density relative to the leaf surface. Both result in thicker leaves. These adaptations, which are seen in the chaparral, are especially important in increasing photosynthesis efficiency in winter, when light intensities are low and limiting. Thus, although the chaparral occurs in a semi-arid environment, light becomes a limiting factor for part of the year, and the vegetation shows responses to light limitation.

PROCESSES AFFECTING PLANT TEMPERATURES

Temperatures are affected by all the processes of energy exchange: solar and infrared radiation, convection, evaporation, and conduction, and affect most of the physiological processes of the plant. Thus all the processes of energy exchange are fundamental in plant-environment interactions.

Plant Temperature

The leaf and stem temperature models link the environment and plant models. The temperature models are adaptions of the energy budget equation for individual leaves (Gates, 1962, 1965). The models estimate the physical processes of solar and infrared radiation absorption, infrared loss from leaves, convectional exchange, and transpirational exchange. The leaf temperature is calculated as the solution balance between the incoming and outgoing fluxes of energy. The leaf temperatures have been tested against measured temperatures in mangroves (Miller, 1972), chaparral (Lawrence, 1975), and Arctic (Tieszen, unpublished data). Measurements of the air boundary layer resistance in natural conditions are few, especially in chaparral where the tightly clustered leaves require calculation on a broader basis, similar to the work of Gates et al. (1965) and Tibbals et al. (1964) for conifers. In the canopy model, energy exchange processes and leaf temperatures are calculated for sunlit and shaded leaves and stems at different levels in the canopy.

Turbulent Transfer

Convectional and evaporational heat exchange depends partly on the air temperature and humidity and on wind speeds at different levels in the canopy.

The model for the turbulent transfer of heat, water vapor, and CO_2 in the canopy has been used in mangrove, Arctic tundra, and chaparral canopies. Models of physical processes affecting canopy microclimate have been proposed by Denmead (1964), Waggoner and Reifsnyder (1968), Waggoner et al. (1969), Murphy and Knoerr (1970, 1972), and Stewart and Lemon (1972) and tested with data from red clover, millet, pine, and corn canopies with reasonable success. Profiles of air temperature and humidity between the top of the canopy and a reference height above the soil surface are calculated with the algorithms proposed by Waggoner and Reifsnyder (1968) and Waggoner et al. (1969) using profiles of net radiation, turbulent exchange, and leaf resistance to water loss as input data. Our model has been tested preliminarily in mangroves and chaparral with fair success (fig. 15) and has been used to calculate ground surface temperatures and soil temperatures in mangroves, chaparral, and Arctic tundra.

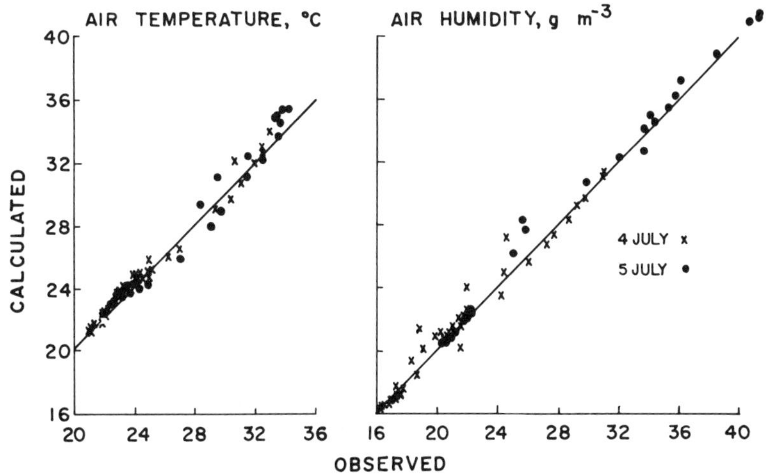

Fig. 15. Air temperatures and humidities calculated with the turbulent transfer model and observed in mangrove scrub forests in June, 1975.

Soil Surface Temperature

Soil surface temperatures affect the air temperatures in the canopy, leaf temperatures, evaporation from the soil, and soil heat conduction.

The soil surface temperature model is a solution of the energy budget equation for the soil surface. The ground or soil surface is defined as the level at which turbulent exchange processes become negligible and conduction becomes the dominant process of heat exchange. The surface temperature model is the link between canopy heat exchange processes and heat conduction in the soil. Most previous models involve only one set of processes and use the surface temperature as one of the boundary conditions. The effect of the canopy on surface temperature is significant in open, natural canopies because of the reciprocal effect of surface temperature on air and leaf temperature, although it may be less important in agricultural crops with dense canopies. In open, natural canopies the canopy-soil interaction cannot be overlooked. When the soil surface temperature is calculated, the resistance of the air boundary layer to the exchange of heat and water vapor and the resistance of the soil to the transfer of water vapor from the effective evaporating surface to the actual ground surface must be known, but these resistances are poorly known (Lemon, 1973; van Keulan, 1975; Ng and Miller, 1977). These relationships have only been measured crudely in mangroves. The surface temperature model has been used in mangroves, chaparral, and Arctic tundra (Miller et al., 1974; Ng and Miller, 1975; Miller and Ng, 1977).

PATTERNS OF WATER UTILIZATION

Background on Mechanistic Models

Plant water relations. The plant water relations model was first developed for the mangroves (Miller, 1974) and used in the alpine tundra (Ehleringer and Miller, 1975a,b), the Arctic tundra (Stoner and Miller, 1975), and the chaparral (Poole, 1974; Miller and Mooney, 1974). The model is essentially that of Honert (1948), modified by Rawlins (1963), and described in nonmathematical terms by Jarvis and Jarvis (1963). It allows for nonequal rates of water uptake and loss and the development of plant water stress. The model requires measurements of relation between leaf resistance or conductance to water loss and leaf water content or water potential, and measurements of the relation between leaf water content and leaf or xylem water potential, and measurements of the root resistance to water uptake and its environmental or plant controls.

The relation of leaf resistance to leaf water content or water potential has been measured in three mangrove species (Miller et al., 1975), eight chaparral and coastal sage species (Poole and Miller, 1975), six Artic tundra species (Stoner and Miller, 1975), six Arctic tundra species (Stoner and Miller), five alpine tundra species (Ehleringer and Miller, 1975a,b), snap beans (Kanemasu and Tanner, 1969), and boreal forest tree species (Jarvis and Jarvis, 1963). The relation can be inferred for conifers from Luposhinsky (1969), heath species from Bannister (1971), and alpine plants from Klikoff (1965). The relation between leaf water content and leaf water potential is discussed by Slatyer (1967) and has been measured on the mangrove, chaparral, tundra, and boreal forest species. A separation of total potential into matric-osmotic and turgor potential components and the relation between leaf potential and xylem potential have been measured in the three mangrove species (Miller et al., 1975).

The root resistances have been measured crudely in Arctic species and have been estimated in simulation for mangroves and chaparral. Kuiper (1964) showed the relation of root resistance to temperature, and Slatyer (1967) and Kramer (1969) discuss the relation of root resistance to the rate of water uptake. The relation of root resistance to temperature, root morphology, and rate of water uptake should be clarified.

Water movement in the soil. The movement of water in the soil is calculated similarly to Hanks and Bowers (1962), Liakopoulous (1965), Amerman (1971), and de Wit and Goudriaan (1974). The soil water model was tested and developed in chaparral and used to estimate evaporation, deep percolation, and transpiration losses (Ng, 1974).

Seasonal Patterns of Water Utilization

Water utilization can be expressed in terms of photosynthesis, to evaluate the costs and benefits of various aspects of water utilization, similar to the evaluation of various aspects of light utilization. Photosynthesis can be calculated as the annual precipitation times the ratio of transpiration to incoming precipitation, times the ratio of photosynthesis to transpiration. Thus,

$$\text{Photosynthesis} = (\text{Precipitation}) \left(\frac{\text{Transpiration}}{\text{Precipitation}} \right) \left(\frac{\text{Photosynthesis}}{\text{Transpiration}} \right)$$

The first ratio will be called here the transpiration efficiency, and relates to the capacity of the plant to capture water. The second ratio is the water use efficiency, or its reciprocal, the transpiration ratio.

Transpiration Efficiency

The transpiration efficiency is usually less than 1.0 because of interception losses, surface runoff, subsurface runoff, and evaporation from the soil surface. Of the incoming precipitation a fraction is intercepted by the vegetation canopy and evaporates. Rowe and Coleman (1951) gave interception losses in chaparral of about 8 percent. The plant crown should hold 0.2 to 0.8 mm of water per leaf area index before water falls or flows from the crown. The crown storage is about 0.3 mm on the south-facing slope and 1.8 mm on the north-facing slope (leaf area indices of 0.6 and 3.4, respectively [Krause, 1975]). This difference is small relative to the size of the storms.

Preliminary data on surface runoff, measured in two storms of 12.8 and 254 mm was 1.4–1.8 percent on the south-facing slope with 70 percent vegetation cover and 0.0 percent on a north-facing slope with complete vegetation cover. Subsurface runoff, calculated from soil moisture flow across the lower two measurement depths (Ng, 1974), was 0.2 to 1.5 mm day^{-1}, varying through the season.

Evaporation from the soil surface of the south-facing slope is calculated to be 109 mm yr^{-1}, varying between 0.02 mm day^{-1} in June, July, August, and September and 0.82 in April (fig. 16). On the north-facing slope soil evaporation is nearly zero. Transpiration was estimated from the plant properties characterizing the water flow through the plant-soil-air system (Ehleringer and Miller, 1975 a,b; Stoner and Miller, 1975; Poole and Miller, 1975) and the annual course of climatic variables. Transpiration was 170 mm yr^{-1} per unit leaf area index, or 102 mm yr^{-1}, on the south facing slope. It was 208 mm yr^{-1} per unit leaf area index, or 713 mm yr^{-1}, on the north-facing slope, assuming that all leaves transpire as in a canopy with a leaf area index of one.

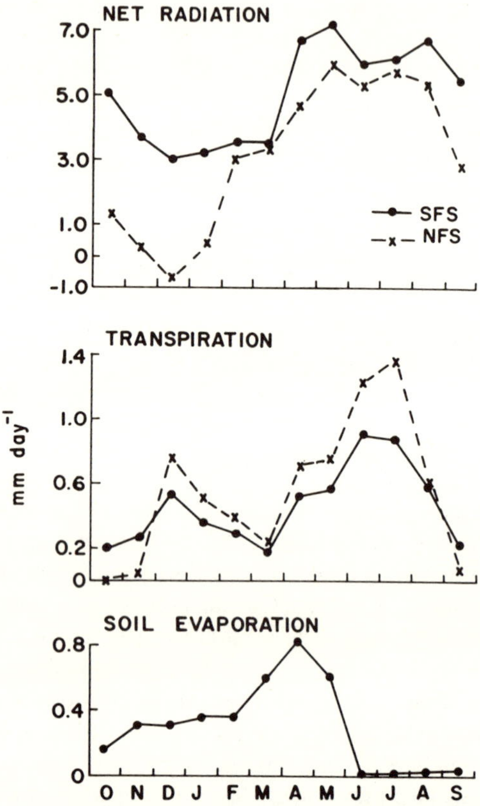

Fig. 16. Seasonal progression of net radiation, transpiration, and soil evaporation, all expressed in mm water equivalents per day, for north facing (NFS) and south facing (SFS) slopes in the chaparral zone. Soil evaporation from the north facing slope is assumed negligible.

Van Keulen (1975) discusses the role of shading and incomplete stomatal opening in the reduction of transpiration per leaf area index as the canopy leaf area index increases. He indicated that transpiration in a canopy with a leaf area index of 3.0 would be only 0.7–0.75 of that expected from fully sunlit leaves. This reduction brings the calculated transpiration close to the normal precipitation of 550 mm. Thus the transpiration efficiency of the south-facing slope is only about 0.19, with losses of 3 mm yr^{-1} from interception, 14 mm yr^{-1} by surface runoff, 320 mm yr^{-1} by subsurface runoff, 109 mm yr^{-1} by soil evaporation, and 102 mm yr^{-1} by transpiration. The transpiration efficiency on the north-facing slope is almost 1.00, with about 18 mm yr^{-1} lost by interception, no surface runoff, no subsurface runoff, and 520 mm yr^{-1} transpiration.

The annual water use estimated for our shrubs is similar to usage in other Mediterranean scrub regions. Kramer (1952) and Pillsbury et al. (1963) showed that regardless of the rainfall total the previous winter, the soil water of chaparral soils in southern California was depleted by the end of the summer. Shachori and Michaeli (1965) reviewed 26 studies of water yield in regions with less than 1,400 mm annual precipitation. Their regression of water yield on precipitation indicated that forested, woodland, and maqui areas with less than 400 mm yr^{-1} lost all the annual precipitation in transpiration, soil evaporation, or interception. They suggested that 400 mm yr^{-1} was the minimum precipitation requirement for woody vegetation development, which corresponded with the geographic distribution of woody vegetation in Israel. Shachori et al. (1967) measured through a 3-yr period an average evapotranspiration of 291 mm (239–378) for maqui shrub and 157 mm (149–165) for pasture or bare areas. Rowe and Coleman (1951) showed that between 380 and 483 mm of water was lost by evapotranspiration from mixed chaparral over a three-year period. Most of the precipitation over this amount was lost by percolation; only 8 percent was lost by interception and evaporation. Between 373 and 490 mm was lost from their *Ceanothus* plot and 457–500 mm from their chamise plot. By contrast 221–424 mm was lost by evaporation from bare soil. Daily evapotranspiration was 1.3 (0.17) mm.

If species respond to increased water utilization, major increases in water use are gained by increased water use through the winter, since subsurface drainage, which occurs during the heavy winter precipitation period, is the major path by which water is lost. Inasmuch as transpiration is proportional to leaf area, evergreen shrubs should be more-efficient water users than deciduous shrubs or grasses, which is substantiated by the common increase in water yields from watersheds when chaparral is converted to grass (Pase and Ingebo, 1965; Pase and Fogel, 1967).

Water Use Efficiency

The water use efficiency is about 4.5 g dry wt per kg H_2O or about 4.5 g dry wt m^{-2} leaf mm^{-1}, based on transpired water (fig. 17). The water use efficiency is higher in winter than in summer (5.5 vs 2.5 g dry wt m^{-2} mm^{-1}) and varies vertically through the canopy. In winter and spring the water use efficiency is highest at the bottom of the canopy and in the late spring and summer, at the top of the canopy (Miller and Mooney, 1974). The pattern of water use efficiency follows, roughly, the pattern of photosynthesis through the canopy. The water use efficiency can be increased in summer-active plants by decreasing leaf width and moderately inclined leaves (fig. 18). In winter-active plants the water use efficiency can be increased by

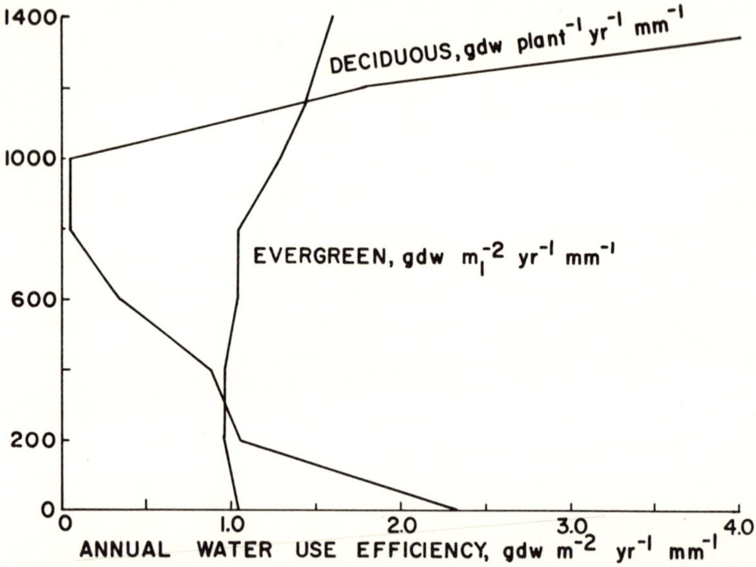

Fig. 17. Annual water use efficiency (production over precipitation) with altitude for deciduous and evergreen shrubs, calculated from production and precipitation estimates discussed in the earlier section on bioclimate.

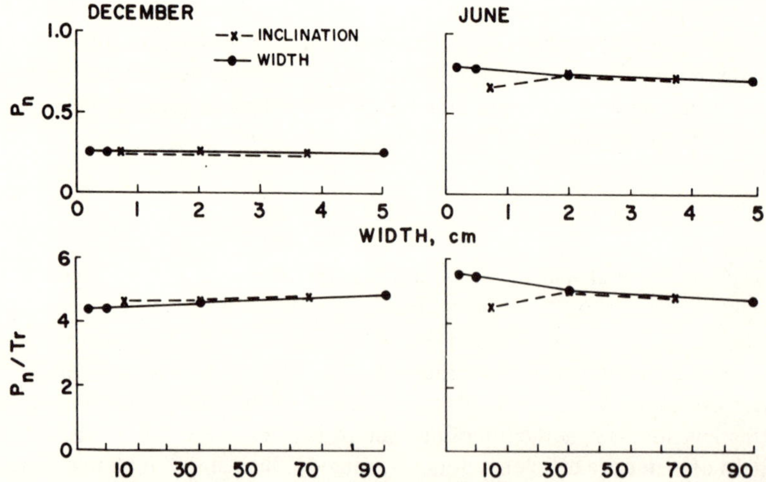

Fig. 18. Net photosynthesis (P_n) in g dry wt m^{-2} leaf day^{-1} and water use efficiency (P_n/T_r) in g dry wt (mg water)$^{-1}$ in December and June for leaves of different widths and inclinations.

increasing leaf width and more horizontal leaf inclinations. Water use efficiency is also increased by decreasing the mesophyll resistance to CO_2 diffusion.

A water use efficiency that includes aspects of the water capture and the net photosynthesis to transpiration ratio can be calculated for the altitudinal transect discussed earlier (fig. 4). This efficiency, obtained by dividing the calculated annual production by the annual precipitation, indicates a gradually increasing efficiency with altitude for evergreen shrubs but a sharp decrease in water use efficiency for deciduous shrubs at mid altitudes. The altitudinal pattern of water use efficiency, which follows the altitudinal pattern of production, reinforces the selection against deciduous shrubs at mid altitudes.

Species Differences in Water Utilization

Aspects of water relations which define the flow of water through the plant were measured on eight chaparral species on south- and north-facing slopes at a coastal and an inland site. The various species differed from one another, but individuals of the same species at different locations were similar. *Arctostaphylos glauca* had the highest conductances, and this species and *Ceanothus greggii* are capable of transpiring at the lowest water potentials (−65 bars). The three *Rhus* species had the lowest conductances and the lowest capability to transpire at low water potentials (−20 bars minimum). The seasonal progression of transpiration indicated that all species show their maximum daily transpiration in June. *Ceanothus greggii* shows some transpiration later in the summer than do the other species. In late summer and early fall transpiration is at a minimum. Dawn and midday water potentials support the relation between leaf conductances and water potential. The dawn plant water potentials are about 40–50 bars lower than soil water potentials in late summer, probably due to decreased absorbing root mass, shrinkage of the roots, and decreased soil conductivity to water flow in dry soils. The hydraulic conductivity of the chaparral soils decreases rapidly with soil water contents below 0.12 $cm^3\ cm^{-3}$. Plants recover from the summer water stress 3–10 days after the fall rains begin (Poole and Miller, 1975). The rapidity of this recovery indicates that soil, rather than plant, factors may contribute to the resistance to water flow.

PATTERNS OF NITROGEN AND PHOSPHORUS UTILIZATION

Similar to light and water utilization, nitrogen and phosphorus use can be considered in relation to photosynthesis and carbon balance. However, the role of nitrogen and phosphorus is less direct than that of light or water. Analogous to solar energy incident on the canopy or to incoming precipitation, the nitrogen potentially available to the plant comes from nitrogen release of decomposing organic matter and nitrogen fixation by living or symbiotic forms, and the phosphorus potentially available is released with decomposition and from weathering bedrock. A fraction of the nitrogen or phosphorus released is taken up by the plant, and a fraction of that taken up is incorporated into leaf biomass. Some is also incorporated in stems or roots. If the synthesis of new leaf biomass requires more nitrogen or phosphorus than is available from uptake, the additional nitrogen or phosphorus may be withdrawn from reserves, thereby suppressing the resynthesis of proteins in mature leaves. Under conditions of nitrogen or phosphorus limitation, older leaves may senesce and die prematurely. The difference between leaf growth and leaf die-back is expressed as a change in leaf biomass, which is associated with a change in photosynthesis, via the interception efficiency in equation (1). The concentrations of nitrogen or phosphorus also affect the conversion efficiencies in equation (1), but much less than the affect of leaf biomass on photosynthesis. Thus,

$$\text{Change in photosynthesis} = (\text{N release} + \text{N fixation}) \left(\frac{\text{N uptake}}{\text{N release} + \text{N fixation}} \right) \left(\frac{\text{Leaf growth}}{\text{N uptake}} \right) \left(1 - \frac{\text{Leaf death}}{\text{Leaf growth}} \right) \left(\frac{\Delta\text{photosynthesis}}{\Delta\text{leaf biomass}} \right).$$

An analogous equation can be written for phosphorus.

Background on Mechanistic Models of Growth and Allocation

Five models of growth and allocation with different levels of sophistication were developed in the tundra research (Miller, 1972; Miller et al., 1975; Miller et al., 1976 a, Miller et al., 1977; Stoner et al., 1978). The models incorporate nitrogen, phosphorus, and calcium in the growth patterns. The model is driven by various climatic factors and seasonal patterns of growth of individual leaves on the tiller, of the leading rhizome, and of the inflorescence. Photosynthesis is calculated with relation between daily photosynthesis and daily mean temperatures, soil moisture content, and irradiance, which were developed from simulation with the canopy photosynthesis model. Root growth is related to soil temperature. The carbon and inorganic nutrient costs associated with the growth of new tissue are calculated from the biochemical composition of mature tissue and the growth respiration costs of creating different biochemical constituents (Penning de Vries, 1972a,b, 1973, 1974; Penning de Vries et al. 1974a).

Seasonal Patterns of Nitrogen and Phosphorus Utilization

Nitrogen release and fixation. Nitrogen release from soil organic matter is related to the effects of temperature and soil moisture on decomposition and to the C:N ratio of the substrate (Beek and Frissel, 1975; Harpaz 1975). In the mediterranean climate, soils are cold and wet in the winter and hot and dry in the summer. The seasonal course of soil temperature and moisture is such that rates of decomposition are probably low throughout the year (fig. 19). Nitrogen release is further decreased by high C:N ratio, on the order of 100, of the fresh litter. About three months of decomposition is required to lower the C:N ratio so that appreciable amounts of nitrogen are released.

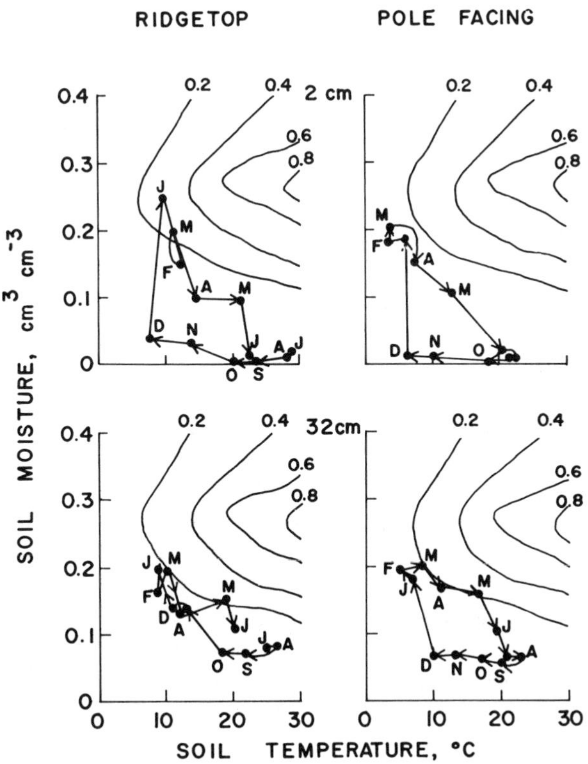

Fig. 19. Seasonal progression of soil temperatures and soil moisture at 2 and 32 cm depths plotted in relation to isolinear showing the fraction of the maximum rate of decomposition expected with different temperatures and moisture contents (from P. C. Miller, unpublished data, and Beck and Frissel, 1973).

Nitrogen fixation by free living forms is not recognized in chaparral, but fixation by root nodules has been shown for *Ceanothus* species and *Quercus dumosa*. The amount of this fixation is unknown, but a rate of about 0.8 g N m^{-2} yr^{-1} seems reasonable (Harpaz, 1975; Penning deVries et al., 1974a,b; Christensen and Muller, 1975). An input of nitrogen in precipitation may also be about 0.8 g N m^{-2} yr^{-1} (Harpaz, 1975). During a fire, 50–70 percent of the nitrogen in the vegetation and litter can be lost to the atmosphere by volatization (Penning deVries, 1974; L. deBano, personal communication). The regrowth of chaparral following a fire is approximately equal to that which can be expected with the annual increments from fixation and precipitation.

Nitrogen taken up/Nitrogen made available. All the nitrogen or phosphorus made available is not taken up because of losses due to leaching or volatilization. This fraction should relate to the absorbing root density, to the ability of the absorbing roots to take up the nutrient from the soil solution or exchange site, and to the ability of the molecule to diffuse to the root surface from the bulk soil. The diffusion ability of NO_3, PO_4, and water are such that absorbing roots should be about 4 mm apart in the soil for complete exploitation of the phosphorus in a few days, 40 mm apart for complete exploitation of nitrogen as nitrate, and 80 mm apart for exploitation of water. These distances are approximate due to the arbitrary subdivision of a continuous system into discrete space and time units, the complexity and the unknown nature of parameters in the soil-chemical diffusion system, and the inadequacy of the data base. Nevertheless, the trends exist. The requirement of a high root density for efficient phosphorus uptake is often met by mycorhizae in some species (Tamm, 1971). The absorbing root densities in chaparral appear adequate to exploit the soil volume for nitrogen and water, but not phosphorus.

Leaf growth/Nitrogen or phosphorus uptake. Primary production in natural vegetation is often limited by nitrogen and/or phosphorus, even in semi-arid regions where water is considered to be in short supply (Helmers et al., 1955; Dabadghao and Shankarnarayan, 1970; Fisher, 1970; Chapin et al., 1974; Harpaz, 1975; van Keulen et al., 1975, van Keulen, 1975). In addition, sclerophylly is thought to result from nitrogen or phosphorus deficiency (Small, 1973; Monk, 1966; Beadle, 1966), affecting plant form and forage quality. Small (1973) suggested that sclerophylly, measured as the ratio of crude fiber to crude protein, increased with nitrogen and phosphorus deficiency as well as with water deficiency, but stated that the mechanism for the development of the changes in leaf form was not clear. Steubing and Alberdi (1973) showed experimentally that sclerophylly increased with decreased phosphorus uptake, whether induced by low phosphorus in the growth solution or by low temperatures. Photosynthesis and nitrogen and phosphorus content were correlated across plant life forms (Skre, 1975), but not necessarily within a life form (Keller, 1971), indicating that in a nutrient-deficient environment growth is suppressed more than photosynthesis.

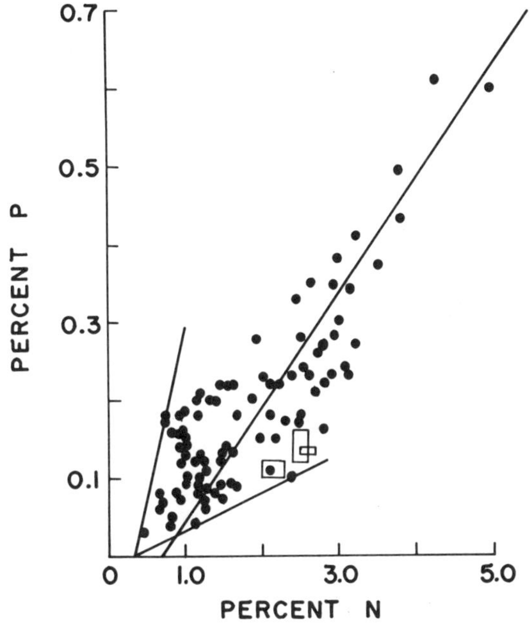

Fig. 20. Nitrogen and phosphorus contents of different plant species. The steepest line is the N:P ratio of nucleoproteins, the shallowest line is the N:P ratio of cytoplasm, the third line is a line for maximum growth. The slopes of the latter two lines were suggested by Kostic et al. (1967), but the intercept has been changed to fit the data points.

Kostic et al. (1967) proposed a preliminary scheme for evaluating the levels of phosphorus relative to nitrogen in perennial ryegrass (fig. 20). They considered that because nitrogen and phosphorus were principally found in the cytoplasm, nitrogen and phosphorus should be in constant proportion in the leaf. However, phosphorus levels should be increased above the level expected on the basis of cytoplasmic proteins because ester phosphates were needed for normal cell functioning and inorganic phosphates were needed for maximum growth.

A compilation of nitrogen and phosphorus contents of a wide variety of plant types (table 2) indicates that the Kostic et al. (1967) formulation, though true in broad form, can be made more precise for naturally occurring plant

TABLE 2

SUMMARY OF N AND P LEVELS IN VARIOUS SPECIES

Species	N	P	Source
Atriplex lentiformis			
1st harvest	2.7	.21	Goodin and McKell, 1970
2d harvest	2.3	.17	
Atriplex polycarpa			
1st harvest	3.1	.24	
2d harvest	2.5	.17	
Valencia and navel			
orange	2.4–2.6	.12–.16	Reuther, 1973
Tung leaves	2.0–2.3	.10–.12	Reuther et al., 1958
Pinus taedo (loblolly)			
Needles, live	.75–.83	.093–.096	Wells and Metz, 1963
Needles, dead	.41–.48	.056–.058	
Calamagrostes holmii	1.44 (3.53)	.12 (.37)	Chapin et al., 1975
Carex aquatilis	2.18 (3.11)	.15 (.23)	
Dupontia fischeri	1.60 (3.24)	.09 (.27)	
Eriophorum augustifolium	2.10 (2.77)	.11 (.26)	
Eriophorum scheuchzeri	2.13 (2.62)	.18 (.23)	
Euzula arctica	1.15	.20	
Poa arctica	1.25 (2.97)	.12 (.28)	
Petosites frigidus	2.26 (3.74)	.22 (.41)	
Salix phlebophylla	2.55	.24	
Salix rotundifolia	1.61	.22	
Saxifiaga cernia	1.00	.18	
Saxifraga punctata	1.58	.22	
Stellaria laeta	2.03	.23	
Salix pulchum	(3.80)	(.49)	
Oil palm, no N & P added	2.5–2.7	.13–.14	Martin, 1974
Oil palm, N & P added	2.5–2.7	.15–.17	
Corn, no P	2.5–2.7	.22–.23	Lutz, 1974
Corn, P added PARP	2.8–3.1	.34–.37	Lutz, 1974
Corn, P added on CSP	2.8–3.0	.38–.40	Lutz, 1974
Cacao	2.4, 2.8	.23, .27	
Peanut (required	2.5	.16	Smith, 1962
P at different	3.0	.20	
levels of N)	3.5	.24	
Acacia	2.5	.10	Unpublished
Sabel grasses	2.8	.165	Unpublished
Straw, sabel	0.25	.02	
Quercusilex			
1st-yr. leaves	1.43	.146	Rapp, 1971
2d-yr. leaves	1.17	.162	
Twigs	.88	.148	
Branches	.54	.150	
Trunks	.22	.074	
Roots	.34	.18	
Gauigue			
Leaves	1.35	.08	
Bois	.54	.03	
Wheat stem mN	.5	.1	Khetawat et al., 1972
N at 40 kg/ha	.5	.07	
N at 80 kg/ha	.6	.025	
N at 160 kg/ha	.5	.05	
Wheat stem ON	1.65	.52	Khetawat et al., 1972
N at 40 kg/ha	1.65	.395	
N at 80 kg/ha	2.1	.37	
N at 160 kg/ha	1.75	.36	
Pine			
SB	.4	.06	
B	.3	.04	

TABLE 2—*Continued*

SUMMARY OF N AND P LEVELS IN VARIOUS SPECIES

Species	N	P	Source
SW	.15	.02	
Rhus ovata	2.54	.18	Mooney et al., 1976
Ceanothus leucodium	2.00	.15	
Quercus agrifolia	1.72	.18	
Ceanothus greggii	1.67	.09	
Quercus dumosa	1.63	.13	
Heteromeles arbutifolia	1.28	.11	
Adenostoma fasciculatum	1.16	.08	
Arctontaphylos glauca	.87	.08	
Trevoa trinervis	2.92	.11	
Kayeguechia oblonga	2.45	.33	
Cryptorarya alba	1.96	.28	
Quercus saponaria	1.42	.22	
Colbiguaya odorifera	1.16	.13	
Eitweja gilliesii	.93	.16	
Sugar cane	.26	.085	Chin, 1972
Olive trees			
1st 6-mo. of new leaf	~2.0–2.5	.10	Govalan, 1973
In winter	>1.8	>.09	
Pine			
Needles	1.02	.10	Koch, 1972
Stembark	.42	.06	
Branches	.32	.04	
Stemwood	.14	.02	
Loblolly pine	1.6	.1–.2	Blain and Epps, 1969
Needles	.95–1.10	.118–.124	White et al., 1970
Inner bark	.26–.30	.062–.067	
Current spruce needles	1.40, 1.36, 1.16, 1.20	.196, .201, .180, .210	Tamm, 1971
Populus nigia			
maxP_s, maxGR	3.0	.3–.35	Keller, 1971
Sitaria			
1v set	1.44	.293	Middleton
1v set	1.32	.302	(Super PO_4 added to all)
Siratra			
1v set	2.89	.210	
1v set	2.83	.203	
Desmodium	2.10	.252	
	2.07	.221	
Yellow mustard	2.96	.34	Hansen, 1971
Pinus taeda			Meade and Pritchett, 1975
Terminal leader			
Current foilage	1.03	0.15	
1-year foliage	.84	0.15	
Upper crown			
1st-order branch			
Current foliage	1.00	0.14	
1-year foliage	.85	.16	
2d-order branch			
Current foliage	.94	.14	
Lower crown			
1st-order branch			
Current foliage	1.00	.13	
1-year foliage	.95	.16	
2-year foliage	.76	.18	
2d-order branch			
Current foliage	1.01	.14	
1-year foliage	.96	.16	
2-year foliage	.77	.17	

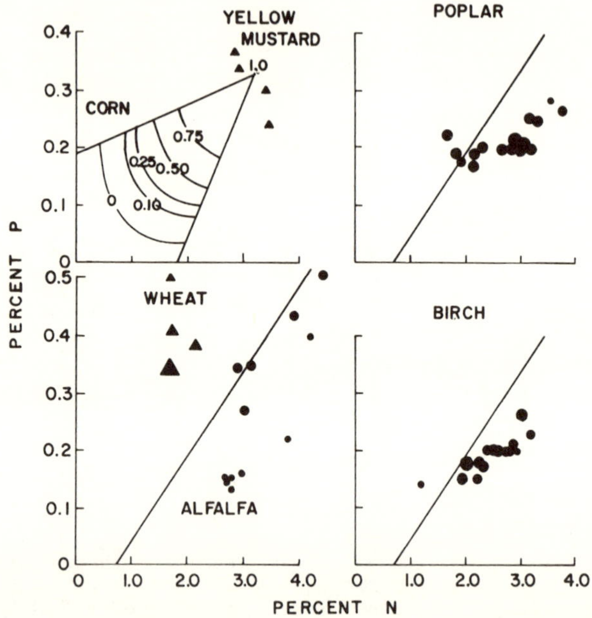

Fig. 21. Indicators of primary production with different nitrogen and phosphorus contents. A. Yellow mustard and corn yield relative to a maximum yield, B. Wheat yield (kg ha^{-1}) and alfalfa yields (10^3 kg ha^{-1}) at 12% moisture, C. Poplar (g dry wt pot^{-1}), D. birch (g dry wt pot^{-1}). Lines are the lines postulated for maximum growth previously.

species and tree crops (fig. 21). In no case surveyed was the phosphorus level as high as suggested for maximum growth, and in many cases phosphorus levels were below the minimal levels necessary for growth. In fact, levels recommended for citrus leaves (Embleton et al., 1973; Reuther et al., 1958) and tung leaves (Reuther et al., 1958) were below the minimal suggested phosphorus levels. Roots and stems were consistently low. However, phosphorus levels in buds and inflorescences were close to the levels suggested for maximum growth.

Low levels of nitrogen and phosphorus commonly occur in species with low rates of growth and photosynthesis. Actively photosynthesizing leaves commonly do not grow, and actively growing leaves commonly do not photosynthesize. In seedlings chlorphyll formation is delayed for several days even under adequate light conditions while rapid growth is taking place (Britton and Goodwin, 1973). In unfertilized plants the nitrogen and phosphorus concentrations are high as leaves begin developing from bud to maturity, but decrease with increasing dry matter, which is largely composed of cellulose, lipids, lignin, and carbohydrates. However, the phosphorus:nitrogen ratio remains about the same. A slight deviation from a constant phosphorus:nitrogen ratio is apparent at low nitrogen and phosphorus levels because the nitrogen content is positive with zero phosphorus. The protein, extensin, which is estimated at 1–2 percent of cell wall constituents (Lyttleton, 1973), would account for positive nitrogen values. If the cell wall constituents are 40 percent of the leaf biomass, at zero phosphorus, nitrogen levels would be 0.5–0.6 percent which is close to the intercept. Roots, stems, and rhizomes had somewhat higher phosphorus, per unit nitrogen, than did leaves and stems. In general the points representing the nitrogen and phosphorus levels of diverse plants fall near the same line at high levels. Mature leaves with low photosynthesis rates have low phosphorus and nitrogen contents and variable phosphorus:nitrogen ratios, probably because the phosphorus and nitrogen levels are dominated by the demands for photosynthesis, which is less sensitive to phosphorus and nitrogen concentrations, than growth. Tung, citrus, oil palm, olive, sahel acacia, and grass had low phosphorus:nitrogen ratios as did many tundra plants from Barrow. Chilean shrubs had high P:N ratios, and Californian shrubs had relatively low P:N ratios.

The similarity of the phosphorus:nitrogen ratios of developing leaves of various species suggests that this ratio may be a requirement for maximum growth. Some experimental support for this hypothesis is gained from studies on corn (Duminel, 1961) wheat (Khetawat et al., 1972), and alfalfa (Lutz, 1973). The highest alfalfa yields occurred with P and N concentrations on the hypothesized line for maximum growth. Yields decreased at the high nitrogen and phosphorus contents, which will be discussed later, and with concentrations lying off the line. Khetawat et al. showed large increases in wheat yields with nitrogen fertilization, without increases in nitrogen concentration. When the wheat yields are plotted against nitrogen and phosphorus concentration, it is apparent that the phosphorus contents are extremely high relative to the nitrogen contents. Fertilization with nitrogen increased yields by 4–5 times. Further nitrogen fertilization might have increased yields further if this analysis is true. Duminel (1961) fitted regression lines to the results of 93 studies of leaf foliage analyses and corn yields. With the regression lines he constructed isopleths of equal yield. The trend of decreasing yield with decreasing nitrogen and phosphorus concentrations has almost the same slope on the line hypothesized for maximum yield.

On the other hand, fertilization of Californian chaparral (Hellmers et al., 1955), which tends to have a low P:N ratio, gave responses to nitrogen, but not to phosphorus, and to both nitrogen and phosphorus. Yellow mustard yields (Hansen, 1971) were low on the center line and increased with lower P:N ratios.

If mature leaves contain relatively high P and N, say greater than 0.25 percent and 2.0 percent, respectively, inhibition of production by some other factors can be suspected. Phosphorus and nitrogen increase production by increasing leaf area, rather than by increasing the rate of photosynthesis per unit of leaf. Total canopy photosynthesis will increase almost linearly with leaf area, at least until light becomes limiting (leaf area indices greater than 2). Photosynthesis increases very little with increased P and N contents of leaves if leaf phosphorus and nitrogen levels are above 0.1 percent and 1.2 percent, respectively. Thus, without other limiting nutrients, the production of leaf biomass dilutes the plant phosphorus and nitrogen until the P and N levels are minimal for photosynthesis. Poplar and birch (van Tuil, 1965) showed maximum growth index nitrogen and phosphorus levels were in a ratio, and decreased growth under the ratio departed from a central ratio, but this line through the concentration giving maximum growth has a different slope than obtained on the other plants. These studies indicate that the concept of a ratio for maximum growth may be valid, but the ratio may not be the same for all species. Leaves or buds with relatively high contents (0.3 percent P and 3.0 percent N) can be expected to be growing. As leaf P becomes less than 0.1 percent and leaf N becomes less than 0.7 percent, leaf function decreases and data points are more scarce. The lower data points are due largely to old leaves on California-Chile shrubs. Fertilization will certainly upset the normal P:N ratio and may produce unwanted side effects on plant growth such as decreasing growth, decreasing frost and drought tolerance, and increasing water use.

SUMMARY OF SOME CONCEPTS IN PLANT ECOLOGY

A fundamental concept in the plant community ecology tradition is that the vegetation in a region, if altered from its natural state, will change with time and return to a state similar to its original undisturbed condition. Vegetation classification has been attempted on the basis of the potential vegetation that would occur in a region without disturbance. Because some disturbances are always present, by natural or man-related causes, controversy developed about the reliability of these stable conditions and the reality of the potential vegetation. Additional controversy raged over the hierarchical organizations of environmental controls, i.e., does climate determine the soils and vegetation, while soils and vegetation interact, or, do climate and soils determine vegetation? The potential vegetation was estimated subjectively, by carefully observing disturbed plant communities occurring simultaneously in similar regional and local settings and then arranging these communities in a sequence assumed to represent a time sequence. When disturbing events could be dated, the successional sequence could be described more accurately.

Recently ecologists in the population-community tradition have developed a "strategy" terminology that invents or postulates goals for which plant populations or communities are selected. From these postulates, characteristics of populations in equilibrium can be predicted. Thus, the strategy thinking is analogous to the "potential vegetation" or climax thinking of plant community ecology. Strategy thinking includes ideas of species evolution and ecosystem development that occurs in the direction of maximizing photosynthesis, maximizing production, or maximizing water use efficiency, (Parkhurst and Loucks, 1972) or of conserving nutrients (Monk, 1966).

The strategy concept of the population-community tradition can be replaced with a quantitative approach from the primary production process tradition, which allows description of the patterns of resource use and the ability to check the consistency of these patterns with those expected on the basis of the goals of maximizing photosynthesis, production, and so forth. Miller and Mooney (1974) concluded that the seasonal pattern of the vertical profiles of primary production and water use efficiency in mediterranean scrub regions of California and Chile was consistent with the seasonal progression of photosynthetic tissue, suggesting a correspondence relating to maximizing production. The correspondence was also reflected in leaf widths and inclinations. They proposed that the distribution of evergreen and deciduous-level plants was related to the length of a period unfavorable for photosynthesis. The analysis presented in this paper adds the additional constraint of temperature into the factors affecting the distribution of evergreen and deciduous forms. Miller et al. (1976) evaluated the potential influence of various factors on photosynthesis of arctic plants and indicated that (1) decreasing the optimal leaf temperatures for photosynthesis lower than those observed, so leaf temperatures were within the range of ambient environmental temperatures, resulted in little increase in photosynthesis: (2) grazing by lemmings could increase production; and (3) light saturation values were influences on daily and annual production.

These results indicate that a carbon balance approach to plant ecology is reasonable. In fact, the idea that succession proceeds toward increasing primary production and a carbon balance has been stated earlier by Clements and Shelford (1939) and Odum and Pinkerton (1955), members of the plant ecology and ecosystem traditions, respectively. On the other hand, Odum (1960) showed that production remained constant while plant species changed in the early stages of the old field succession in Georgia, and Crocker and Major (1955) showed that the climax species were present in early stages of succession on glacial remains and that succession was related to the rate of growth of the species rather than a sequencing of species invasion and establishments.

One can also examine other strategy concepts from a quantitative approach. The reproduction cost of sexual vs. asexual plant reproduction are in a ratio of about 30 to 1 from a carbon balance point of view. In nutrient-deficiency situations, primary production of a species with inherently high turnover rates is lower than that of species with inherently low turnover rates because the nutrient deficiency limits the total leaf biomass and total attainable canopy photosynthesis, while the high turnover rate causes a high rate of respiration loss due to growth.

Thus, many of the problems discussed subjectively and qualitatively in the literature of plant ecology can be analyzed in a rigorous quantitative framework, much of which already exists in the literature of other disciplines.

ACKNOWLEDGMENTS

The research that this paper is based on was carried out on NSF grant GB-27161 as part of the U.S. IBP Origin and Structure of Ecosystems project, NSF grant DEB 75-19491 on Resource Utilization and Allocation by Vegetation in Mediterranian Scrub: A Study of Convergent Ecosystems, NSF grant 29345 as part of the U.S. IBP Tundra Biome, and AEC contract AT(04–3)-807. Much of the synthesis was accomplished while the author was on a NATO fellowship in the Netherlands. I thank Drs. F. W. T. Penning de Vries, C. T. deWit, and J. Goudriaan for their stimulating discussions while I was in the Netherlands. Mr. W. A. Stoner and Ms. S. A. Barkley provided much help in modeling and critical analysis of ideas and data. Drs. D. Billings, F. S. Chapin, III, J. Kummerow, M. Leck, J. McKendrick, D. Mauriello, W. Oechel, and L. L. Tieszen contributed ideas and data in the course of developing these models. In the final analysis, only the mistakes are my own.

LITERATURE CITED

Acock, B., J. H. M. Thornley, and J. Warren Wilson. 1970. Spatial variation of light in the canopy. *In* Prediction and measurement of photosynthetic productivity. Pp. 91–102. Centre for Agricultural Publishing and Documentation, Wageningen.

Amerman, C. R. 1971. Numerical solution of the flow equation. *In* D. Hillel, Soil and water. Pp. 241–54. Academic Press, New York.

Anderson, M. C. 1966. Stand structure and light penetration. II. A theoretical analysis. J. Appl. Ecol. 3:41–54.

Anderson, M. C., and O. T. Denmead. 1969. Shortwave radiation on inclined surfaces in model plant communities. Agron. J. 61:867–72.

Bannister, P. 1971. The water relations of heath plants from open and shaded habitats. J. Ecol. 59:51–64.

Bates, T. E. 1971. Factors affecting critical nutrient concentrations in plants and their evaluation: A review. Soil Sci. 112:116–30.

Beadle, N. C. W. 1966. Soil phosphate and its role in molding segments of the Australian flora and vegetation, with special reference to xeromorphy and schlerophylly. Ecology 47:992–1007.

Beek, J., and M. J. Frissel. 1973. Simulation of nitrogen behavior in soils. Centre for Agricultural Publishing and Documentation, Wageningen.

Bieleski, R. L. 1973. Phosphate pools, phosphate transport, and phosphate availability. Annu. Rev. Plant Physiol. 24:225–52.

Birkebak, R. C., and R. Birkebak. 1964. Solar radiation characteristics of tree leaves. Ecology 45:646–49.

Blair, R. M., and E. A. Eppa. 1969. Seasonal distribution of nutrients in plants of seven browse species in Louisiana. U.S. Dept. Agric. South-Forest Exp. Stn. Annu. Res. Pap. 50–51. 35 pp.

Bodtkin, D. B. 1975. A functional approach to the niche concept in model communities. Paper delivered at NSF Conference on the Future of Systems Ecology, Utah State University, 19–21 Feb. 1975.

Braun-Blanquet, J. 1951. Pflanzensoziologie, Grundguge der Vegetation—spunde. 2d ed. Springer-Verlag, Vienna.

Brichambaut, G. P. de, and C. C. Wallen. 1963. A study of argroclimatology in semi-arid and arid zones of the near east. World Meterol. Organ. Tech. Note 56. 64 pp.

Bronkman, R., and W. Bennert. 1971. Chemical composition of plants of the field layer. Preliminary Report, p. 57–60. *In* H. Ellenberg (ed.), Intergrated experimental ecology. Ecological Studies Analysis and Synthesis, vol. 2. Springer-Verlag, Berlin.

Brunt, D. 1932. Notes on radiation in the atmosphere. Int. Qtly. J. Roy. Meterol. Soc. 58:389–420.

Caldwell, M. M., and O. A. Fernandez. 1975. Dynamics of Great Basin shrub root systems. *In* N. F. Hadley (ed.), Environmental physiology of desert organisms. Pp. 38–51. Dowden, Hutchinson, and Ross, Stroudsburg, Pa.

Cartier, P. 1969. A model for CO_2 assimilation in the leaf. *In* Prediction and measurement of photosynthetic productivity. Pp. 307–15. Centre for Agricultural Publishing and Documentation, Wageningen.

Chapin, F. S., B. H. McCowan, L. L. Tieszen, M. L. Allessio, and M. C. Lewis. 1977. Plant growth and nutrient allocation. *In* J. Brown, F. Bunnell, P. C. Miller, S. MacLean, and L. L. Tieszen (eds.), U.S. Tundra Biome terrestrial synthesis volume. Dowden, Hutchinson & Ross, Stroudsburg, Pa. (in press).

Chapin, F. S., L. L. Tieszen, and K. van Cleve. 1974. Seasonal patterns of aboveground inorganic nutrient content of vascular plants at Barrow, Alaska. U.S. Tundra Biome Data Report 74–38. 94 pp.

Chin, N. Y. 1972. Growth studies on sugarcane. II. Nutrient uptake. Malaysian Agric. J. 48:357–84.

Christensen, N. L., and C. H. Muller. 1975. Effects of fire on factors controlling plant growth in Adenostoma chaparral. Ecol. Monogr. 45:29–55.

Clements, F. E. 1916. Plant succession. Carnegie Institution, Washington, D.C.

Clements, F. E. 1928. Plant succession and indicators. H. W. Wilson Co., New York.

Clements, F. E., and V. E. Shelford. 1937. Bio-ecology. John Wiley and Sons, New York. 698 pp.

Cochemé, J. 1968. FAO/UNESCO/WHO agroclimatology survey of a semi-arid area of West Africa south of the Sahara, *In* Agroclimatological methods. Pp. 235–48. UNESCO, Paris.

Cochemé, J., and P. Franquin. 1967. An agroclimatology survey of a semi-arid area in Africa south of the Sahara. World Meteorol. Organ. Notes 86 and 210, and World Meterol. Bur. Tech. Pap. 110. 109 pp.

Crocker, R. L., and J. Major. 1955. Soil development in relation to vegetation and surface age, Glacier Bay, Alaska. J. Ecol. 43:427–48.

Dabadghao, P. M., and K. M. Shankarnarayan. 1970. Studies of Iseileme, Sehima, and Heteropogon communities of the Sehima-Dichanthium zone. *In* M. J. T. Norman (ed.), Proc. XI Int. Grassland Cong. Pp. 36–38. Univ. Queensland Press, Queensland.

Davidson, J. L., and J. R. Phillip. 1958. Light and pasture growth. *In* Climatology and microclimatology. Pp. 181–87. UNESCO, Paris.

Dement, W. A., and H. A. Mooney. 1974. Seasonal variation in the production of tannins and cyanogenic glucosides in the chaparral shrub, *Heteromeles arbutifolia*. Oecologia 15:65–76.

Denmead, O. T. 1964. Evaporation sources and apparent diffusivities in a forest canopy. J. Appl. Meterol. 3:383–89.

De Wit, C. T. 1965. Photosynthesis of leaf canopies. Agric. Res. Rep. 663, Inst. for Biol. Chem. Res. Field Crops Herb., Wageningen.

De Wit, C. T., and Goudriaan, J. 1974. Simulation of ecological processes. Centre for Agricultural Publishing and Documentation, Wageningen. 159 pp. (Simulation Monographs series)

Donald, C. M. 1961. Competition for light in crops and pastures. *In* Soc. Expt. Biol. Symp. XV. Pp. 282–313. Academic Press, New York.

Drosdoff, M. 1954. Plant analysis and fertilizer problems, pp. 181–90. Institut de Recherches pour les Huites et Oleogineux, Paris.

Drude, O. 1913. Die Okologie der Pfanzen. Die Wissenschaft 50.

Duminel, L. 1961. Nitrogen and phosphorus composition of corn leaves and corn yields in relation to critical levels and nutrient balance. Soil Sci. Soc. Am. Proc. 25:295–98.

Duncan, W. C., R. S. Loomis, W. A. Williams, and R. Hanau. 1967. A model for stimulating photosynthesis in plant communities. Hilgardia 38:181–205.

Ehleringer, J. R., and P. C. Miller. 1975*a*. A simulation model of plant water relations and production in the alpine tundra, Colorado. Oecologia 19:177–93.

Ehleringer, J. R., and P. C. Miller. 1975*b*. Water relations of selected plant species in the alpine tundra, Colorado. Ecology 56:370–80.

Elton, C. 1958. The ecology of invasions by animals and plants. Methuen and Co., Ltd., London. 181 pp.

Embleton, T. W., W. W. Jones, C. K. Labanauskas, and W. Reuther. 1973. Leaf analysis as a diagnostic tool and guide to fertilization. *In* W. Reuther (ed.), The citrus industry. Vol. II: Production technology. Pp. 184–205. University of California Press, Berkeley.

Fisher, M. J. 1970. The effects of phosphorus and water stress on townsville luceine (*Stylosanthes humilis* H. B. K.). *In* M. J. T. Norman (ed.), Proc. XI Int. Grassland Cong. Pp. 481–83. University of Queensland Press, Queensland.

Gaastra, P. 1963. Climatic control of photosynthesis and respiration. *In* L. T. Evans (ed.), Environmental control of plant growth. Pp. 113–40. Academic Press, New York.

Gates, D. M. 1962. Energy exchange in the biosphere. Harper and Row, New York. 151 pp.

Gates, D. M. 1965. Energy, plants, and ecology. Ecology 46:1–13.

Gates, D. M., E. C. Tibbals, and F. Kreith. 1965. Radiation and convection for ponderosa pine. Am. J. Bot. 52:66–71.

Goodall, D. W. 1962. Quantitative aspects of plant distribution. Biol. Rev. 27:194–245.

Greig-Smith, P. 1964. Quantitative plant ecology. Butterworth, London. 256 pp.

Griesebach, A. 1872. Die Vegetation der Erde nach ihrer klimatischen Anornung, pp. 603–35. Ein Abriss der vergleichenden Geographie der Pflanzen. W. Engelman, Leipzig.

Hamilton, E. L., and P. B. Rowe. 1949. Rainfall interception by chaparral in California. Calif. Dept. of Nat. Resources, Div. of Forestry. 43 pp.

Hanes, T. L. 1971. Succession after fire in the chaparral of southern California. Ecol. Monogr. 41:27–52.

Hanks, R. J., and S. B. Bowers. 1962. Numerical solution of the moisture flow equation into layered soils. Soil Sci. Soc. Am. Proc. 26:530–34.

Hansen, H. 1971. Growth, physiological and seed quality experiments with yellow mustard (*Sinopis alba*). Pp. 417–22. *In* R. M. Samish (ed.), Recent advances in plant nutrition, vol. 1. Gordon and Breach Science Publishers, New York.

Harpaz, Y. 1975. Simulation of the nitrogen balance in semi-arid regions. Ph.D. diss., Hebrew University, Jerusalem.

Harper, J. L. 1967. A Darwinian approach to plant ecology. J. Ecol. 55(2):247–70.

Harper, J. L., and J. White. 1974. The demography of plants, p. 419–464. *In* R. F. Johnston (ed.), Annual review of ecology and systematics. Annual Reviews Inc., Palo Alto, Calif.

Harrison, A. T. 1971. Temperature related effects of photosynthesis in *Heteromeles arbutifolia* M. Roem. Ph.D. diss., Stanford University, Palo Alto.

Hellmers, H., J. F. Bonner, and J. M. Kelleher. 1955. Soil fertility: A watershed management problem in the San Gabriel Mountains of southern California. Soil Sci. 80:189–97.

Honert, T. H. van den. 1948. Water transport in plants as a catenary process. Disc. Faraday Soc. 3:146–53.

Horn, H. S. 1971. The adaptive geometry of trees. Princeton University Press, Princeton, N.J.

Idso, S. B., and C. T. de Wit. 1970. Light relations in plant canopies. Appl. Opt. 9:177–84.

Jarvis, P. G., and M. S. Jarvis. 1963. The water relations of tree seedlings. IV. Some aspects of the tissue water relations and drought resistance. Physiol. Plant. 16:501–16.

Jenny, H. 1946. Arrangement of soil series and types according to functions of the soil forming factors. Soil Sci. 61:375–91.

Kanemasu, E. T., and C. B. Tanner. 1969. Stomatal diffusion resistance of snap beans. II. Effect of light. Plant Physiol. 44:1542–46.

Keller, T. 1971. Gaseous exchange—a good indication of nutritional states and fertilizer responses of forest trees. *In* R. M. Samish (ed.), Recent advances in plant nutrition, vol. 2. Gordon and Breach Science Publishers, New York.

Kenworthy, A. L. 1961. Plant analysis and fertilizer problems. Publ. 8, Amer. Inst. Biol. Sci., Washington, D. C. 454 pp.

Kheatawat, G. K., C. P. Vinayak, and S. P. Seth. 1972. Effect of different levels of nitrogen on the uptake of N, P, and K by wheat (Sonora-64). J. Indian Soc. Soil Sci. 20(1):27–30.

Klikoff, L. 1965. Photosynthetic response to temperature and moisture stress of three timberline meadow species. Ecology 46:516–17.

Koch, P. 1972. Utilization of the southern pines. Vol. 1: The raw material. USDA Forest Service. AH No. 420.

Kostić, M., W. Dijkshoorn, and C. T. de Wit. 1967. Evaluation of the nutrient status of wheat plants. Neth. J. Agric. Sci. 15(1967):267–80.

Kramer, P. J. 1952. Plant and soil water relations on the watershed. J. For. 50:92–95.

Kramer, P. J. 1969. Plant and soil water relationships: a modern synthesis. McGraw-Hill, New York. 482 pp.

Krause, D. 1975. Xeromorphic structure and soil moisture in the chaparral. M. S. thesis, San Diego State University, San Diego, Calif.

Kuiper, P. J. C. 1964. Water uptake of higher plants as affected by root temperature. Meded. Landbouwhogesch. Wageningen 63:1–11.

Lambert, J. R., and F. W. T. Penning de Vries. 1973. Dynamics of water in the soil-plant atmosphere system: a model named Troika. *In* E. A. Hadas et. al., Ecological studies, analysis and synthesis, vol. 4. Pp. 257–73. Springer-Verlag, Berlin.

Lawrence, W. T. 1975. A radiation model for chaparral canopies. M. S. thesis, San Diego State University, San Diego, Calif.

Lemon, E. R. 1967. The impact of the atmospheric environment of the integument of plants. Int. J. Biometeor. 3:57–69.

Lessander, R., and J. Gary. 1975. Photosynthesis and Wachstum du Clätter von jungan Zuchurrübenpflanzen in Abhangigkeit von der P. Ernährung. Arch. Acker. u. Pflanzentau u. Bodenkid, Berlin 19(8):545–54.

Liakopoulos, A. C. 1966. Theoretical prediction of evaporation losses from groundwater. Water Res. Res. 2:227–40.

Lommen, P. W., C. R. Schwintzer, C. S. Yocum, and D. M. Gates. 1971. A model describing photosynthesis in terms of gas diffusion and enzyme kinetics. Planta 98:195–220.

Luposhinsky, W. 1969. Stomatal closure in conifer seedlings in response to leaf moisture stress. Bot. Gaz. 130:258–63.

Lutz, J. A., Jr. 1973. Alfalfa response to P and K fertilization. J. Indian Soc. Sci. 21(1):63–69.

Lutz, J. A., Jr. 1974. Growth, chemical composition and yield of no-tillage corn as affected by N, source of P and rate and source of K. J. Indian Soc. Sci. 22 (1):60–65.

Lyttleton, J. W. 1973. Proteins and nucleic acids. *In* G. W. Butler and R. H. Bailey (eds.), Chemistry and biochemistry of herbage. Pp. 63–100. Academic Press, New York.

McAlpine, J. R. 1970. Estimating pasture growth periods and droughts from simple water balance models. *In* Proc. XI Int. Grassland Cong. Pp. 484–98. University of Queensland Press, Queensland.

MacArthur, R. 1970. Species packing and competitive equilibrium for many species. Theor. Pop. Bio. 1:1–11.

McCree, K. 1970. An equation for the rate of respiration of white clover plants grown under controlled conditions. *In* Prediction and measurement of photosynthetic productivity. Pp. 221–29. Proceedings IBP/PP Technical Meeting, Trebon, Pudoc, Wageningen, 1970.

McCree, K. J., and J. H. Troughton. 1966. Prediction of growth rate at different light levels from measured photosynthesis and respiration rates. Plant Physiol. 41:559–66.

Martin, G. 1974. Contribution to the study of phosphate fertilization of oil-palm in para state (Brazil). Phosphorisan Agric. No. 63, pp. 13–23.

Mead, D. J., and W. L. Pritchett. 1975. Fertilizer movement in a slash pine ecosystem. I. Uptake of N and P and N movement in the soil. Plant and Soil 43:451–65.

Miller, P. C. 1972. Bioclimate, leaf temperature, and primary production in red mangrove canopies in south Florida. Ecology 53:22–45.

Miller, P. C. 1973. A model of temperatures, transpiration rates, and photosynthesis of sunlit and shaded leaves in vegetation canopies. *In* Plant response to climatic factors. Pp. 427–34. Proc. Uppsala Symposium, 1970. UNESCO, Paris.

Miller, P. C. 1975. Simulation of water relations and net photosynthesis in mangroves in south Florida. *In*. C. E. Walsh, S. C. Snedaker, and H. G. Teas, eds. Proc. First International Symposium on Mangrove Biology and Management, Honolulu, Hawaii. 2:615–31. University of Florida, Gainesville, Florida.

Miller, P. C., D. Brandbury, N. Thrower, V. La Marche and E. Hajek. 1977. Past and present environment. *In* H. H. Mooney (ed.), Convergent evolution in Chile and California Mediterranean climate ecosystems. Pp. 27–72. Dowden, Hutchinson & Ross, Stroudsberg, Pa.

Miller, P. C., J. R. Ehleringer, B. Hynum, and W. A. Stoner. 1974. Digital simulation of potential reforestation problems in the Rung Sat Delta, Viet Nam. *In* H. T. Odum, M. Aell, M. Brown, J. Zucchetto, C. Swallows, J. Browder, T. Ahlstrom, and L. Peterson, The effects of herbicides in south Vietnam. Part B: Working papers. Models of herbicide, mangroves, and war in Vietnam. Pp. 100–164. National Academy of Sciences–National Research Council, Washington, D.C.

Miller, P. C., and E. Hajek. 1977. Macroclimate of central Chile and southern California. *In* N. J. W. Thrower and D. E. Bradbury, eds. The Chile-California mediterannean scrub atlas. Dowden, Hutchinson & Ross, Inc., Stroudsburg, Pa.

Miller, P. C., J. Hom, and D. K. Poole. 1975. Water relations of three mangrove species in south Florida. Oecol. Plant. 10: 355–67.

Miller, P. C., and H. A. Mooney. 1974. The origin and structure of American arid-zone ecosystems. The producers: Interactions between environment, form, and function. *In* Proc. First International Congress of Ecology. Pp. 201–9. The Hague, Netherlands. (Also to appear *In* C. T. de Wit and G. W. Arnold (eds.), Simulation monographs. Centre for Agricultural Publishing and Documentation, Wageningen.

Miller, P. C., W. A. Stoner, and L. L. Tieszen. 1976a. A model of stand photosynthesis for the wet meadow tundra at Barrow, Alaska (submitted to Ecology).

Miller, P. C., W. A. Stoner, L. L. Tieszen, M. Allessio, B. McCown, S. T. Chapin, and G. Shoven. 1975b. A model of carbohydrate, nitrogen and phosphorus allocation and growth in tundra plant production.

Miller, P. C., and L. Tieszen. 1972. A preliminary model of processes affecting primary production in the arctic tundra. Arct. Alp. Res. 4:1–18.

Miller, P. C., B. D. Collier, and F. Bunnell. 1975. Development of ecosystem modeling in the Tundra Biome. *In* B. Patten (ed.), Systems analysis and simulation in ecology, vol. III. Pp. 95–115. Academic Press, New York.

Miller, P. C., and E. Ng. 1976. Root:shoot ratios in shrubs in southern California and central Chile. (Submitted to Madroño.)

Monk, C. D. 1966. An ecological significance of evergreeness. Ecology 47:504–5.

Monsi, M., and T. Saeki. 1953. Uber den Lichtfaktor in den Pflanzengesellschaften und seine Bedeutung fur die Stoffproduktion. Jap. J. Bot. 14:22–52. (English trans. W. Woodard and P. C. Miller, mimeo.)

Monteith, J. L. 1973. Principles of environmental physics. Edward Arnold, London. 241 pp.

Mooney, H. A. 1972. The carbon balance of plants. Ann. Rev. Ecol. Syst. 3:315–46.

Mooney, H. A. (ed.). 1977. Convergent evolution in Chile and California mediterranean climate ecosystems. Dowden, Hutchinson & Ross, Stroudsberg, Pa. (in press).

Mooney, H. A., and C. Chu. 1974. Seasonal carbon allocation in *Heteromeles arbutifolia*, a California evergreen shrub. Oecologia 14:295–306.

Mooney, H. A., and R. I. Hays. 1973. Carbohydrate storage cycles in two Californian Mediterranean-climate trees. Flora 162:295–304.

Muller, C. H. 1966. The role of chemical inhibition (Allelopathy) in vegetational composition. Bull. Tor. Bot. Club 93(5):332–51.

Muller, C. H., R. B. Hanawalt, and J. K. McPherson. 1968. Allelopathic control of herb growth in the five cycles of California chaparral. Bull. Tor. Bot. Club 95:225–31.

Murphy, C. E., and K. R. Knoerr. 1970. Modeling the energy balance proceses of natural ecosystems. U.S. IBP (Int. Biol. Program) Anal. Ecosyst. Program Eastern Deciduous Forest Biome Subproject Res. Rep. Final 1969-1970, Oak Ridge National Laboratory. 168 pp.

Murphy, C. E., and K. R. Knoerr. 1972. Modeling the energy balance processes of natural ecosystems. U.S. IBP (Int. Biol. Program) Anal. Ecosyst. Program Eastern Deciduous Forest Biome Subproject Res. Rep. EDFB-IBP 72–10, Oak Ridge National Laboratory. 141 pp.

Ng, E. 1974. Soil moisture relations in chaparral. M.S. thesis. San Diego State University, San Diego, Calif.

Ng, E., and P. C. Miller. 1975. A model of the effect of tundra vegetation on soil temperatures. *In* G. Weller and S. A. Bowling (eds.), Climate of the arctic. Pp. 222–26. 24th Alaska Science Conference, Fairbanks, Alaska, August 15–17, 1973. Geophysical Institute, University of Alaska, Fairbanks, Alaska.

Ng, E., and P. C. Miller. 1977. Validation of a model of the effect of tundra vegetation on soil temperatures. Arct. Alp. Res. 9:89–104.

Nobel, P. S., L. J. Varagoza, and W. K. Smith. 1975. Relation between mesophyll surface area photosynthetic rate and illumination level during development for leaves of *Plectranthus parvislorus* Henckel. Plant Physio. 55:1067–70.

Odum, E. 1960. Organic production and turnover in old field succession. Ecology 41(1):34–48.

Odum, H. T., and R. C. Pinkerton. 1955. Time's speed regulator. Amer. Sci. 43:331–43.

Omar, M. H. 1968. Potential evapotranspiration in a warm arid climate, pp. 347–353. *In* Agricolimatological methods. Proc. Reading Symposium. UNESCO, Paris.

Parkhurst, D. F., and O. L. Loucks. 1972. Optimal leaf size in relation to environment. J. Ecol. 60:505–37.

Parsons, D. J. 1973. A comparative study of vegetation structure in the Mediterranean scrub communities of California and Chile. Ph.D. diss., Stanford University, Palo Alto, Calif.

Parsons, D. J., and A. R. Moldenke. 1974. Convergence in vegetation structure along analogous climatic gradients in the Mediterranean climate ecosystems of California and Chile. U.S. -IBP, Origin and Structure of Ecosystems Tech. Rept. 74–11. 18 pp.

Pase, C. P., and M. M. Fogel. 1967. Increasing water yield from forest, chaparral, and desert shrub in Arizona, p. 753–764. *In* Proc. Int. Conf. of Water for Peace.

Pase, C. P., and R. A. Ingebo. 1965. Burned chaparral to grass: early effects on water and sediment yields from two granitic soil watersheds in Arizona. Proc. Arizona Watershed Symp., Sept. 22, 1965. Tempe, Ariz.

Penman, H. L. 1948. Natural evaporation from open water, bare soil and grass. Proc. Roy. Soc. Series A, 193: 120–45.

Penning de Vries, F. W. T. 1972a. A model for simulating transpiration of leaves with special attention to stomatal functioning. J. Appl. Ecol. 9:57–77.

Penning de Vries, F. W. T. 1972b. Respiration and growth. *In* A. R. Rees, K. E. Cockshull, D. W. Hand, and R. G. Hurd (eds.), Crop processes in controlled environments. Pp. 327–47. Academic Press, London.

Penning de Vries, F. W. T. 1973. Substrate utilization and respiration in relation to growth and maintenance in higher plants. Ph.D. diss., Agricultural University Wageningen.

Penning de Vries, F. W. T. 1974. Substrate utilization and respiration in relation to growth and maintenance in higher plants. Neth. J. Agric. Sci. 22:40–44.

Penning de Vries, F. W. T. 1975a. The cost of maintenance processes in plant cells. Ann. Bot. 39:77–92.

Penning de Vries, F. W. T. 1975b. Use of assimilates in higher plants in photosynthesis and productivity in different environments. IBP, vol. 3. Cambridge University Press, Cambridge.

Penning de Vries, F. W. T., A. H. M. Brunsting, and H. H. van Laar. 1974. Products, requirements and efficiency of biosynthesis: a quantitative approach. J. Theor. Biol. 45:339–77.

Penning de Vries, F. W. T., C. E. Murphy, Jr., C. G. Wells, and J. R. Jorgensen. 1976. Similation of nitrogen distribution in time and space in even-agral of loblolly pine plantations and its effect on productivity. IBP Contribution #156. *In* Proc. Symp. Mineral Cycling in Southern Ecosystems. Augusta, Ga., 1–3 May 1974.

Pillsbury, A. F., J. F. Osborn, and R. E. Peleshek. 1963. Residual soil moisture below the root zone in southern California watersheds. J. Geophys. Res. 68:1089–91.

Pillsbury, A. F., J. F. Osborn, and R. E. Peleshek. 1963. Residual soil moisture below the root zone in southern California watersheds. J. Geophys. Res. 68:1089–91.

Poole, D. K. 1974. Water relations of selected species of chaparral and coastal sage communities. M.S. thesis, San Diego State University, San Diego, Calif.

Poole, D. K., and P. C. Miller. 1975. Water relations of selected species of chaparral and coastal sage communities. Ecology 56:1118–28.

Poole, D. K., P. C. Miller, and W. Lawrence. Aspects of water relations of five Chilean shrubs (in prep.).

Possingham, J. V. 1971. Some affects of mineral nutrient deficiencies on the diloroplasts of higher plants. *In* R. M. Samish (ed.), Recent advances in plant nutrition, vol. 1. Pp. 155–64. Gordon and Breach Science Publishers, New York.

Prevot, P., and M. Ollognier. 1957. Plant analysis and fertilizer problems. *In* Institut de Recherches pour les Huites et Oleagineux, Paris. 410 pp. Pp. 177–95.

Rapp, M. 1971. Cycle de la matiére organique et des éléments minéraux dans quelques écosystéms Méditerranéens. Editions du Centre National de la Recherches Scientifique, Paris.

Raunkiaer, C. 1934. The life forms of plants and statistical plant geography. Clarendon Press, Oxford.

Rawlins, S. L. 1963. Resistance to water flow in the transpiration stream. *In* I. Zelitch (ed.), Stomata and water relations in plants. Pp. 69–84. Conn. Agric. Exp. Sta. Bull. 664.

Reuther, W. 1973. Climate and citrus behavior. *In* W. Reuther (ed.), The citrus industry. Vol. 3: Production technology. Pp. 281–33. Univ. of Calif. Division of Agricultural Sciences.

Reuther, W. T., T. W. Embleton, and W. W. Jones. 1958. Mineral nutrition of tree crops. Ann. Rev. Plant Physiol. 9:175–206.

Roberts, S. W. 1974. The morphology and solar radiation environment of two Mediterranean shrubs. M.S. thesis, San Diego State University, San Diego, Calif. 125 pp.

Rowe, P. B., and R. A. Colman. 1951. Disposition of rainfall in two mountain areas of California. USDA Forest Service Tech. Bull. 1018. 84 pp.

Schimper, A. F. W. 1898. Pflanzingiographie auf physiologischer Grundlage. Jena.

Shachori, A. Y., and A. Michaeli. 1965. Water yields of forest, maquis and grass covers in semi-arid regions: a literature review. *In* F. E. Eckardt (ed.), Methodology of plant ecophysiology. Pp. 467–77. Proc. of the Montpellier Symposium. UNESCO, Paris.

Shachori, A., D. Rosenzweig, and A. Poljakoff-Mayber. 1967. Effect of Mediterranean vegetation on the moisture regime. *In* W. E. Sopper and H. W. Lull (ed.), International symposium on forest hydrology. Pp. 291–311. Pergamon Press, New York.

Skre, O. 1975. CO_2 exchange in Norwegian tundra plants studied by infrared gas analyzer technique. *In* F. E. Wielpolaski (ed.), Fennoscandian tundra ecosystems. Part I: Plants and microorganisms. Pp. 168–83. Springer-Verlag, New York.

Slatyer, R. O. 1967. Plant water relationships. Academic Press, New York. 366 pp.

Slatyer, R. O. 1970. Comparative photosynthesis, growth and transpiration of two species of Atriplex. Planta 93:175–89.

Small, E. 1973. Xeromorphy in plants as a possible basis for migration between arid and nutritionally deficient environments. Bot. Notiser 126:534–39.

Smith, P. F. 1962. Mineral analysis of plant tissues. Ann. Rev. Plant Physiol. 13:81–108.

Stanhill, G. 1961. The accuracy of meteorological estimates of evapotranspiration in arid climates. J. Inst. Water Eng. 15(7):

Steubing, L., and M. Alberdi. 1973. The influence of phosphorus deficiency on the scherophylly. Oecol. Plant. 8(3):211–18.

Stewart, D. W., and E. R. Lemon. 1972. The energy budget at the earth's surface: a simulation of net photosynthesis of field corn. ECOM Atmospheric Sciences Lab. Fort Huachuca, Ariz. Int. Rep. 69–3.

Stoner, W. A., and P. C. Miller. 1975. Plant water relations in the wet coastal tundra, Barrow, Alaska. Arct. Alp. Res. 7:109–24.

Stoner, W. A., P. C. Miller, S. P. Richards, and S. A. Barkley. 1976. Internal nutrient cycling as related to plant life form—a simulation approach. *In* Proc. Environmental Chemistry and Cycling Processes Symp. 28–30 April 1976, Augusta, Ga.

Stoner, W. A., P. C. Miller, and L. L. Tieszen. 1978. A model of plant growth and phosphorus allocation for *Dupontia fischeri* in coastal, wet meadow tundra. *In* L. L. Tieszen (ed.). Vegetation and production ecology of an Alaskan arctic tundra. Springer-Verlag, New York.

Tamm, C. D. 1971. Optimum nutrition of trees in field experiments—a long term approach in forest nutrition research. *In* R. M. Samish (ed.), Recent advances in plant nutrition, vol. 1. Pp. 623–56. Gordon and Breach Science Publishers, New York.

Thrower, N., and D. E. Bradbury (eds.). 1978. The Chile-California mediterranean scrub atlas. Dowden, Hutchinson & Ross, Stroudsberg, Pa.

Tibbals, E. C., E. K. Carr, D. M. Gates, and F. Kreith. 1964. Radiation and convection in conifers. Am. J. Bot. 51:529–38.

Tieszen, L. L., P. C. Miller, M. C. Lewis, J. C. Mayo, F. S. Chapin, III, and W. C. Oechel. 1978. Processes of primary production in the tundra. *In* J. J. Moore (ed.), The ecology of tundra and elated habitats. Cambridge University Press, Cambridge (in press).

van Keulen, H. 1975. Simulation of water use and herbage growth in arid regions. Centre for Agri. Pub. and Docum., Wageningen, the Netherlands. 176 pp.

van Keulen, H., N. G. Seligman, and J. Goudriaan. 1975. Availability of anions in the growth medium of roots of an actively growing plant. Neth. J. Agric. Sci. 23 (1975):131–38.

Waggoner, P. E., G. M. Furnival, and W. E. Reifsnyder. 1969. Simulation of microclimate in a forest canopy. For. Sci. 15:37–45.

Waggoner, P. E., and W. E. Reifsnyder. 1968. Simulation of temperature, humidity, and evaporation profiles in a leaf canopy. J. Appl. Meterol. 7:400–409.

Wallén, C. C. 1968. Agroclimatological studies in the Levant. *In* Agroclimatological methods. Pp. 225–33. Proc. of the Reading Symposium. UNESCO, Paris.

Wallén, C. C., and G. P. de Brichambaut. 1962. A study of agroclimatology in semi-arid and arid zones of the near east. FAD Rome, Italy. 185 pp.

Warming, E. 1895. Plantesamfund. Grundtraek af den okologiske Plantegeografi. Kjobenhavn.

Wells, C. G., and L. J. Metz. 1963. Variation in nutrient content of loblolly pine needles with brasm, age, soil, and position in the crown. Soil Sci. Soc. of Am. Proc. 27(1):90–93.

White, J. D., C. G. Wells, and E. W. Clark. 1970. Variations in the inorganic composition of inner bark and needles of loblolly pine with tree height and soil series. Canad. J. Bot. 48(6):1079–84.

Williams, C. B. 1964. Patterns in the balance of nature and related problems in quantitative ecology. Academic Press, New York. 324 pp.

Yim, Y., H. Ogawa, and T. Kira. 1969. Light interception by stems in plant communities. Jap. J. Ecol. 19:233–38.

R. G. WIEGERT

Population Models: Experimental Tools for Analysis of Ecosystems

8

INTRODUCTION

Analysis versus Synthesis

The theme of this symposium volume, ecosystem analysis, implies, according to my dictionary, an examination to distinguish the component parts of the ecosystem separately and in their relation to the whole. An important component of the ecosystem is the population, a unit most often composed of individuals of the same species.[1] Although the changes caused by temperature variation, predation, and reproduction are observable at the level of the individual, the net effect of such physical and biological parameters at the population level is the prime determinant of the character of an ecosystem. Indeed, some important evolutionary and ecological processes affecting the ecosystem (change in gene pool, succession, etc.) can be discussed *only* with reference to population change. Thus the analysis of an ecosystem profits from a clear understanding of the properties of the component populations.

Conversely, an understanding of the properties of components of systems permits synthesis, the reverse of the analytic process, whereby knowledge of the individual components of the system is pieced together to compose statements about the behavior of the whole. In particular, information accumulated about populations in field or laboratory may permit hypotheses about the behavior of ecosystems under a wide variety of environmental conditions.

One current approach to such a goal is through simulation models of ecosystems (Wiegert, 1975a). To be used in this manner, simulation models must comprise population submodels of sufficient ecological quality (=real-

R. G. Wiegert, Department of Zoology, University of Georgia, Athens, Georgia 30602.

ity) to be useful theoretical tools in their own right. That is, the population model, given the information about available resources and a measure of losses, could be used alone to investigate effects of environmental manipulations. Although this has been one goal of population ecology for decades, the models used have fallen short of the necessary quality. The purpose of this paper is to examine briefly the shortcomings of previous models and to suggest some useful alternatives. I further show how these alternative population models, when coupled to form very simple food chains, can be used as experimental tools to formulate both qualitative and quantitative hypotheses about the behavior of simple ecosystems.

POPULATION MODELS

Characteristics

All models, whether verbal, diagrammatic or mathematical, are abstractions of the real world. The ecologist first formulates the model verbally, outlining the known characteristics of the population and, most importantly, the objectives and limitations of the desired model. Limitations are defined by those factors that are omitted as unimportant to the objectives. For example, if population models do not explicitly consider the effects of the abiotic environment as it acts to change parameters, then the predictions of such models are limited to a narrow range of environmental conditions and cannot be expected to respond to perturbations of temperature, moisture, and so on.

Models may document changes in the gene pool responding to selective pressures (population genetic models). Alternatively, models may include competitive interactions of organisms and their resources, predicting changes in numbers, biomass, or energy (population growth models).

This paper deals exclusively with models of population dynamics, although the influence of both genetic and environmental changes can be implicitly modeled by making the parameters variables instead of constants. Explicit representation of changes in genotype incorporated into models of population dynamics may eventually unite the objectives of population genetics and ecology (see, for example, Charlesworth, 1973; Lewontin, 1976).

Models of population may be deterministic or stochastic. The former have certain advantages of simplicity, but assume large population size to surmount (or average out?) the effects of variability in the behavior of individuals. If the assumption of large numbers is met and in addition the variation caused by environmental fluctuations is small relative to the size of the interaction coefficients, then stability conclusions based on the use of deterministic models agree with those obtained with the stochastic form (May, 1971). This is not to say that deterministic models cannot produce variable results. Indeed, under certain conditions the behavior of deterministic models is indistinguishable from that of completely random models (May, 1974), a point discussed at some length later in this paper.

Realistic, Precise Models

Levins (1966) characterized models as: (1) those that sacrificed generality for the advantages of reality and precision, (2) those that emphasized generality and reality but sacrificed precision, and (3) those that were unrealistic but very general and precise.

In the first category are models such as those of Watt (1961) and Holling (1963, 1965). Watt constructed models of insect pest dynamics using difference equations (the density at time $t + 1$ is a function of density at time t). These models employed phenological as opposed to chronological time and were built up of submodels (and sub-submodels) of various processes (mortality, fecundity) affecting population density. The result was a model that incorporated a large number of causal pathways arrived at deductively together with parameter values inductively derived from regression tests of measured data. Such a model is excellent for predicting the dynamic changes of a specific population; i.e., it is realistic and precise, but is relatively poor for constructing hypotheses about the general behavior of populations.

Similarly, Holling (1963, 1965), modeling predator-prey interactions, sought to break the interaction into all its relevant component parts, the ultimate analysis of a simple system, so to speak.

Models of the Watt and Holling type are good predictive tools and provide much useful insight into the relative importance of the various parts of the host-parasite and predator-prey interactive sequences. But the wealth of experimental-observational detail required to implement such models is difficult and expensive to obtain. The specific nature of the many causal pathways modeled precludes any general application of the results.

Generalized, Realistic Models

A second class of model according to Levins (1966) contains those in which precision is sacrificed for generality and reality. This approach is characteristic of Levins's own work as well as that of MacArthur (MacArthur and Levins, 1965) and Rosenzweig (Rosenzweig and MacArthur, 1963). Such models, because they are concerned with qualitative rather than quantitative phenomena, can be more flexible, even graphical. Conclusions are expressed in ordinal rather than interval scale of measurement; something is greater than or less than another value, but the degree of difference is not specified. A major argument in favor of this approach is the robustness of the conclusions when the latter are derived from consideration of several models differing in detail but with common biological assumptions (Levins, 1966). Quantitative results are thought to be only important in testing the results. In many instances, however, the quantitative values of parameters in identical models may cause very different results, and the model may be complex enough to defy simple graphical interpretation. It was this kind of problem that led very early in the history of population ecology to the development of the third class of model described by Levins (1966).

Generalized, Precise (Classical) Models

The classical models of population dynamics, those developed before 1935 (according to Royama, 1971) all share the quality of generality coupled with precision, i.e., they engender general statements about the quantitative relationships within and between populations, but do so at the expense of realistic causal interactions.

An extensive literature describes these population models. Useful reviews are found in Pielou (1969), May (1973a), and Maynard Smith (1974). A detailed critical monograph on predator-prey configurations is that of Royama (1971). The models all employ a single variable representing population size, and the relationships (at least in the original form) are expressed as differential equations.[2]

Growth and subsequent control of density in the classical models are represented by the interaction of universal population processes. For example, a property of all living populations is an inherent capacity for autocatalytic (exponential) growth. When resources are unlimited and losses are minimal,

$$dx/dt = \rho \cdot x \qquad (1)$$

where: ρ is a maximum specific rate of growth (the intrinsic rate of increase) with dimensions time^{-1}. Such a model is of limited application, because populations seldom exhibit uncontrolled growth for long, if ever. Therefore, the earliest population models contained a term describing the effect of competition for limited resources. The most famous of these earlier classical models is the Verhulst-Pearl logistic model,

$$dx/dt = \rho \cdot x \; (1 - x/\gamma) \qquad (2)$$

where γ is the maximum density of carrying capacity.

Equation (2) produces the familiar sigmoid growth curve; the population grows at close to rate ρ when x is close to zero and asymptotically approaches a zero rate of growth as x approaches γ. This logistic model may be derived in at least three ways (Pielou, 1969). The simplest uses the definition of negative feedback term, $f(x)$, for positive Δx and a linear relationship, thus $f(x) = (\rho - x/\gamma)$. The assumption of a linear-feedback term is the simplest possible and is thus justified in the absence of biological information to the contrary. The frequent success of the logistic equation, despite its other weaknesses, in simulating real population growth is probably a testimony to the commonness of such linear feedback relationships. Of course, the feedback term of equation (2) is only one of many possible ways that the density of a population may affect its growth. Other feedback terms that have been proposed (from May, 1973a) are those by Gompertz (1825)

$$\rho \cdot \ln (\gamma/x), \tag{3}$$

Smith (1963)

$$\rho(\gamma - x)/(\gamma - \phi x), \tag{4}$$

Schoener (1973)

$$\rho[(\gamma/x) - 1], \tag{5}$$

Rosenzweig (1971)

$$\rho[(\gamma/x)^{-\eta} - 1], \tag{6}$$

and Goel, Maitra, and Montroll (1971)

$$\rho[1 - (x/\gamma)^{\eta}]. \tag{7}$$

($1 \geq \eta > 0$ for Eqs. 6 and 7.) ϕ and η are constant coefficients.

All these stabilize growth (May, 1971), although the trajectory of population density toward the limit will be different.

When two (or more) populations utilize a common resource, the model must have feedback control functions incorporating the additional competitor(s). The earliest models of this type used a modification of the simple linear logistic term, the result being the familiar Lotka-Volterra two species competition equations:

$$dx_1/dt = \rho_1 x_1[1 - (x_1 + \beta_1 x_2)/\gamma_1] \tag{8}$$

$$dx_2/dt = \rho_2 x_2[1 - (x_2 + \beta_2 x_1)/\gamma_2] \tag{9}$$

where the β values are the coefficients of competition.

A simple analysis of such equations shows that the persistence (=coexistence) of such a population system will depend on the relative magnitude of the β and γ values (Slobodkin, 1961). Plotting x_1 against x_2, if the lines representing the zero isoclines (growth of $x_1=0$ and growth of $x_2=0$) do not intersect, then the species with greater intercepts on the x_1 and x_2 axes will always win. If, however, the zero isocline lines do intersect, the equilibrium (point of intersection) will be stable or unstable depending on whether $\gamma_1/\beta_1 > \gamma_2$ and $\gamma_2/\beta_2 > \gamma_1$ (stable) or vice versa (unstable). In the case of the unstable equilibrium, the winner will be determined by the intial values of the two species.

Competition for limited resources is one way in which the growth of a population can be reduced to zero. Predation (or parasitism), by increasing losses, can accomplish the same thing. Again, the classical models of this type were developed independently by Lotka and Volterra (see Lotka, 1924; Volterra, 1931, cited in Caswell, 1972). Regarding the predator-prey interaction as simply a matter of the density of prey affecting the growth of the predator (x_2) and the density of the predator affecting growth of the prey (x_1), they wrote:

$$dx_1/dt=x_1(\rho_1-\phi_2x_2) \tag{10}$$

$$dx_2/dt=x_2(-\lambda_2+\phi_1x_1) \tag{11}$$

The biological basis for these Lotka-Volterra predator-prey equations rests on the relative effects of feeding by the predator acting in concert with the losses of the predator and the inherent growth rate of the prey. Neither predator nor prey is constrained by intraspecific density dependent controls. The model postulates a continuous rise in predator consumption as prey density goes up, i.e., no satiation of feeding by individual predators.

The dynamic behavior of predator-prey populations modeled with the Lotka-Volterra equations has been analyzed by Pielou (1969). The model always produces oscillating predator-prey numbers with neutral stability, i.e., the amplitude of the fluctuation is determined by the initial densities. This result can be shown by assuming that small amplitude fluctuations are approximated by ellipses (Pielou, 1969). Neutrally stable models are not, however, satisfying representations of ecological reality (May, 1973a). Given the potential feedback controls that can be developed by populations, persistent systems would be expected to show either stable points or else limit cycles. Indeed, in nature even a neutrally stable system would not exhibit constant amplitude of oscillations because of the change in the starting densities following random perturbations. Constancy of oscillation could only be expected from a relationship involving a limit cycle, where the amplitude of the fluctuations is independent of starting densities.

Various "improvements" of the Lotka-Volterra predator-prey model have been introduced, most commonly via some form of self-damping of prey (MacArthur, 1970) or of the predator (Dixon and Cornwell, 1970). These improvements do not, however, address those basic difficulties discussed in the next section; they are simply combinations of the competition and predator-prey classical models.

A New Perspective

Of the three general types of model discussed by Levins (1966), the classical models of competition and predation, because they offer both generality and precision, are the best for generating specific hypotheses and also for developing coupled, predictive simulation models of ecosystems (see Wiegert, 1975a,b). But before this utility can be realized fully, their most obvious failings need to be corrected. In particular we need to see if these drawbacks, which concern mainly the departures from reality, can be corrected without too great a loss of generality or precision.

The unrealistic aspects of a model such as the Lotka-Volterra were, to a large extent, required because of inadequate computational aids at the time of its formulation. In an age when pencil and paper calculation was required, precise general models had to consist of simple, easily manipulatable equations. Numerical solution of sets of simultaneous differential equations was a laborious process. Viewed in this light, the reluctance to add complexity to the models in their general form was understandable. The development of analogue and digital computers removed the restriction on complexity. The recent improvements in speed and cost of digital interactive terminal computing facilities have placed within the easy reach of ecologists a powerful new tool for exploring the dynamics of populations, and may be as great a stimulus to the development of ecological theory as was the electron microscope to advances in cellular morphology.

The failure of the classical models to represent realistically the dynamics of interacting populations stems from three structural defects: (1) the models are closed, i.e., they do not represent flows of material or energy between populations; (2) the models do not separately represent the controlling effects of scarcity of material resources as opposed to the effects of direct competition for available space; and (3) the models do not provide for the possibility of threshold effects, i.e., the appearance of behaviorial or physiological responses which would substantially alter the form of the relationship between the population and its environment.

Closed models do not need to ensure consistency between the inflows and outflows of populations connected trophically. Thus in the Lotka-Volterra predator-prey model, for example, the loss term to predation from the prey need have no direct quantitative correspondence to the "prey capture" term in the predator growth equation. Such models are not suited for coupling into models of simple systems with more than two links in a food chain. This is the least damaging and the most easily corrected of the major failings of the classical models. Open flow models are a desirable preliminary to the correction of the other two structural defects of the classical models of population.

The most serious flaw in the early models is the failure to represent, separately, renewable material resources and the nonrenewable resource comprising living space (area or volume). Organisms competing for a material resource affect each other indirectly through the effect of each on the standing crop and rate of supply of the resource. Organisms competing for space affect each other directly through behavioral interactions that may change the feeding rate, affect fecundity, or vary the nonpredatory mortality. Because one type of resource is renewable and the other is not, their separation in population models is vital. Yet few have attempted to solve this problem. In his last book MacArthur (1972) developed a "recipe" for calculating the value of *k*, regarding it as the limiting or equilibrium density of populations competing solely for a material renewable resource. Schoener (1973) recognized the importance of the two different types of resource and constructed models that separate "exploitative" from "interference" competition. The Schoener models did not explicitly represent flows; material transfers were represented as change in density of individuals.

Rogers and Hassell (1974) pointed out that searching behavior of predators or parasites is responsive to spatial and temporal changes in the host or prey population as well as to their own density and that realistic models must take these into account.

A number of models of insect population growth incorporate terms to separate density-dependent "interference" or competition for space from the "exploitative" use of renewable resources (see Hassell and Rogers, 1972; Hassell and May, 1973). Wiegert (1973) introduced an open energy flow model of a plant-herbivore trophic link in which competition for space (interference) was separated from competition for a material resource. This model provided explicit biological definitions for the parameters governing feedback control, and was structured in a manner that allowed any number of different feedback functions to be developed, depending on the biologies of the interacting populations. The model also proposed to solve a third major deficiency of all early models by providing for threshold behavior on the part of populations reacting to limitations caused by interference and competitive exploitation of a scarce resource. For example, the logistic equation generates a negative feedback effect of density whenever the population is greater than zero. Thus the familiar S curve of population growth represents the lower bound of a family of curves of which the upper bound is exponential growth that becomes limited only when the density reaches the equilibrium level. Specifying a threshold density level where the effect of crowding on growth rate is first noticed is an example of the kind of reality that needs to be incorporated into models of population to effect a productive compromise between generality, realism, and precision.

Such modification of the classical equations of population dynamics by representing flows instead of numbers, separation of exploitation from interference, and introducing threshold behavior has been used to reexamine the theory of competition (Wiegert, 1974) and to simulate the dynamics of food chains and ecosystems (Wiegert, 1975a,b).

In the present paper I develop the rationale for using such models as subjects for experimentation with a view to further developing ecological theory. The basis of the generalized realistic model is first reviewed, then used to construct a predator-prey model that complements the competition model of Wiegert (1974). The latter is modified slightly to form an open flow model; then both models are subjected to a range of enrichments and time delays. Finally, the use of such models as experimental tools similar to laboratory microcosms is discussed.

A SINGLE-SPECIES MODEL

Model Structure

General characteristic. The general model is based on three commonly accepted observations about living organisms.

First, populations undergo autocatalytic growth when provided with optimum levels of resource, both renewable (material) and nonrenewable (space). Under these conditions the growth equation should reduce to a simple exponential, exhibiting a maximum specific rate[3] of ingestion and growth ($\rho = \tau - \lambda$), ρ being the intrinsic rate of increase, τ the maximum specific rate of ingestion, and λ the minimal specific rate of loss.

Both τ and λ are defined for optimum conditions, but they are genetically determined parameters for the given environment. Thus, τ may often be a variable, changing with genotype and physical factors. Although both τ and λ will be assumed constant in the particular experiments reported in this paper, the structure of the model easily accepts them as variables.

The specific rate of loss (λ) is here restricted to losses that are functions of the population. This is a reasonable assumption when conditions are optimal, but is included only to simplify the discussion and is not a requirement of the model. Often, for example, λ would be a function of predation or of the available resources without in any way compromising the experimental approach put forward in this paper.

Second, because of their potential for exponential growth, living organisms tend to press against their resources, and thus negative feedback mechanisms for controlling growth must exist. In short, populations may exhibit either steady state or limit cycle behavior in which the long-term cumulative mean density change approaches zero. If such limits do not exist in the long run, the system cannot persist; one or more populations will become extinct.

Third, because they require matter/energy for survival, populations deprived of any ingestion under otherwise optimum conditions for growth must decline according to a rate of loss, λ, or any other operative mortality factors included in the model.

A general flow diagram is shown in figure 1 using the (slightly modified) notation of Forrester (1961). A model of x_j that will satisfy the foregoing three observations, in the differential equation form is:

$$dx_j/dt = x_j\{[\tau_{ij} - \tau_{ij}f(x_i) - (\tau_{ij} - \lambda_j)f(x_j)]_+ - \lambda_j\} \qquad (12)$$

where

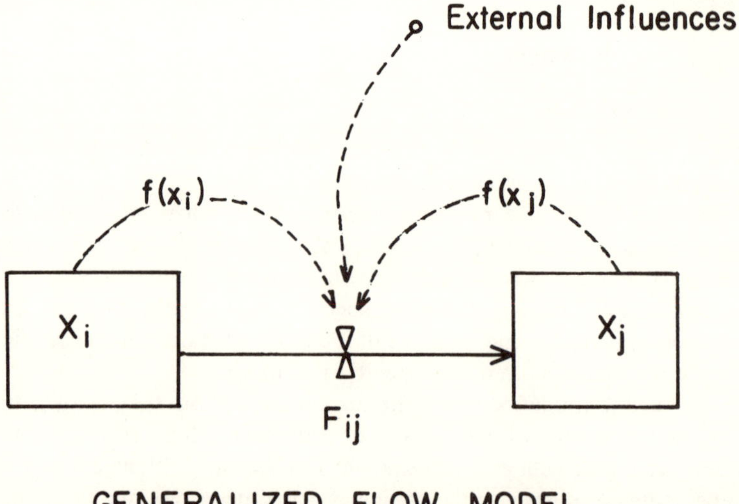

GENERALIZED FLOW MODEL

Fig. 1. A generalized flow model of two compartments showing the possible sources of control. External influences are not considered further. Compartment χ_i assumed to receive input from a source; χ_j assumed to have output to a sink.

τ_{ij} = maximum specific rate of ingestion

λ_j = minimum specific rate of loss

$f(x_i)$ = feedback control of ingestion due to scarcity of material resources

$f(x_j)$ = feedback control of ingestion due to scarcity of space

and

$$[\cdot]_+ = \begin{cases} 0 \text{ if } [\cdot] \leq 0 \\ [\cdot] \text{ if } [\cdot] > 0 \end{cases}$$

Equation (12) presents the simplest form of the general model, one that will be used in the subsequent simulation experiments. Applications of this approach to ecosystem modeling where other terms such as egestion, respiration, and excretion are used can be found in Wiegert (1973, 1975b) and in Wiegert et al. (1975). The subscript plus notation in Eq. 12 assures that the population (x_j) can never ingest negative quantities. Equation 12 satisfies the ecological constraints itemized above: (1) it permits growth at the maximum rate ($\tau - \lambda = \rho$) when all resources are optimum (feedback functions = 0), (2) it specifies $dx_j/dt = 0$ when ingestion is just balanced by losses, and (3) it permits decrease in density at specific rate λ_j when ingestion is reduced to zero.

The modeling problem now reduces to one of finding ecologically acceptable forms for the feedback control functions $f(x_i)$ and $f(x_j)$.

Competition for a material resource. Consider the scarcity of a material renewable resource and its effect on the ingestion by a population, i.e., exploitive competition. Many organisms respond to variation in the standing stock or density of a material resource by showing an effect on ingestion (and thus growth) when the resource first reaches some threshold saturation density (α_{ij}). As the density of the resource continues to decrease a refuge level may be encountered ($\gamma_{ij} \geq 0$) where the resource is no longer available and thus ingestion is precluded. The simplest way to represent this relationship is to assume that a unit change in the resource will have a constant effect on ingestion by the consumer of the resource.

$$f(x_i) = [(\alpha_{ij} - x_i)/(\alpha_{ij} - \gamma_{ij})]_+ \qquad (13)$$

Equation (13) assumes no change in the behavior or searching ability of the consumer as the resource becomes scarce. Note that equation (13) is constrained such that the feedback term ≥ 0.

Many other forms of this function could be postulated. Two discussed in Wiegert (1975c) are cases where the predator increases its searching effort rapidly as prey first become scarce or alternatively, the predator rapidly abandons the prey after scarcity first becomes apparent. The form of equation (13) is used in the experiments reported in this paper.

Competition for space. The effects of competitive interference can begin in theory as soon as organisms are dense enough that individuals become aware of the presence of other individuals of their own or competing species. The actual threshold response density (α_{jj}) is usually greater than this minimal density below which no communication and thus no competitive interference can take place. At higher densities the ingestion and growth rates of the population will continually decrease until the latter reaches zero. At this point the maximum asymptotic density (γ_{jj}) has been reached.

A somewhat related phenomenon, the Allee effect, requires a minimum density for viability; this is easily illustrated in those social organisms where a minimum density is required for effective courtship behavior and breeding. Even for solitary sexual organisms, there is obviously some minimum density below which the probability of male and female gametes' being brought together is so low that the population declines to zero. The Allee effect is not specifically used in the models discussed in this paper, but could be easily incorporated as another density-responsive feedback term, one that was equal to unity above a threshold density, then declined to zero at some lower minimum survival density. This will be discussed in more detail elsewhere (Wiegert, ms in prep.). For a graphical example of the Allee effect, see May and Oster (1976).

The asymptotic density (γ_{jj}) is deferred solely on the basis of competitive interference in the struggle for space; the supply of renewable material resources is assumed to be optimal and thus exploitative competition does not occur. The simplest function to represent this type of feedback term is:

$$f(x_j) = [(x_j - \alpha_{jj})/(\gamma_{jj} - \alpha_{jj})]_+ \qquad (14)$$

where α_{jj} and γ_{jj} are as defined above.

Note that except for the response threshold density (α_{jj}) the mathematical form of this function is equivalent to the feedback term of the Verhulst-Pearl logistic equation. Thus when this form is used for $f(x_j)$, and if $f(x_i)$ and α_{jj} are zero, equation (13) reduces to the logistic whenever $x_j \leq \gamma_{jj}$ (Wiegert, 1974).

Equation (14) assumes a linear relationship between the changes in x_j and dx_j/dt. Other relationships are often required in order to represent a particular ecological interaction realistically. For example, in a model of an algal population the effect of a unit change in algal density was inversely related to the algal density (Fraleigh and Wiegert, 1975); bacteria on the surface of particles required the feedback control to be a function of both x_i and x_j (Wiegert et al., 1975).

In addition to the model equations, the dynamics of a population model are affected greatly by the manner in which spatial and temporal factors are incorporated into its structure. With respect to the persistence of all components of a model, spatial heterogeneity tends to decrease the probability of extinction and time lags tend to increase this probability. Before developing the general multi-species model with which experiments are to be performed, some attributes of spatio-temporal heterogeneity need to be discussed, at least insofar as they influence the dynamics of single-species population models.

Spatial-Temporal Considerations

Spatial heterogeneity. Any real environment is far from a physically uniform place. Temperature, nutrient concentrations, wind, soil types, and so on, all create a mosaic of physical habitats in even the most monotonous land- or seascape. Whether the patches are discrete or connected by a continuous gradient of the variable(s) in question is of relatively little consequence, although assuming a continuum may convey certain advantages in analysis (Levin, 1976). This patchy nature of the environment fosters a similar spatial heterogeneity of the distribution of organisms. But a population model usually deals in averages—of the population, its resources, or its enemies. Furthermore, one of these resources is space itself; and if it is not uniformly occupied, that fact should be conveyed to the feedback control terms governing competition for space.

One aspect of spatial heterogeneity is automatically conveyed by the refuge term in equation (13). Organisms protected by the "refuge" are in a different sort of space, by definition, than are the vulnerable prey. But suppose the situation is not one of physical restrictions on where the predator càn go or which prey it can consume. Instead, assume that prey are all vulnerable when found but are clustered in groups throughout the environment. Such a situation can quickly lead to extinction if the predator is easily able to find all groups. But the classic experiments of Huffaker (1958, 1963) with prey and predator mites showed that provision of sufficient physical islands for prey could lead to persistent coexistence by reducing the chances of discovery of prey by the predator. More recently, Luckenbill (1973, 1974) using *Paramecium* and *Didinium,* showed that the same effect (at least for a while) could be obtained by simply slowing down the movements of the predator with methyl cellulose, in this way reducing the chances of prey discovery. In fact, if both predator and prey movements could be slowed enough, relative to the prey density, a patchy but persistent system should be obtained. The predators would exterminate prey from a given area, reproduce, then die out. Their descendants would move on, colonize and exterminate a new area, die out, and so on. Meanwhile, the prey would reoccupy the bare environment and increase in numbers until again discovered by the predator.

Incorporating spatial heterogeneity of this type into population models can be solved by constructing and running separate models. But the problem has only recently been recognized as a central one in population and ecosystem modeling. Smith (1972), in an important early paper on this topic, discusses many of the ways in which spatial heterogeneity may "stabilize" communities, i.e., render them persistent. He suggests the use of frequency distributions as a way of incorporating this effect of heterogeneity into models. Gridding or zoning the environment is another way of handling heterogeneity. This latter technique usually requires a separate model, although if the heterogeneity applies to only one state variable, subdividing that variable into several states with transfer rates between them will often suffice. Wiegert (1975b) employed this technique in modeling the effects of temperature heterogeneity in a mat of thermal algae. Allen (1975) suggests handling the patchiness by setting up transition matrices. An excellent review of this topic is that of Levin (1976).

Temporal heterogeneity. Specifically with reference to populations, temporal heterogeneity requires the dynamics at time t to be a function of the conditions at time $t-\varepsilon$, $(\varepsilon > 0)$ as well as density at time t.[4] Some kinds of time delay are inherent in the makeup of all populations. Thus, the fecundity, growth rate, and mortality of all organisms are to a greater or lesser degree dependent on the past nutritional and physical history of the organism. Furthermore, any factor that affects fecundity cannot exert an influence on population density before the onset of the next breeding period in any individual. This raises the immediate question, How important is separate consideration of age classes in models of population? Clearly, the dynamics of large numbers of individuals can be projected by models considering population density as a single variable. However, as age class distribution changes, so will average fecundity per individual as well as average survival. Changes in population genotype may also occur, further confounding the issue. Finally, a given age class may itself exhibit varied fecundity, depending on its prior reproductive history (Hubbell, unpublished preprint). Obviously, the particular population and the objectives of the population model will have to be taken into account in deciding how to represent the population density variable in the model.

One type of model incorporates the population age structure explicitly in the variables. Developed by Leslie (1945, 1948), these models employ a matrix of demographic parameters (age-specific birth and survival rates) that is premultiplied by a vector of population incorporating the numbers in each age class. The Leslie matrix models have the advantage of separating age classes in the population, but they are basically convenient algorithms for calculating population growth, not a single model in the sense discussed here. Indeed, Leslie (1948) provided a number of examples in which different interaction equations could be used to represent the effects of density-limited growth, predator mortality, and so forth. Williamson (1959) and Usher (1972) used the Leslie matrix approach as the basis of a general model of population interaction and extended Leslie's results by showing that the population age class could be expanded to a population matrix in which genetic variables such as sex and polymorphisms could be incorporated to model selection and nonrandom mating.

The discrete nature of many population processes has often led to the adoption of difference equations in models of population. Kuenzel (1975) discusses this point in the context of a general discussion of models of population.

Nonoverlapping generations and chaos. Difference equations are clearly appropriate for modeling populations with nonoverlapping generations. Annual insects and plants provide the largest number of species in this group, but many other species with longer lifetimes, from periodical cicadas to bamboo (Evans, 1976), fall into this category. For all these species, the population of generation $t+1$ is a function of the density of generation t (although complicated time delays may be present, stretching back several generations).

But even the simplest deterministic difference equations may introduce unexpected complexity into simulations of population growth. May (1974) used a difference form of the logistic equation to explore the consequences of increasing τ_{max} when modeling populations with nonoverlapping generations and found the ultimate result to be "chaos." The models produced oscillating and finally aperiodic (chaotic) cycles in density as the maximum specific growth rate (ρ_{max}) used exceeded 2.69,[5] a value less than that possessed by the majority of annual species. That such a purely deterministic model could generate such unpredictable behavior is indeed surprising. Of immediate concern is the question, Can we expect all simple, deterministic difference equation models of populations with nonoverlapping generations to generate such behavior when ρ_{max} per generation is greater than 2.69? Or is this apparent restriction on ρ_{max} just a further example of the biological unreality of the logistic model?

The Verhulst-Pearl logistic model (equation [2]) permits populations to grow at close to the maximum geometric rate when unchecked ($x << \gamma$). Because of the discrete time step involved in difference equations, the density at time t can often exceed the carrying capacity even though $x(t-1) < \gamma$.[6] The negative feedback function built into the logistic model requires that the population density in the following $(t+1)$ generation be reduced by the factor exp $(\rho - \rho x(t)/\gamma)$. Thus as the growth rate ρ_{max} becomes larger, leading to possibly greater and greater overshoot of the equilibrium density in any one generation, the depression in the following generation will approach zero ever more closely. The "aperiodic" or "chaotic" patterns of May (1974) show such proportionally low populations during the intervals between population highs that in the real world the situation would represent successive extinctions and reinvasions. This point is also noted by May and Oster (1976).

Many (perhaps most) populations with nonoverlapping generations and relatively high growth rates per generation do not exhibit this kind of negative feedback. Increase above the equilibrium density in one generation is often followed in the next generation by a tendency to decrease toward that equilibirum density, not toward zero. If the negative feedback operated to bring density back exactly to equilibrium, no oscillations would result. For ρ_{max} values $>>2.69$ per generation, decreases that fell short of this ideal could result in (a) an asymptotic return to equilibrium over the next several generations, (b) a return via damped oscillations, or (c) extinction, the result predicted using the feedback term in the logistic equation. Which of these three behaviors a population exhibits is not a function of the maximum specific growth rate, but is instead a property of the kind of mortality and/or emigration that causes the population to decline after exceeding the equilibrium density. This fact was recognized by Hassell et al. (1976). They charted the behavior of simulations (stable equilibrium cycles or chaos) as a function of two parameters, ρ_{max} and the slope, of the density-dependent relationship in the more realistic population model developed by Hassell (1975). When the slope is small $(\rightarrow 1)$, indicating "contest' rather than "scramble" competition, the model provides a stable equilibrium point for any ρ_{max}.

A general model of population growth should have provision for representing a variety of negative feedback functions; it should not be constrained to a single one, as is the logistic model. Furthermore, as population density approaches the equilibrium density, the negative feedback terms that control ingestion, and thus growth, should not be used to control mortality after overshoot has occurred, as does the model of Hassell (1975). Ingestion and mortality are two different processes that deserve to be separately represented in any model.

In the model of equation (12) mortality is governed by the parameter λ_j. independent of ρ. It is instructive to see how the model fares when expressed in different form and used to treat the same range of population behavior analyzed by May (1974) and Hassell, Lawton and May (1976).

Neglecting for the moment the feedback term involving material resources (the logistic model does not distinguish between the two kinds of resources), equation (12) becomes

$$dx_j/dt = x_j\{[\tau_{ij} - (\tau_{ij} - \lambda_j)f(x_j)]_+ - \lambda_j\} \tag{15}$$

Assuming (as in the logistic model) no density response threshold, equation (13) becomes

$$f(x_j) = x_j/\gamma_{jj} \tag{16}$$

Then, using the identity $\tau_{ij} = \rho_{ij} + \lambda_j$, equations (15) and (16) are combined to give

$$dx_j/dt = x_j\{[\rho_{ij} - \frac{\rho_{ij}x_j}{\gamma_{jj}} + \lambda_j] - \lambda_j\} \tag{17}$$

which in difference form is

$$x_j(t+1) = x_{j(t)}\exp\{[\rho_{ij} - \frac{\rho_{ij}x_j(t)}{\gamma_{jj}} + \lambda_j]_+ - \lambda_j\}. \tag{18}$$

Note, when

$$\lambda_j = \rho_{ij}(x_{j(t)}/\gamma_{jj} - 1), \tag{19}$$

Equation (18) reduces to the logistic whenever $x_{j(t)} \leq \gamma_{jj}$. But as $x_j(t)$ exceeds γ_{jj}, the mortality term determining the population decrease in the next generation cannot exceed λ_j. The model now permits examination of the effects of differing mortality functions (λ_j) interacting with various values of the maximum specific growth rate (ρ_{ij}).

For those few species with high ρ_{max} per generation that commonly invade newly opened areas, overexploit the resources for a few generations, and then become extinct, a loss term as in equation (19) may be suitable. The long-term population dynamics would then exhibit aperiodic fluctuations, the period depending on a return of the system to a receptive condition for reinvasion.

However, if a species tends to decrease toward the equilibrium or carrying capacity following overshoot, a more realistic loss function might be

$$\lambda_j = (\ln x_{j(t)}/\gamma)_\delta \tag{20}$$

When $\delta = 1$, the density in the generation following an overshoot of γ will be exactly γ (fig. 2a). Values of δ different from 1.0 will produce a variety of behavior for any given ρ_{max},[7] ranging from an asymptotic approach to γ from above (fig. 2b), damped oscillations (fig. 2c), two-generation cycles (fig. 2d), and multi-generation cycles (fig. 2e), leading finally to aperiodic (chaotic) cycles at some point between $5 < \beta \leq 15$ (fig. 2f).

The reason for the aperiodicity in the model (and the reason why this represents extinction in the real world) is apparent upon considering the magnitude of the decline after the second peak in figure 2f. From a peak of slightly more than ten times the equilibrium density γ_{jj}, the population decreases in the next generation to a low of 7×10^{-15} times γ_{jj}, a total decline of more than fifteen orders of magnitude. It takes six generations for the population to recover to the point of appearing on the graph and seven generations to reach a new peak. Probably no population known could withstand such a proportional decline without becoming locally extinct.

Hassell et al. (1976) fitted existing data from the literature to the model of Hassell (1975) and found only one data set falling into the region defining chaotic solutions. Nevertheless, the possibility of such behavior needs to be kept in mind as more and more detailed and realistic models of population are constructed.

Thus the aperiodic (chaotic) behavior of the deterministic model is not a necessary result of high ρ_{max} per generation as implied by May's (1974) analysis but is instead caused by the interaction of ρ_{max}, which determines the level of overshoot of γ_{jj} in any one generation, with the mortality function (λ_j) that determines the degree of depression of the population in the generation following an overshoot.

Continuous versus discrete time delays. The time delay associated with the case of nonoverlapping generations and represented so well by difference equations (equation [18]) is, however, somewhat of a side issue because of its specialized nature. Caswell (1972) has considered the general effect of time delays in both predator and competition models. Two kinds of delays are possible, continuous and discrete. A population whose behavior is conditioned by the average population size extending to some finite time in the past is said to show the effects of a continuous time lag. Volterra himself was among the first to study the effects of such time delays (Caswell, 1972). The credit for exploring the first population models with discrete time delays is generally given to Hutchinson (1948), who incorporated a time lag of the form $t - \varepsilon$ into the logistic model of growth of a single species. May (1973b) has discussed the mathematical consequences of introducing time lags into population models. He points out the curious and interesting circumstance that introducing continuous time lags, which would seem to complicate matters, in fact simplifies the mathematical analysis! Both Caswell (1972) and May (1973b) regard the continuous time lag as ecologically more realistic, although the only evidence cited is the paper by Caperon (1969). Whether this

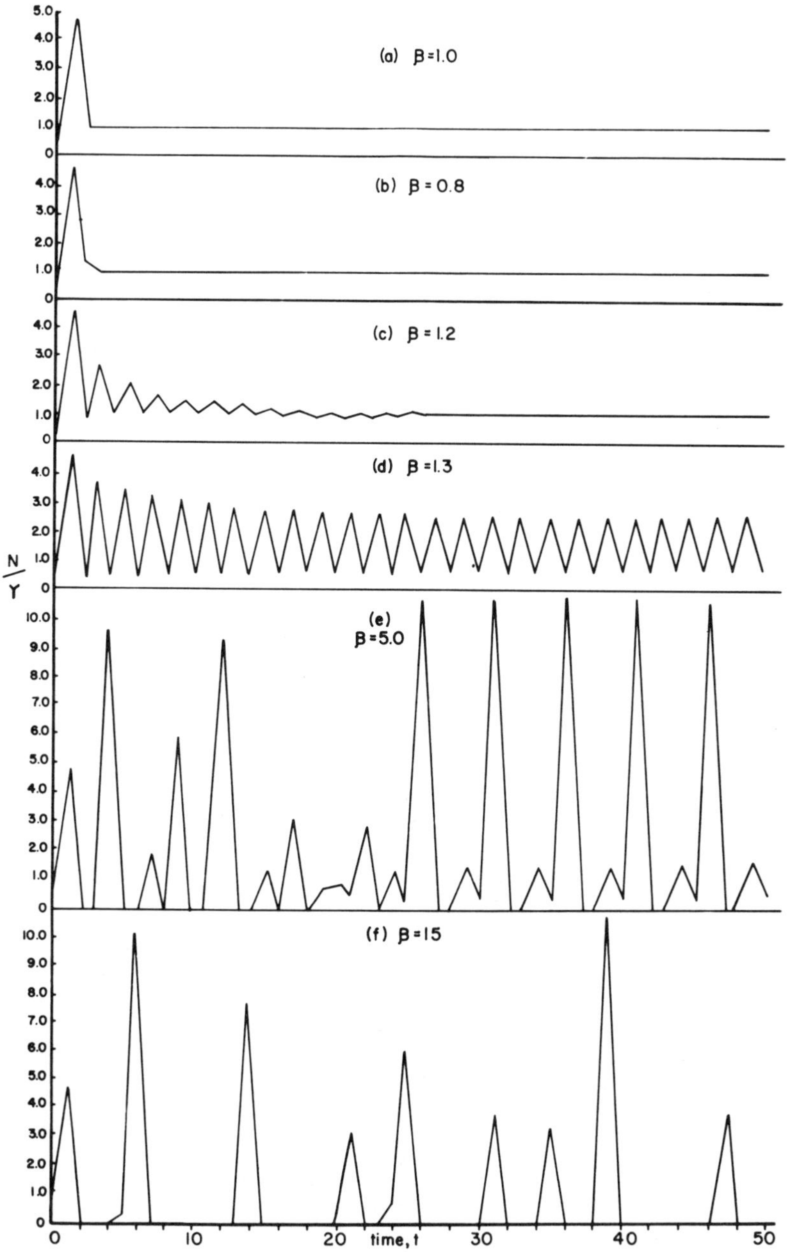

Fig. 2. The range of behavior exhibited by the difference form of a model simulating the dynamics of nonoverlapping generation population growth. In this example γ_{max} per generation equals 5.0 and the type of mortality generated by densities exceeding the carrying capability is governed by the parameter β (see text).

view is accepted is a question of what biological process is being modeled. A concentration of food may trigger reproduction at a particular moment sometime later irrespective of the prey conditions in the intervening time. If the period of increased prey concentration necessary to elicit the response is short compared to the intervening time before the response occurs, then the discrete time lag may be adequate. In many, perhaps most, cases the continuous time lag will be more realistic. For example, the past nutritional history of an organism over a reasonably long period usually plays a large role in determining its behavior in the present.

The evaluation of both continuous and discrete time delays of this type requires that population models be incorporated into a simple systems context, either as a competition or as a predator-prey model. The next step is thus to develop differential equations (12–14) into models of interacting populations suitable for experimental manipulation by varying the level of enrichment and the magnitude of time delay.

MULTI-SPECIES FLOW MODELS

Model Structure

Competition model. A three-compartment flow model of two competing populations and a resource is illustrated (fig. 3). The resource compartment

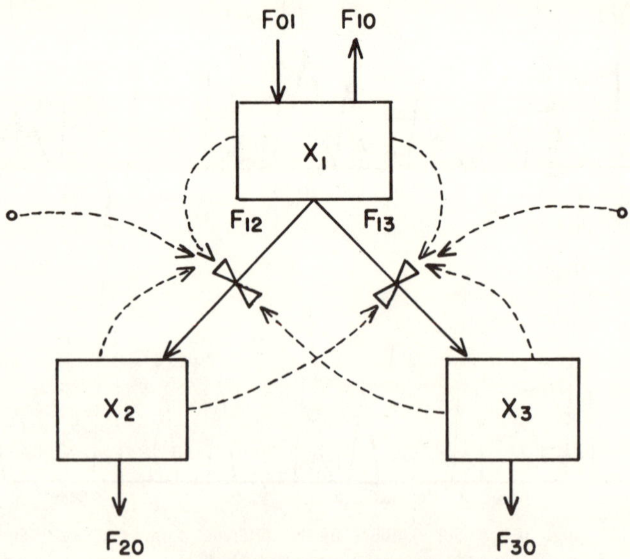

Fig. 3. The form of the three-comaprtment competition model showing the origins of control functions.

(x_1) has a constant inflow F_{01} and loses both to consumers x_2 and x_3 and to the surroundings (F_{10}). The latter loss (as well as F_{20} and F_{30}) is determined by the specific rate λ and takes the simple linear form:

$$F_{j0} = \lambda_j \cdot x_j \tag{21}$$

Figure 3 shows an open energy flow model. The model could be transformed easily to a partially closed elemental cycling model by the addition of flows F_{21} and F_{31}. These changes would add some complexity but would not alter the general conclusions reached through simulation study of this model.

Flows F_{12} and F_{13} are given (from equation [12]) by

$$F_{ij} = x_j[\tau_{ij} - \tau_{ij} f(x_i) - (\tau_{ij} - \lambda_j) f x_j, x_k)]_+ . \tag{22}$$

The feedback control $f(x_i)$ is that of equation (13). The control exerted by competition for space $f(x_j, x_k)$ was derived from equation (14) by the insertion of a competition coefficient to represent the interspecific effect of the presence of the competitors on each other. Thus,

$$f(x_j, x_k) = [(x_j + \beta_k x_k - \alpha_{jj})/(\lambda_{jj} - \alpha_{jj})]_+ \tag{23}$$

where β_k is the proportional effect of a unit of species k on ingestion by members of species j relative to the effect of the same unit on ingestion by members of species k.

Predation model. A simple three-compartment predator-prey model comprises a linear food chain with a resource, prey, and predator (fig. 4). The loss terms F_{10}, F_{20}, and F_{30} were represented as in the competition model. Flows F_{12} and F_{23} were written as equation (22) with the inclusion of feedback terms from both resource and self, according to equations (13) and (14) respectively. No interspecific competition for space was included in the predator-prey model.

Simulation experiments on these models were made with a CDC Cyber 74 computer using a simple Euler integration routine with integration interval of 0.1 unit of time. (Some preliminary studies with the models showed no significant change in stability conclusions or stable equilibrium point for intervals of 0.01 and 0.001.) The two Fortran programs are given in Appendix 1. Note that parameters α, β, λ, and τ are represented in the program by A, B, G, L and T.

Effects of Enrichment

Destabilization. Discussion of schemes to increase the productivity of natural ecosystems by enrichment is becoming more common, but there have been few field experiments directed to this topic and even fewer theoretical

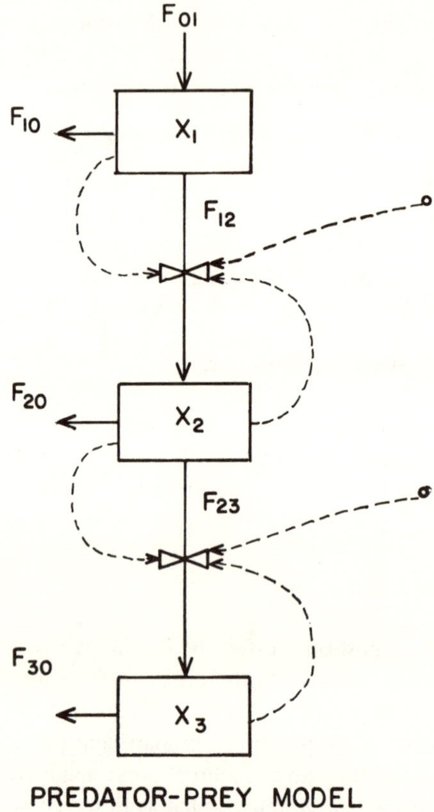

Fig. 4. The form of the three-compartment predation model showing the origins of control functions.

treatments. Rosenzweig (1971) analyzed the behavior of six simple 2-species predator-prey models and concluded that sufficient input to the free resource would destabilize (cause extinction of one or more components) all of them, given the assumption that the time scale was short enough so that evolutionary change would be negligible. Although some controversy arose about the interpretation of a field experiment, Rosenzweig's conclusions remained substantially intact for the particular 2-species predator-prey systems used in his experiments (see McAllister et al., 1972; Rosenzweig, 1972). The difficulty is that his models were all slightly modified versions of the Lotka-Volterra predator-prey models with a self-damping term added to the prey compartment. The question is, How badly is the generality of the conclusions compromised by the ecological unreality of the models?

When proposing a realistic general theory of competition (Wiegert, 1974), I speculated that the outcome of enriching a 2-species (three-compartment) competition model would depend on the parameter values, including the competition coefficients. I now examine the effects of such enrichment in detail, both for the competition model and for the predator-prey model described in the previous section. Of particular interest are the steady state responses of these models to successive, small, permanent increments of enrichment.

Enrichment of the competition model. Each of the two consumers in the competition model (fig. 3) has its ingestion controlled by two feedback terms (equation [22]). For a steady state solution to be possible, at least one of these must be positive for each species. Thus there are nine possible sets of steady state equations, depending on the value, positive or zero, of the feedback control terms governed respectively by the state of material resources and by available space. The nine possible combinations are listed in table 1. Case 1 designates a steady state in which the density of each competitor is controlled solely by scarcity of material resources. This condition requires a low level of enrichment (F_{01} small). Case 9, where control is exerted solely by intra- and interspecific competition for space, ocurs when enrichment has proceeded to the point of saturating the ability of *both* competitors to process material (F_{01} large). The free resource represented by F_{01} may of course be regarded as either a nutrient or a prey population. Wiegert (1974) pointed out that a case 1 steady state is not only improbable but, even if the parameter values were such as to make it possible, the steady state would have an infinite number of solutions with respect to the relative densities of species 1 and species 2.

TABLE 1

NINE POSSIBLE STEADY STATE CASES
FOR A THREE-COMPONENT COMPETITION MODEL

	Mode of Limitation			
	Consumer 1 (x_2)		Consumer 2 (x_2)	
Case	Material $f(x_1)$	Space $f(x_2, x_3)$	Material $f(x_1)$	Space $f(x_3, x_2)$
1	+	0	+	0
2	+	0	+	+
3	+	+	+	0
4	+	+	+	+
5	0	+	+	+
6	+	+	0	+
7	0	+	+	0
8	+	0	0	+
9	0	+	0	+

NOTE. A + entry indicates limitation of the ingestion of the species at steady state by scarcity of the designated resource. The material resource is designated x_1. See text and Appendix A for discussion of the feedback control functions.

Random fluctuations in the environment would cause the species with the greater potential growth rate near equilibrium to ultimately cause the extinction of its competitor.

Coexistence of both competitors depends on particular values of the competition coefficients (β_j) and the maximum densities (γ_{jj}). If these values are chosen to simulate the condition for an unstable equilibrium, i.e., one in which the competitive outcome depends on relative starting densities (Slobodkin, 1961), then level of enrichment also plays a role in determining the outcome of competition.

Simulating to steady state with a model using high values of β (Appendix, table 8) produced the results of table 2. At very low levels of enrichment ($F_{01} = 1$) neither species could survive. When F_{01} was increased to 1.5 units per interval of time, species 2 survived, control being imposed by resource scarcity. The density of x_2 is less than the threshold value (α_{22}) that determines the onset of competition for space, and the value of x_1 is less than the threshold value (α_{12}) that determines when x_1 is considered "scarce."

As F_{01} is increased to 2, an abrupt shift in competitive ability occurs, and species 3 survives, being limited by both scarcity of material resources and competition for space. As enrichment continues, the steady state density of species 3 continues to rise until the point when resource scarcity is not a factor. At this point the competitive advantage again shifts to species 2, which has a greater tolerance for crowding (see G_{22} Appendix, table 8). In summary, species 2 is tolerant of free resource levels lower than those at which ingestion by species 3 is reduced below maintenance levels. At intermediate levels of enrichment the combined tolerances of species 3 to crowding and nutrient

TABLE 2

THE EFFECT OF VARYING DEGREES OF ENRICHMENT (INCREASE IN FLOW TO RESOURCE COMPARTMENT) ON THE COMPETITIVE ADVANTAGE OF SPECIES 2 AND 3

	Competition Model						
	Steady state stc.			x_2		x_3	
F_{01}	x_1	x_2	x_3	Res.	Space	Res.	Space
1.	10	0	0				
1.5	12	7	0	+	0		
2.	15	0	11			+	+
2.5	19	0	13			+	+
3.	23	0	15			+	+
3.5	27	0	17			+	+
4.	31	0	19			+	+
4.5	36	23	0	+	+		
5.	40	25	0	0	+		
5.5	45	25	0	0	+		

NOTE. Values represent the steady state resulting from simulation with starting densities of $x_2 = 2$ and $x_3 = 5$.

TABLE 3

THE EFFECT OF CHANGING THE MAXIMUM CROWDING TOLERANCE OF SPECIES 3 ON COMPETITIVE ABILITY UNDER VARYING LEVELS OF ENRICHMENT OF THE RESOURCE (x_1)

Competition Model G_{33} from 21 to 25							
	Steady state stc.			x_2		x_3	
F_{01}	x_1	x_2	x_3	Res.	Space	Res.	Space
1	10	0	0				
1.5	12	8	0	+	0		
2	14	0	11			+	+
2.5	18	0	14			+	+
3	22	0	16			+	+
3.5	26	0	19			+	+
4	29	0	21			+	+
4.5	33	0	24			+	+
5	38	0	25			0	+
5.5	43	0	25			0	+

NOTE. All other parameter values as in table 2.

levels gives it the competitive advantage, but its upper limit of density permitted by space alone is too low to enable it to survive the competition at high enrichment levels.

The dependence of this enrichment effect on the parameter values and starting densities is demonstrated easily. The same simulation runs with the same starting densities but with γ_{33} changed from 21 to 25 (to equal γ_{22}) causes species 3 to always win at any enrichment level (table 3) other than when the resource is very scarce ($F_{01} = 1.5$). Now the species have clear

TABLE 4

THE EFFECT OF CHANGING THE RELATIVE STARTING DENSITIES OF SPECIES 2 AND 3 ON COMPETITIVE ABILITY UNDER VARYING LEVELS OF ENRICHMENT OF THE RESOURCE (x_1)

Competition Model x_2/x_3 from .4 to 2.4							
	Steady state stc.			x_2		x_3	
F_{01}	x_1	x_2	x_3	Res.	Space	Res.	Space
1	10	0	0				
1.5	12	8	0	+	0		
2	16	11	0	+	+		
2.5	20	13	0	+	+		
3	24	16	0	+	+		
3.5	28	18	0	+	+		
4	32	20	0	+	+		
4.5	36	23	0	+	+		
5	40	25	0	0	+		
5.5	45	25	0	0	+		

NOTE. All other parameter values as in table 2.

differences. Species 2 competes better for scarce resource, and species 3 does better under crowded conditions where space is at a premium because, although its maximum growth rate is lower ($\tau_{12}=0.27$, $\tau_{13}=0.2$), it has the advantage in initial density. If instead of raising γ_{33} from 21 to 25, the initial ratio of x_2/x_3 is changed from 0.4 to 2.4, species 2 will always win as resource input is increased from the lowest level (1.5) at which survival of either species is possible (table 4).

The parameter values used in this model of an unstable equilibrium situation were arbitrarily chosen, although all are biologically reasonable rates. Thus examples discussed here are in no direct way representative of any real situation. But this phenomenon could explain the sudden replacement of a dominant species (of phytoplankton in a lake, for example) by a species previously very rare or absent as a consequence of an increase in enrichment. In the latter case a means of initial colonization must be assumed. But the important point here is that such a phenomenon can occur, theoretically at least, without *any change in the genetically determined coefficients of growth or competitive ability*.

If the maximum density threshold (γ_{jj}) and competition coefficients are changed so that a stable coexistence is possible (see Slobodkin, 1961), then enrichment of the system has a markedly different effect. The parameter values used in the runs from which figure 5 was compiled are given in Appendix table 8. Species 2 and 3 were purposefully made very different with respect to their tolerances for crowding and scarcity of resources, and in the struggle for space each has a much greater competitive effect on itself than on its competitor (β values <1.0). When F_{01} is very low, species 2 survives but not species 3. As the resource input is increased past the point where both species survive, each increases its stable equilibrium densities until species 2 has reached its maximum density and is entirely free from any effects of resource scarcity. At this point consumer 3 can still increase, and it does so more rapidly with continued enrichment, reducing the steady state density of consumer 2 through direct competitive interference. When both consumer species reach maximum mutual crowding densities, i.e., no effect of resource scarcity on either, increased enrichment shows up solely as an increase in the standing crop of the free resource. This latter phenomenon I have termed a "food chain block," but it is more appropriately discussed in the next section dealing with the predator-prey food chain model.

The enrichment of a competition-based system can thus have effects quite different from those expected from such an outwardly simple perturbation. Depending on the values of the biological parameters involved, we might expect a switch in the winning species when the system "normally" exhibits an unstable equilibrium and at least a major change in the relative densities of species that under "normal" conditions coexist.

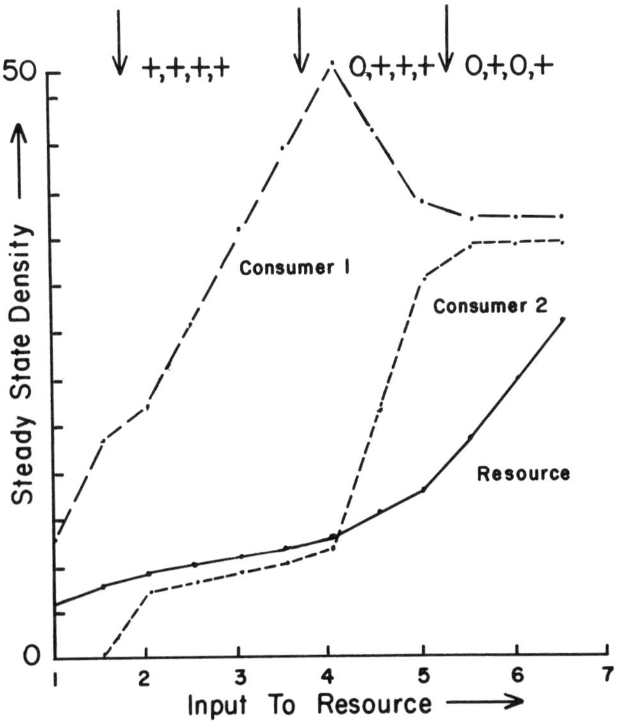

Fig. 5. Effect of progressive enrichment of the resource compartment on the dynamics of two competing populations, culminating in a "food chain block."

Enrichment of the predator-prey model. More possible sets of steady state equations are possible for a three-component model arranged in the predator-prey configuration shown in figure 4. This is because the prey population can, under appropriate conditions, be so severely limited by predatory mortality that it experiences neither limitation by scarcity of its material resources ($f(x_1)=0$) nor limitation due to intraspecific competition for space ($f(x_2)=0$). Thus instead of the number of possible sets of steady state equations being 3^n, where n is the number of food chain links (table 1 and fig. 3), the predator-prey model has a total possible number of sets of steady state equations[8] equal to $4 \times 3^{n-1}$ or 12 for a model with 2 food chain links. In the predator-prey model of figure 4, the parameters governing intraspecific competition for space determine the extent of squabbling between predators (Rosenzweig, 1971). Absence of interspecific competition for space between predator and prey is assumed.

Case 1 of table 5 represents a steady state in which prey are limited solely by scarcity of a material resource and the predator is controlled solely by

TABLE 5

TWELVE POSSIBLE STEADY STATE CASES FOR A THREE-COMPONENT PREDATOR-PREY MODEL

	MODE OF LIMITATION			
	Prey (x_2)		Predator (x_3)	
Case	Resource $f(x_1)$	Space $f(x_2)$	Resource $f(x_2)$	Space $f(x_3)$
1	+	0	+	0
2	0	0	+	0
3	+	+	+	0
4	0	+	+	0
5	+	0	+	+
6	0	0	+	+
7	+	+	+	+
8	0	+	+	+
9	+	0	0	+
10	0	0	0	+
11	+	+	0	+
12	0	+	0	+

NOTE. A + entry indicates limitation of the species at steady state by scarcity of the designated resource. The material resource of the prey is designated by x_1. The material resource of the predator is the prey (x_2). See text and Appendix A for discussion of the feedback control function.

scarcity of prey. As enrichment is added (F_{01} increases) the system must ultimately be driven toward the case 12 situation where control at steady state is exerted entirely through competition for space. The possibility of intermediate cases depends on the particular parameter values used.

If the predator capture (ingestion) rate is made high ($\tau_{2.3}=0.5$, Appendix, table 8), and if the input to the resource compartment is systematically increased, the resulting change in the equilibrium densities seems at first to follow a somewhat more predictable pattern than does the competition model. Figure 6 shows that as the level of enrichment climbs, the system passes from a condition where predators cannot survive (density of prey maintained below the refuge level) to a condition where the predator controls the prey at a constant density and much of the added enrichment shows up as an increase in the predator density. This illustrates the response noted by Rosenzweig (1972) where the predator isocline is a vertical line on the phase-plane graph of predator versus prey, i.e., the predators are not limited by intraspecific competition for space. The increase in the resource concentration relaxes the control on ingestion by prey, and the turnover rate of the latter is greatly increased. For rates of resource input (F_{01}) greater than 4, the predator population is allowed to exceed the threshold response density of intraspecific competition for space ($\alpha_{3.3}$). This limitation allows the prey steady state density to increase, with the net result of an even greater rate of increase in the steady state density of the predator. Finally, with predator growth rate high, the system at higher levels of enrichment enters a phase where predator density controls prey density at low levels but is itself fixed by the density

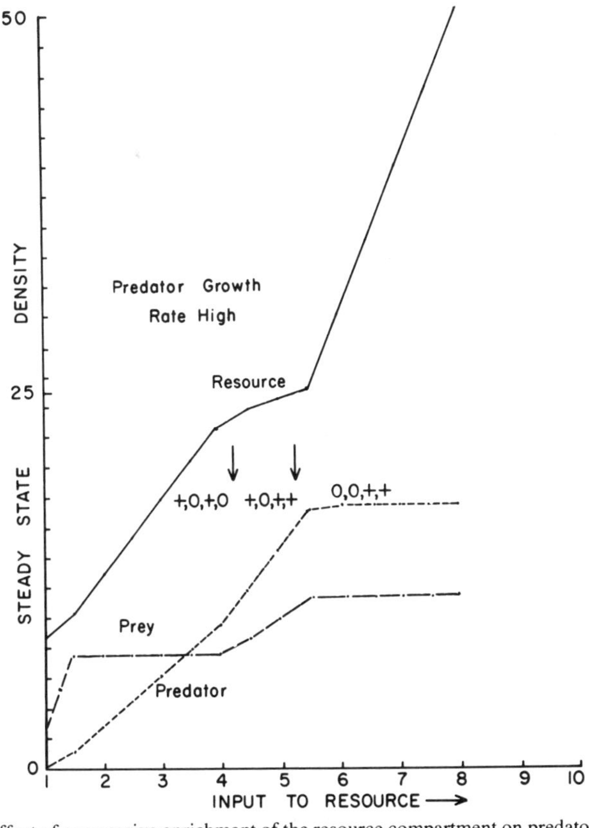

Fig. 6. Effect of progressive enrichment of the resource compartment on predator-prey populations where growth potential of the predator is high.

maximum (γ_{33}). Added enrichment can only show up as an increase in the steady state density of free resource. This is another example of the "food chain block" noted in the competition model. This situation may be expected as a consequence of enrichment whenever the assimilative capacity of a population is saturated, either because it is held to a constant low level by predation or because it has reached the maximum density permitted because of intraspecific interference. In any real ecosystem a situation as shown in figure 6 would not persist. Some species would move to utilize the free resource that accumulates because of the inability of the consumers to keep up with the supply. Possibly a food chain block may be responsible for the increased prominence of previously rare species when ecosystems are enriched by pollution. Both this mechanism and the competitive displacement noted in figure 5 could account for the often unexpected results caused by the enrichment of natural ecosystems.

At first the final result of figure 6 was puzzling, because I expected to see case 12 (both prey and predator free of the effects of scarcity of material resources) emerge upon continued enrichment. The failure of this situation to develop was at first explained as a consequence of the high predator growth rate rendering the prey incapable of realizing their maximum potential steady state density under even high levels of enrichment.

In a second series of experiments, the predator ingestion rate ($\tau_{2.3}$) was changed from 0.51 to 0.35 and a new series of runs made at various levels of enrichment (fig. 7). Much the same pattern of change as in figure 6 emerged as enrichment rates were increased to 7.4. As the level of input passed 7.4, however, the model showed a distinct change in the relative position of the steady state densities and the control situation rapidly passed from a prey population free of resource and competitive control through one controlled by

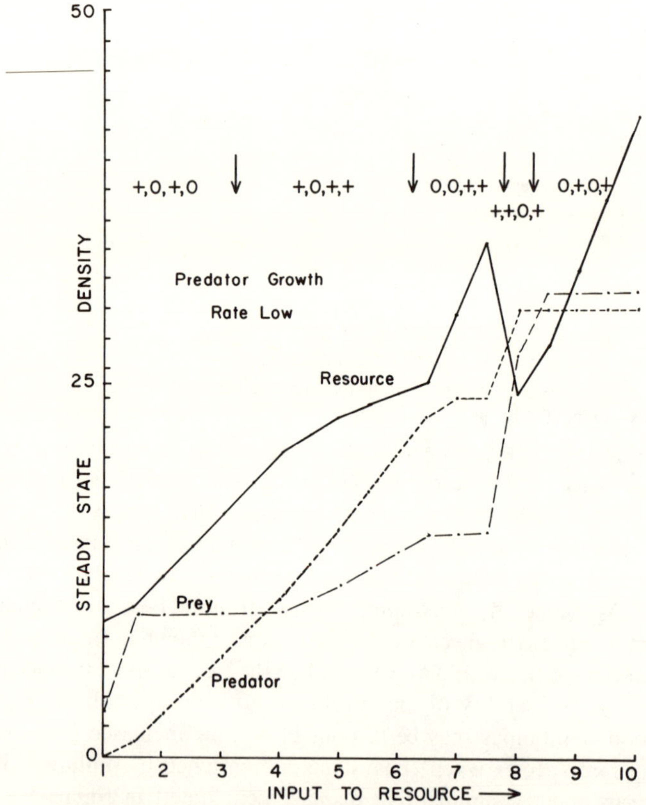

Fig. 7. Effect of progressive enrichment of the resource compartment on the predator-prey populations where growth potential of the predator is low.

resource and intraspecific competition to, finally, the case 12 situation, where both predator and prey reached their density maxima determined solely by space available. There was initially a large decrease in the steady state density of free resource and a consequent increase in the predator and prey. Subsequent increases in degree of enrichment ($F_{01} > 8.5$) produced a steady rise in the accumulated free resource and no change in the steady state densities of predator or prey.

This result is the same no matter what the relative starting densities of predator, prey, and resource, provided only that they are not so disparate as to cause extinction of prey before the oscillations can be damped to the steady state. However, in experimenting again with the high predator growth rate model (table 6), I discovered that at enrichment levels of 9.5 or above the system had *two possible steady states* and that the realized steady state of the system did in fact depend on the initial conditions. Thus, after one of the two possible steady states had been reached, a sufficiently great perturbation of this steady state could cause a shift to the second one. Starting a run with predator greater than prey, for example, produced the steady state of prey 11 and predator 17, the latter controlling the former at low levels relative to available material resources and space. Small perturbations of this steady state will not affect it. For example, following a doubling of the prey population the original steady state is reestablished quickly. However, if a drastic reduction of predators is made ($x_3 = 1$) or if prey (x_2) is increased almost threefold, to 35, then the resultant steady state is one in which both predator and prey have reached their maximum densities. In other words, the prey has escaped from predator control and subsequently become limited by density, where the density maintained is high enough so that the predators, themselves constrained by a density limitation, can never make sufficient inroads on the prey to regain control.

TABLE 6

THE EFFECT OF PERTURBING THE HIGH PREDATOR GROWTH RATE MODEL WITH HIGH ENRICHMENT

	Predation Model Predator Growth Rate High (.51) Enrichment High (9.5) Steady State			
	Res.	Pred.	Pred.	
	64	11	17	0, 0, +, +
Perturb. 1		Reduce Predator Density to 1		
	38	31	30	0, +, 0, +
Perturb. 2		Increase Prey Density to 35		
	38	31	30	

NOTE. Original steady state has both prey and predator held to low density. Either of two perturbations will cause the system to assume the second of the two steady state possibilities with both prey and predator at higher density.

This situation is conceivable in a real system, although to my knowledge it has never been experimentally demonstrated or documented in the field. Indeed, many ecologists may support the view of Patten (1975), who denies the reality of more than one steady state, saying that ecosystems with more than a single forced steady state are structurally unstable. Yet the predator-prey model discussed here, although very much simplified, does utilize realistic interactions and ecologically definable parameters, and the solutions are certainly stable. The two steady states each have a perfectly straightforward explanation. Further analytical work with this model should help predict when this behavior may be expected. Note that the model does not imply multiple steady states of a single set of differential equations, for, with the change in the feedback control terms from a 0,0,+,+ to a 0,+,0,+ situation, we have a new set of steady state equations, i.e., the rules of the game are changed. This capability is the single most important contribution of the modeling approach outlined in the present paper.

Time Delays versus Enrichment

Delay in competitive effects. The deterministic models discussed so far all exhibit either a stable or unstable point equilibrium or they are unstable. No limit cycles are observed. This is because they are tightly constrained by the imposition of terms reflecting intra- or interspecific competition. The introduction of time delays tends to destabilize ecological interactions; thus an examination of the interaction of time delays with level of enrichment in the two simple models provides further insight into the ecological consequences of enriching simple model systems.

In the competition model exhibiting coexistence (β values low), two levels of enrichment were selected: $F_{01}=2.5$ and $F_{01}=6.5$ (Appendix B). These are referred to as low and high respectively. Time delays of a discrete nature were introduced into the model using arrays to store previous values of the standing stock of x_1, x_2, and x_3. Once each interval of time (every 10 iterations) the array values were changed (these are the SX values of the programs reproduced in Appendix A). Thus the fluxes and feedbacks in the model employing time delays were calculated using, for 10 iterations, the value of a variable that appeared 1, 2, 3, etc. time intervals earlier, the actual interval being set by the parameter TLAG.

Although continuous time delays are thought to be ecologically more realistic (see earlier discussion), I here use discrete intervals, first because the degree of unreality introduced is slight and unlikely to change drastically the outcome of a simulation and, second, because the objective in this paper is to pinpoint that critical interval length that led to instability. This would have been more difficult with continuous delays. Continuous time delays will be introduced in future experimentation with the models presented here.

When a time delay was introduced into the competition model with low levels of enrichment, oscillations of increasing amplitude were produced, culminating in the "extinction" of the resource (fig. 8). The graph of the simulations was stopped at this point. In fact, because a constant F_{01} continues to feed the free resource compartment, the system continues cycling indefinitely, the resource standing crop ranging from about 11 to 0, with an over-all period in excess of 100 days. The initial large fluctuation is caused by the initially high standing crop of resource. However, this initial value does not influence the later behavior of the simulation.

If the input to the resource component is increased to 6.5 (high enrichment), the model will withstand a time delay of 6 without the depletion of the resource to zero at any time (fig. 9). The amplitude of the fluctuations is increased as the time delay is made greater, but the greater rate of input to the free resource compartment keeps the standing crop above zero even at the low point of the cycle. Because the free resource never reaches zero, there is no resetting as happens in the low enrichment model. Thus the short-term cycles

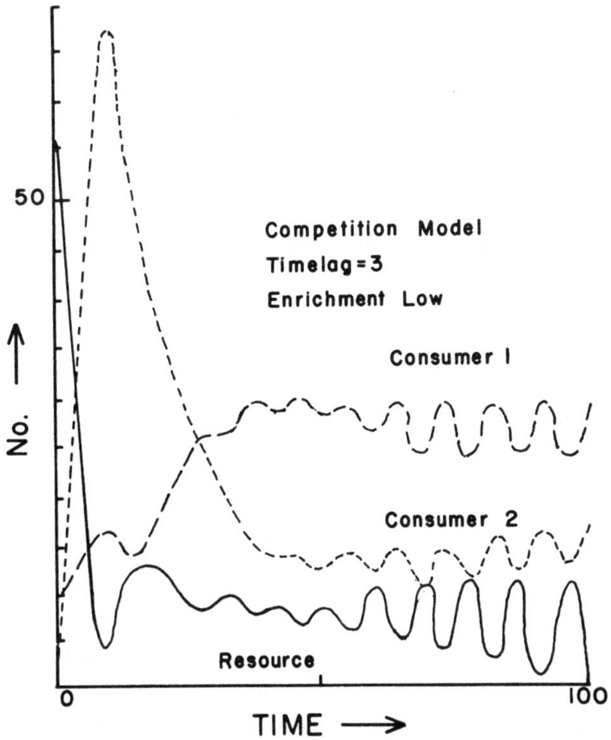

Fig. 8. Effect of a time delay of three units on the behavior of competitors in an environment with a low level of enrichment.

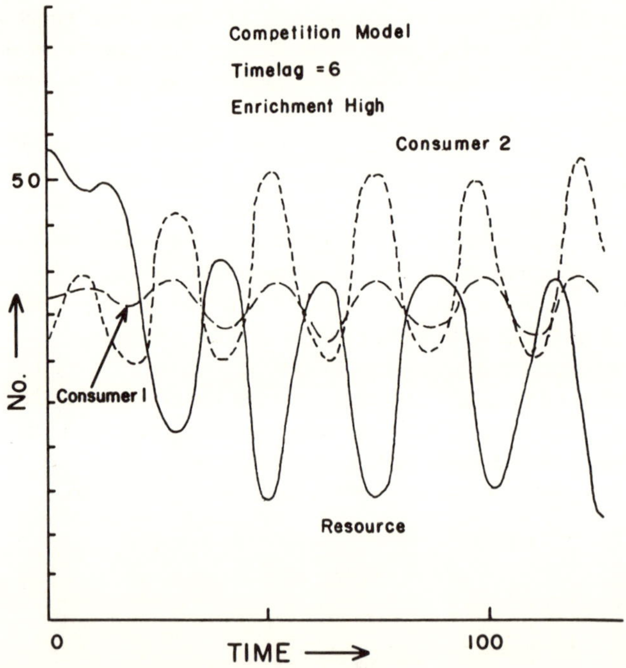

Fig. 9. Effect of a time delay of six units on the behavior of competitors in an environment with a high level of enrichment.

are all that remain in the high-enrichment simulation. They do not progressively increase, and there is no long-term secondary periodicity in the system. The over-all effect of enrichment in the competition model with coexistence possible is to stabilize the system by increasing its toleration of time delays.

Delay in effects of predation. If the predator-prey model with a high predator ingestion rate ($\tau_{2,3} = 0.51$) is subjected to a time delay of length 3, the system remains stable in the sense of all components persisting, but it enters an undamped cycle (fig. 10) that persists unchanged through at least 1,000 time units, when the simulation was stopped. If the time delay is increased to 4, continuously growing oscillations are induced and the system crashes upon extinction of the prey (fig. 11). Note that in figure 11 the larger time delay has displaced the peak of the predator population further to the right of the prey peak than in figure 10, thus causing the growing oscillations.

If the ingestion rate of the predator is lowered to 0.35, all other parameters remaining the same, the oscillations grow at a much slower rate (fig. 12); but the end result is the same: the prey and predator eventually become extinct.

If, however, the enrichment level is increased to 9.5 with low predation and a time lag of 4, the system stablilizes very quickly (fig. 13). This is because

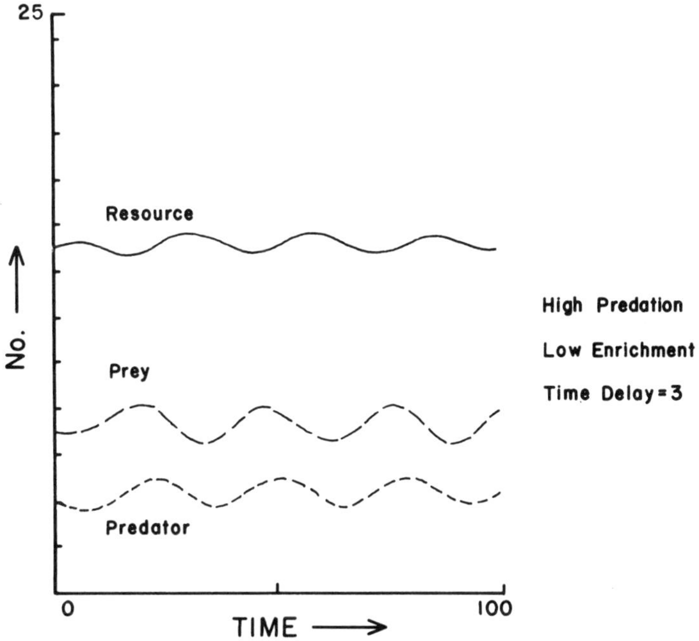

Fig. 10. Effect of a time delay of three units on the behavior of predator and prey in an environment with a low level of enrichment and a high potential for growth by the predator.

the increased flow of resource through the system has raised the density of the predator population to an average level where the constraining action of intraspecific competition among predators for space now plays a role. This reduces greatly the amplitude of the fluctuations and permits damping. In the simulation of figure 12, by way of contrast, the oscillations never become large enough for the density feedback control to become positive. Thus in the predator-prey model, both decreasing the rate of predator ingestion and increasing the degree of enrichment at the free resource level tended to increase the stability of the system. The same result would have been realized, of course, by decreasing the tolerance of the predator for crowding, i.e., by lowering α_{33} and/or γ_{33}.

Rosenzweig (1971) cautioned against schemes for enrichment because he found it led eventually to destabilization of simple model systems of predator and prey. The results here do not contradict these findings but rather add some further criteria that must be considered when evaluating the advisability of any such plans. Rosenzweig did not consider systems with both intra- and interspecific competition for space, density-dependent limitation of the predators in his models or time delays. These factors serve to increase the possible outcomes derived from graphical analysis of any particular case (see Levine,

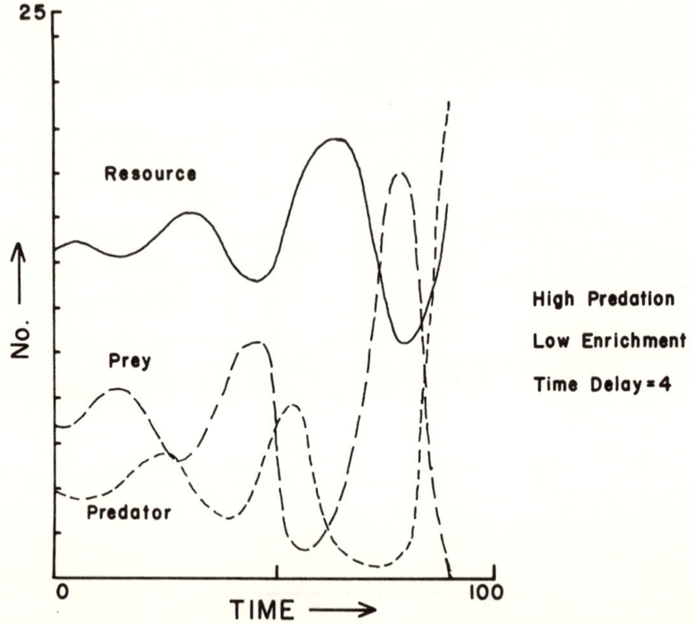

Fig. 11. Effect of a time delay of four units on the behavior of predator and prey in an environment with a low level of enrichment and a high potential for growth by the predator.

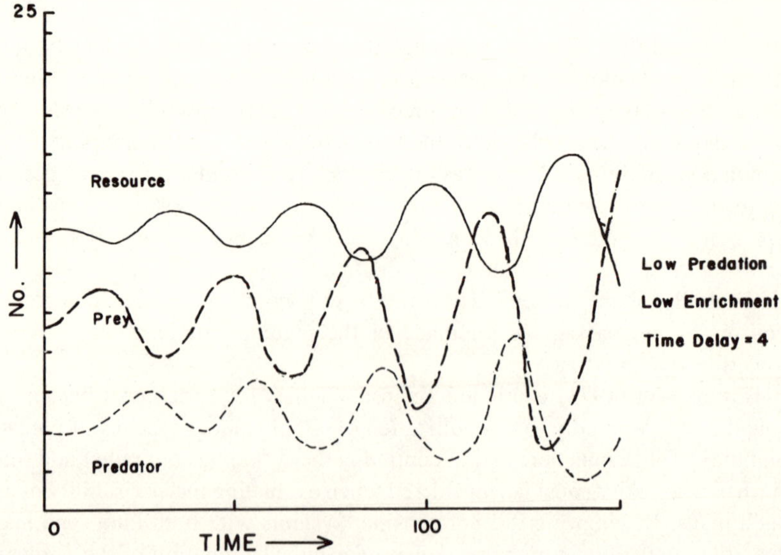

Fig. 12. Effect of a time delay of four units on the behavior of predator and prey in an environment with a low level of enrichment and a low potential for growth by the predator.

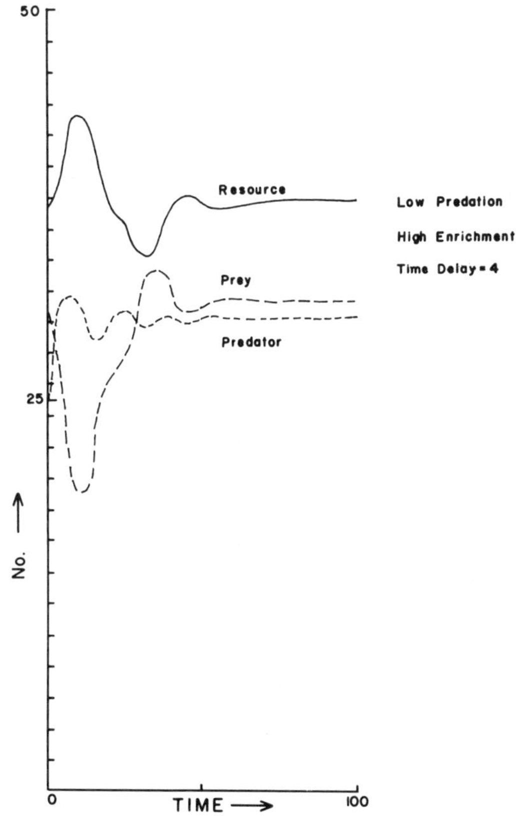

Fig. 13. Effect of a time delay of four units on the behavior of predator and prey in an environment with a high level of enrichment and a low potential for growth by the predator.

1975). Rosenzweig (1972), in reply to his critics, did reemphasize the possible stabilizing effect of density-dependent control of predators. The present results confirm this with quantitative predictions from some realistic models that are also of wide generality in application.

Results from experimentation with models of the kind described in this paper can be employed in constructing hypotheses regarding the behavior of real ecological systems. L. B. Slobodkin once commented to me that the field ecologist deals with systems existing in nature, the laboratory ecologist deals with all possible systems, and the theoretical ecologist deals with all conceivable ecosystems. The difficulties of theoretical studies have been the definition of "conceivable" systems and the problem of obtaining enough insight into their possible modes of behavior to justify an initial set of axioms that could be tested and from which, if supported, subsequent laws could be deduced. Given the computational aids presently available to ecologists, simple models set up to obey realistic ecological constraints provide the starting point for simulation experiments that may offer a useful complement to laboratory experiment with simple microcosms. Defining the behavior of interesting "conceivable" population systems in realistic, yet precise and sufficiently general, terms will permit a more efficient directed search for possible existence of their counterparts in the laboratory and field. Thus the model can become an important experimental tool for the analysis of ecosystems.

ACKNOWLEDGMENTS

Many colleagues and students have provided invaluable criticism and comment during the development of ideas presented in this paper. For particular comments on earlier versions of it, I thank J. Imberger, C. Montague, D. Smith, and R. L. Wetzel. Support for the research and preparation of the paper was received from N.S.F. Grant GB 21255 and DES 7520842.

1. Although other kinds of population may often be recognized, they must be qualified, as: nonbreeding, multi-species, etc.

2. The consequences of using the difference form in the case of nonoverlapping generations will be examined in detail later.

3. Such parameters are often erroneously called instataneous rates. Instantaneous rates refer to the derivative of a function. Thus in the common equation of exponential growth $dx/dt = \rho x$, the term ρx is the instantaneous rate with dimensions org.xtime^{-1} whereas the parameter ρ is the *specific* rate of growth with dimensions time^{-1}.

4. For the human population, or populations under direct control of man, the dynamics at time t could be $f(t+\varepsilon)$, $(\varepsilon > 0)$. Thus the rapid decrease in birthrate in industrialized Western countries in recent years could have been the result of planning (or trying to avoid) something in the future; Managed populations of domesticated animals or plants are usually manipulated in anticipation of future economic conditions.

5. Note that these ρ_{max} values are expressed per generation. Thus any organism that only doubles per generation, e.g., a microorganism, has a value of ρ_{max} of only .69 per generation and would not show the aperiodic cycling even though, compared over identical spans of absolute time, the microorganism has a much higher potential growth rate than does an annual general insect.

6. The same behavior and identical consequences are observed for the differential form of the logistic when time delays are introduced.

7. For example, the ρ_{max} value used in fig. 2 was 5.0, the highest used by May (1974), a value for which the logistic model predicts only aperiodic cycles.

8. The number of possible sets of steady state *equations* does not imply an equal number of steady state *solutions* that are biologically valid. Most of the sets of equations can only be satisfied by negative values of one or more state variables. The number of possible steady state *solutions* is discussed in the text.

LITERATURE CITED

Allen, J. C. 1975. Mathematical models of species interactions in time and space. Amer. Nat. 109(967):319–42.

Caperon, J. 1969. Time lag in population growth response of *Isochrysis galbana* to a variable nitrate environment. Ecology 50:188–92.

Caswell, H. 1972. A simulation study of a time lag population model. J. Theor. Biol. 34:419–39.

Charlesworth, B. 1973. Selection in populations with over-lapping generations. V. Natural selection and life histories. Amer. Nat. 107:303–11.

Dixon, K. R., and G. W. Cornwell. 1970. A mathematical model for predator and prey populations. Res. Popul. Ecol. 12:127–36.

Evans, G. C. 1976. A sack of uncut diamonds: the study of ecosystems and the future resources of mankind. J. of Ecology 64:1–39.

Forrester, J. W. 1961. Symbols for flow diagrams. *In* Industrial Dynamics. Pp. 81–85. MIT Press, Cambridge, Mass.

Fraleigh, P. E., and R. G. Wiegert. 1975. A model explaining successional change in standing crop of thermal blue-green algae. Ecology 56(3):656–64.

Goel, N. S., S. C. Maitra, and E. W. Montroll. 1971. On the Volterra and other nonlinear models of interacting populations. Rev. Mod. Phys. 43:231–76.

Gompertz, B. 1825. On the nature of the function expressive of the law of human mortality. Phil. Trans. 115:513–85.

Hassell, M. P. 1975. Density dependence in single species populations. J. Anim. Ecol. 44:283–95.

Hassell, M. P., J. H. Lawton, and R. M. May. 1976. Patterns of dynamical behaviour in single-species populations. J. Anim. Ecol. 45:471–86.

Hassell, M. P., and R. M. May. 1973. Stability in insect host-parasite models. J. Anim. Ecology 42:693–726.

Hassell, M. P., and D. J. Rogers. 1972. Insect parasite responses in the development of population models. J. Anim. Ecology 41:661–76.

Holling, C. S. 1963. An experimental component analysis of population processes. Mem. Ent. Soc. Can. 32:22–32.

Holling, C. S. 1965. The functional response of predators to prey density and its role in mimicry and population regulation. Mem. Ent. Soc. Can. 45:1–60.

Huffaker, C. B. 1958. Experimental studies on predation: dispersion factors and predator-prey oscillations. Hilgardia 27(14):343–83.

Huffaker, C. B., K. P. Shea, and S. G. Herman. 1963. Experimental studies on predation: complex dispersion and levels of food in an Acarine predator-prey interaction. Hilgardia 34(9):305–30.

Hutchinson, G. E. 1948. Circular causal systems in ecology. Ann. N.Y. Acad. Sci. 50:221–46.

Kuenzel, N. T. 1975. A differential-difference equation model of population dynamics in a natural life system of *Trialeurodes packardi* (Homoptera: Aleyrodidae) on *Impatiens pallida* (Balsaminaceae). Ph.D. thesis, Cornell University, Ithaca, N.Y.

Leslie, P. H. 1945. The use of matrices in certain population mathematics. Biometrika 33:183–212.

Leslie, P. H. 1948. Some further notes on the use of matrices in population mathematics. Biometrika 35:213–45.

Levin, S. A. 1976. Population dynamic models in heterogeneous environments. Ann. Rev. Ecol. Syst., vol. 7. In press.

Levine, S. H. 1975. Discrete time modeling of ecosystems with applications in environmental enrichment. Math. Biosciences 24:307–17.

Levins, R. 1966. The strategy of model building in population biology. Amer. Sci. 54(4):421–31.

Lewontin, R. C. 1976. Fitness: genetic aspects of survival. Presented at Third Ann. Biosciences Colloq., Anal. of Ecol. Syst., Ohio State University.

Lotka, A. J. 1924. Elements of physical biology. William and Wilkins, Baltimore. (Reissued as Elements of mathematical biology, Dover, 1956).

Luckinbill, L. S. 1973. Coexistence in laboratory populations of *Paramecium aurelia* and its predator *Didinium nasutum*. Ecology 54(6):1320–27.

Luckinbill, L. S. 1974. The effects of space and enrichment on a predator-prey system. Ecology 55(5):1142–47.

MacArthur, R. H. 1970. Species packing and competitive equilibrium for many species. Theoret. Pop. Biol. 1:1–11.

MacArthur, R. H. 1972. Geographical ecology. Harper and Row, New York.

MacArthur, R. H., and R. Levins. 1965. Competition, habitat selection, and character displacement in a patchy environment. PNAS 51:1207–10.

McAllister, C. D., R. J. LeBrasseur, and T. R. Parsons. 1972. Stability of enriched aquatic ecosystems. Science 175:562–64.

May. R. M. 1971. Stability in multi-species community models. Math. Biosci. 12:59–79.

May, R. M. 1973a. Stability and complexity in model ecosystems. Princeton University Press, Princeton, N.J.

May, R. M. 1973b. Time-delay versus stability in population models with two and three trophic levels. Ecology 54(2):315–25.

May, R. M. 1974. Biological populations with non-overlapping generations: stable points, stable cycles, and chaos. Science 186:645–47.

May, R. M., and G. F. Oster. 1976. Bifurcations and dynamic complexity in simple ecological models. Am. Nat. 110:573–99.

Maynard-Smith, J. 1974. Models in ecology. Cambridge University Press. Cambridge.

Patten, B. C. 1975. Ecosystem linearization: An evolutionary design problem. Am. Nat. 109:529–39.

Pielou, E. C. 1969. An introduction to mathematical ecology. Wiley-Interscience, New York.

Rogers, D. J., and M. P. Hassell. 1974. General models for insect parasite and predator searching behavior. Interference. J. Anim. Ecology 43:239–53.

Rosenzweig, M. L. 1971. Paradox of enrichment: distabilization of exploitation ecosystems in ecological time. Science 171:385–87.

Rosenzweig, M. L. 1972. Stability of enriched aquatic ecosystems—reply. Science 175:564–65.

Rosenzweig, M. L., and R. H. MacArthur. 1963. Graphical representation and stability condition of predator-prey interactions. Am. Natur. 97:209–23.

Royama, T. 1971. A comparative study of models for predation and parasitism. Res. Pop. Ecol. Suppl. no. 1. 91 pp.

Schoener, T. W. 1973. Population growth regulated by intraspecific competition for energy or time: some simple representations. Theor. Pop. Biol. 4:56–84.

Slobodkin, L. B. 1961. Growth and regulation of animal populations. Holt, Rinehart, and Winston, New York.

Smith, F. E. 1963. Population dynamics in *Daphnia magna* and a new model for population growth. Ecology 44:651–63.

Smith, F. E. 1972. Spatial heterogeneity, stability, and diversity in ecosystems. Trans. Conn. Acad. Arts and Sci. 44:307–35.

Usher, M. B. 1972. Developments in the Leslie Matrix model. *In* Mathematical models in ecology. Pp. 29–60. J. N. R. Jeffers (ed.). Blackwell Science Publ., Oxford.

Volterra, V. 1931. Leçons sur la théorie mathematique de la lutte pour la vie. Lauthier-Villars, Paris.

Watt, K. E. F. 1961. Mathematical models for use in insect pest control. Can. Ent., Suppl 19.

Wiegert, R. G. 1973. A general ecological model and its use in simulating algal-fly energetics in a thermal spring community. *In* Insects: studies in population management. Pp. 85–102. P. W. Geier, L. A. Clark, D. J. Anderson, and H. A. Nix (eds.). Ecol. Soc. Aust. (Memoirs 1), Canberra.

Wiegert, R. G. 1974. Competition: a theory based on realistic, general equations of population growth. Science 185:539–42.

Wiegert, R. G. 1975a. Simulation models of ecosystems. Ann. Rev. Ecol. Systematics 6:311–38.

Wiegert, R. G. 1975b. Simulation modeling of the algal-fly components of a thermal ecosystem: effects of spatial heterogeneity, time delays and model condensation. *In* Systems analysis and simulation in ecology. B. C. Patten (ed.) Vol. 3. Pp. 157–81. Academic Press, New York.

Wiegert, R. G. 1975c. Mathematical representation of ecological interactions. *In* Ecosystem analysis and prediction. S. A. Levin, (ed.). Pp. 43–53. Proc. SIAM-SIMS Conf., Alta, Utah. Philadelphia Soc. Industrial and Appl. Math.

Wiegert, R. G., R. R. Christian, J. L. Gallagher, J. R. Hall, R. D. H. Jones, R. L. Wetzel. 1975. A preliminary ecosystem model of coastal Georgia *Spartina* marsh. *In* Estuarine research, Vol. 1. Pp. 583–601. Academic Press, Inc.

Wiegert, R. G., and R. L. Wetzel. (In press.) Simulation experiments with a 14-compartment model of a *Spartina* salt marsh. *In* Marsh-estuarine simulation. R. Dame, ed. Univ. of S. Carolina Press, Columbia, S.C.

Williamson, M. H. 1959. Some extensions of the use of matrices in population theory. Bull. Math. Biophysics 21:13–17.

ADDENDUM

While this paper was in press, I changed the way in which the two feedback controls (material resources and space) are incorporated into the growth equation. In eqs. 12 and 22 the detrimental effects of scarcity of a material resource and of interference due to crowding are assumed additive, and thus are subtracted from maximum ingestion. Although this sometimes occurs, a stronger argument can be made for the usual result to be multiplicative. Thus, consider a predator population facing shortage of prey such that it averaged twice as long to capture each prey than under optimal conditions ($f(x_i)=.5$). If, simultaneously, crowding of predators had reached the point where half the time normally spent hunting was spent squabbling with other predators, such that $(\tau_{ij}-\lambda_j)f(x_j)=.5$, then net ingestion would be one fourth of the maximum: $F_{ij}=x_j\tau_{ij}[f(x_i)\cdot(\tau_{ij}-\lambda_j)f(x_j)]_+$ not zero as when $F_{ij}=x_j\tau_{ij}[1-f(x_i)-(\tau_{ij}-\lambda_j)f(x_j)]_+$

This change is an improvement in the ecological reality of the model but does not change the fundamental stability or responses to enrichment and time delays. These equations have been used in a model of carbon cycling in the coastal Georgia salt marsh (Wiegert and Wetzel in press). The feedback control terms must now be written so that under optimal conditions they equal one (rather than zero) and vice versa. Thus eqs. 13 and 14 become:

$$f'(x_i)=1-f(x_i)=[(x_i-\alpha_{ij})/(\alpha_{ij}-\gamma_{ij})]_+$$

and

$$f'(x_j)=1-f(x_j)=[(\alpha_{jj}-x_j)/(\gamma_{jj}-\alpha_{kk})]_+$$

or more generally,

$$f(x_i)=\left[1-\left(\frac{\alpha_{ij}-x_i}{\alpha_{ij}-\gamma_{ij}}\right)_+\right]_+$$

and

$$f(x_j)=\left[1-\left(\frac{x_j-\alpha_{jj}}{\gamma_{jj}-\alpha_{jj}}\right)_+\right]_+$$

The latter two formulations ensure that both $f(x_i)$ and $f(x_j)$ remain within $0\leq f(x)\leq 1.0$. This is important when $f(x_i)$ is used to evaluate feeding preference (Wiegert and Wetzel, in press) or in the case of $f(x_j)$, when a separate limitation function and/or a separate mortaility rate are to be used whenever density (x_j) exceeds carrying capacity (γ_{jj}).

APPENDIX A

Fortran IV programs for the competition model (I) and the predator-prey model (II). Each of these programs incorporates the provision for a time delay. If simulation runs with no time delay are to be made, the value IGST is set to zero and TLAG is set equal to 1.

Program I

```
PROGRAM COMPLAG (INPUT,OUTPUT,A,TAPE5=INPUT,TAPE6=OUT-
PUT,TAPE7=A)
DIMENSION SX1(50),SX2(50),SX3(50)
INTEGER TLAG
3 FORMAT (1H ,8F8.2)
2 FORMAT (1H0,*DAY, X1,X2,X3,FF12,FF22,FF13,FF33*)
4 FORMAT (1HO,*TLAG=*,15)
F1=10
T12=.27
T13=.2
RL1=.1
RL2=.04
RL3=.05
G12=6
G22=25
A12=40
A22=8
G13=5
G33=21
A13=35
A33=10
B2=1.2
B3=1.2
DELT=.1
PRINT=5
JCOUNT=IFIX(PRINT/DELT+.5)
TLAG=1
LGST=10
ITIME=100000
1 ICOUNT=0
DAY=0.0
X1=50
X2=5
X3=5
DO 9 J=1,50
SX1(J)=X1
SX2 (J)=X2
9 SX3 (J)=X3
```

```
WRITE (7,2)
WRITE (7,4) TLAG
WRITE (7,3) DAY,X1,X2,X3,FF12,FF22,FF13,FF33
DO 10 I=1,ITIME
FF12=DIM(A12,SX1 (1))/(A12=G12
FF22=DIM(SX2 (1)+B3*SX3 (1) ,A22)/(G22-A22)
FF13=DIM(AL3,SX1 (1))/(A13-G13
FF33=DIM(SX3 (1)+B2*SX2 (1) ,A33)/(G33-A33)
F01=F1
F10=RL1*X1
F12=T12*X2*DIM(1. ,FF12+ (1.-RL2/T12)*FF22)
F13=T13*X13*DIM (1.,FF13+(1.-RL3/T13)*FF33)
F20=RL2*X2
F30=RL3*X3
DX1=F01-F10-F20
DX2=F12-F20
DX=F13-F30
DIFF1=X1
DIFF2=X2
DIFF3=X3
X1=DIM(X1+DX1*DELT,0.)
X2=DIM(X2+DX2*40DELT,0.)
X3=DIM(X3+DX3*DELT,0.)
DAY=DAY+DELT
LAG=LAG+1
IF (LAG.LT.LGST) GO TO 14007
LAG=0
DO 13 J=1,TLAG
SX1(J)=SX1 (J+1)
SX2 (J)=SX2 (J+1)
SX2 (J)=SX2 (J+1)
13 SX3 (J)=SX3 (J+1)
SX1 (TLAG)=X1
```

Program II

```
PROGRAM PREDLAG (INPUT,OUTPUT,A,TAPE5=INPUT,TAPE6=OUTPUT,
TAPE7=A)
DIMENSION SX1 (50), SX2 (50), SX3 (50)
INTEGER TLAG
3 FORMAT (1H ,8F8.2)
2 FORMAT (1H0,*DAY,X1,X2,X3,FF12,FF22,FF23,FF33*)
4 FORMAT (1HO,15)
F1- 8
T12- .27
T23=.35
RL1=.1
RL2=.04
RL3-.15
```

```
G12=6
G22=50
A12-25
A22=20
G23=3
G33=30
A23=18
A33=10
B2=0
B3=0
ICOUNT=0
DELT=.1
PRINT=5
JCOUNT=IFIX (PRINT/DELT+.5)
TLAG=1
LGST=10
ITIME=3000
1 ICOUNT=0
DAY=0.0
X1=50
X2=2
X3=5
DO 9 J=1,50
SX1 (J)=X1
SX2 (J)=X2
9 SX3 (J)=X3
WRITE (7,2)
WRITE (7,3) DAY, X1,X2,X3,FF12,FF22,FF23,FF33
DO 10 I=1,ITIME
FF12=DIM(A12,SX1 (1))/(A12-G12)
FF22=DIM(SX2 (1)+B3*SX3 (1) ,A22)/(G22-A22)
FF23=DIM (A23,SX2(1))/(A23-G23)
FF33=DIM(SX3 (1)+B2*SX2 (1),A33)/(Gee-A33)
F01=F1
F10=RL1*X1
F12=T12*X2*DIM(1.,FF12+(1.-RL2/T12)*FF22)
F23=T23*X3*DIM (1.,FF23+(1.-RL3/T23)*FF33)
F20=RL2*X2
F30=RL3*X3
DX1=F01-F10-F12
DX2=F12-F20-F23
DX3=F23-F30
DIFF1=X1
DIFF2=X2
DIFF3=X3
X1=DIM(X1+DX1*DELT, 0.)
X2=DIM(X2+DX2*DELT,0.)
X3=DIM (X3+DX3*DELT, 0.)
DAY=DAY+DELT
LAG=LAG+1
```

```
IF(LAG.LT.LGST) GO TO 14
LAG=0
DO 13 J=1,TLAG
SX1 (J)=SX1 (J+1)
SX2 (J)=SX2 (J+1)
13 SX3 (J)=SX3 (J+1)
SX1 (TLAG)=X1
```

APPENDIX B

Model Parameter Values of Predator-Prey Model

	Fl (F01)	τ1 (,2)	τ2 (,3)	Rλ1 (1)	Rλ2 (2)	Rλ3 (3)	γ12 (1,2)	γ22 (2,2)	α12 (1,2)	α22 (2,2)	γ13 (1,3)	γ33 (3,3)	α13 (1,3)	α33 (3,3)	β2 (B2)	β3 (B3)
Competitor model: effect of enrichment, B values high	var.	.27	.2	.1	.04	.05	6	25	40	8	5	21	35	10	1.2	1.2
Competitor model effect of enrichment, B values low[1]	var.	.2	.5	.1	.05	.05	2	55	10	2	5	50	15	12	.4	.5
Predator-prey model:			T23 (2.3)								G23 (2,3)		A23 (2,3)			
effect of enrichment, pred. growth rate high[2]	var.	.27	.51	.1	.04	.15	6	50	25	20	3	30	18	10	. . .	. . .

Note. Symbols are those used in Fortran program; equivalent symbols used in equations in text are given in parentheses.

[1]All other competitor models have identical parameter values except: high enrichment Fl = 6.5; low enrichment Fl = 2.5.

[2]All other predator-prey models have identical parameter values except: predator growth rate low T23 = .35; enrichment low Fl = 2.5; enrichment high Fl = 9.5.

DAVID J. HORN AND ROBERT V. DOWELL

Parasitoid Ecology and Biological Control in Ephemeral Crops

9

Of practical interest to all ecologists is the relationship of ecological models to problems of ecosystem management. Recent years have seen the development of a spate of descriptive and analytical models that purport to describe events following introductions of parasitic insects to control agricultural pests. In this paper we review some of these models and their application to the use of parasitic wasps, or parasitoids, in practical biological control of field crop pests.

Biological control has been variously defined, but it is basically the use of natural agents, usually parasitoids, pathogens, or predators, to increase mortality of a pest. Debach (1964) defined *partial* biological control as an increase in host mortality, resulting in lower average pest population densities in subsequent generations. *Complete* biological control is reduction in host population density below the economic injury level (Stern et al., 1959). (For detailed discussion of economic injury levels, see Headley's [1972, 1975] reviews.) Complete biological control implies a permanent reduction in host density with low-amplitude oscillations (Fig. 1).

Biological control in practice has proceeded, and continues to proceed, on a largely empirical basis. As practiced by personnel of the U.S. Department of Agriculture, the University of California Biological Control Division, and the Commonwealth Institute of Biological Control, "classical" biological control involves the following: (1) foreign exploration for natural enemies of exotic introduced pests; (2) quarantine culture of promising natural enemies to determine their precise identity and to learn something of their biology; (3)

David J. Horn and Robert V. Dowell, Department of Entomology, Ohio State University, Columbus, Ohio 43210.

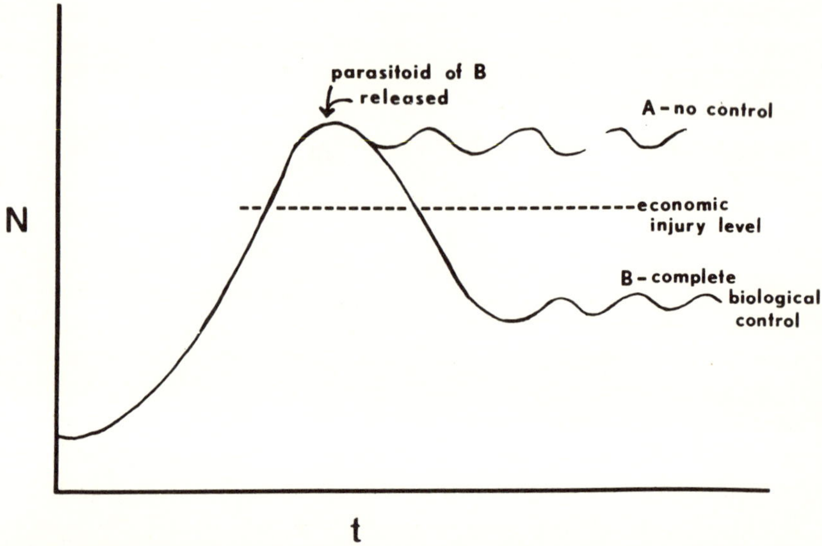

Fig. 1. Relationship of biological control to economic injury level.

release of the most promising (or the most easily obtainable) natural enemies in areas of pest occurrence; and (4) evaluation of control, often merely the empirical conclusion that any observed decline in pest density represents success. This method, here outlined, assumes that the pest is introduced from a foreign country. Pimentel (1963) and others have suggested the importation of exotic parasitoids whose normal hosts are closely related to native insect pests.

Successful instances of complete biological control have been in the minority. Huffaker (1971) and DeBach (1974) documented cases resulting in partial and complete success. Very few of the partial, and *none* of the complete, successes were on ephemeral agricultural crops. The low success rate has been attributed to a variety of ecological and agricultural reasons, nearly all of which have ignored the genetic dimension. Introductions of parasitoids nearly always involve a small number of insects, drawn from a small, often laboratory-reared, genetic pool. Lewontin (1965), Remington (1968), and Levins (1969) have called attention to the possibility that introduction of a very few, genetically homogeneous individuals leads to a high probability of extinction.

Several ecological questions have arisen in the application of biological control. Most important are: (1) Is there a way to predict the "best" agent of biological control?, and (2) Is it better to introduce the single "best" agent or a series or sequence of biological control agents that attack a single pest species? There has been continual debate on question (2). DeBach (Doutt and

DeBach, 1964) and others have championed the introduction of a series of parasitoids, whereas Turnbull and Chant (1961) and Watt (1965) argued that competition so generated may be detrimental. The arguments of each are empirical, with examples selected to support each case. An analytical look at models, followed by current analysis of actual biological control, may shed light on these questions in agricultural systems.

MODELS

A biological model is "a reconstruction of nature for the purpose of study" (Levins, 1968). Models of population ecology have attempted to simplify reality by abstracting appropriate facets that, one hopes, retain the essential reality of the system under study. Inevitably, a single model cannot be simultaneously precise, general, and real, but a useful model should lead ultimately to a testable hypothesis. Many models of host-parasitoid interactions have been proposed, and some of these do offer hypotheses concerning the practice of biological control, particularly the question of single- versus multiple-species introductions. We consider such models in two categories; (a) verbal models, largely empirical, *post facto* general descriptions; and (b) analytical models, mathematically rigorous, quantitative and predictive, based largely on laboratory studies.

EMPIRICAL, VERBAL MODELS

Empirical Biological Control

DeBach (1974) best echoes the sentiments of legions of biological control workers in describing the empirical attributes of the "best" parasitoid from analysis of many successes and failures. Price (1975) recognized basic limitations in parasitoid biology that might limit their potential as biological control agents. The following are important parameters of successful biological control agents:

1. Parasitoid and host populations must be synchronized in space and time.
2. Parasitoids should be host-specific, unless alternate hosts are important in assuring parasitoid survival when the primary host is rare or absent.
3. Parasitoids must be adaptable to a wide range of physical conditions inhabited by the host, though many species are specialized to search for specific host stages in a narrow range of habitats.
4. Dispersal ability exceeding that of the host is desirable. Sweetman (1936) early pointed out that the rather high rate of success in biological control of scale insects was probably due to superior dispersal.

5. Ideally, reproductive potential of parasitoids should exceed that of the host. This depends on the availability of suitable refugia and food for adult parasitoids.
6. Agents of biological control should be efficient in locating and successfully parasitizing hosts. Searching efficiency depends on available time for searching, on interference among searching parasitoids, and on host defenses, especially encapsulation of parasitoid eggs by hemocytes within the host.

According to DeBach, successful biological control is most likely when these conditions are met. The biological characteristics that may prevent a parasitoid from meeting DeBach's criteria are often adaptations that are eminently suited to parasitoids' survival (Price 1975); they simply are at variance with the economic objectives of biological control. This very general description of parasitoid biology does provide a set of useful guidelines applicable to the search for promising agents of biological control.

Adaptive Strategies

To analyze the organization of the parasitoid communities from which agents of biological control are selected, Force (1972) and Price (1973) have suggested consideration of adaptations, especially those centered around reproductive biology, for several parasitoid species using the same host. Force (1972), following Pianka's (1970) extension to MacArthur and Wilson's (1967) genetical theory, categorized parasitoids as "r-strategists" or "K-strategists" (table 1). According to this view, r-strategists are typical of unstable habitats in space and time whereas K-strategists are more likely to be denizens of stable, undisrupted areas. Force (1972) and Price (1973) perceived the evolution of parasitoid complexes to proceed generally from r- to K-strategists, that species of superior competitive ability would, by definition, eventually "outcompete" others were the habitat to remain stable. Force (1970) found that the parasitoid with the highest reproductive and dispersal

TABLE 1

QUALITATIVE COMPARISON OF r- AND K-STRATEGY
(Adapted from Force 1972)

	r-strategist	K-strategist
Fecundity	High	Low
Handling time	Short	Long
Competition	Inferior	Superior
Dispersal	Rapid	Slow
Habitat	Unstable	Stable
Host density	High	Low

abilities was the competitive loser, and dominated in newer host infestations in disturbed habitats rather than older areas. Force (1972) suggested that K-strategists might be most numerous in endemic environments from which exotic pests have come, but that these species might not possess the best colonizing, dispersing, and reproductive abilities. Agricultural environments, especially field crops, are temporary in time and space, so r-strategists might have the best chances of establishment in a biological control program (Force 1974). Ehler and van den Bosch (1974) found that the predators with the highest r were most successful in utilizing resources in California cotton fields. The adaptive strategy model points out, in a way more amenable to analysis, the same characteristics that DeBach (1974) feels are those of a "successful" biocontrol agent.

ANALYTICAL, QUANTITATIVE MODELS

A number of simple yet tractable analytical models have been devised to describe the interactions between hosts and parasitoids in quantitative terms.

Nicholson-Bailey Model

Nicholson and Bailey (1935) were the first to apply to insect parasitoids the generalized predation models of Lotka and Volterra. Essentially, the Nicholson-Bailey model is:

$$N_{(t+1)} = F N_t e^{-aP_t}$$

$$P_{(t+1)} = N_t(1 - e^{-aP_t})$$

where:

N_t = host population density at time t
P_t = adult parasitoid population density at time t
F = host population rate of increase
a = area of discovery, the average area searched by a parasitoid in its lifetime.

This model describes simple coupled oscillations in which N and P oscillate out of phase. Although simple, it assumes several things that are not true for most parasitoids, to wit: Searching is random, and search area (a) remains a constant for each species regardless of host density. Alternate hosts are not considered; parasitoids are not even limited by their capacity to produce eggs. Perhaps the major descriptive failing of the Nicholsonian model is that it does not permit existence of more than a single parasitoid species. The equilibrium is unstable and generates successively larger oscillations resulting in extinction, whether or not an additional parasitoid species is added (Hassell and May, 1973). Most insect hosts have several associated parasitoid species.

Holling Model

Holling (1959) devised a model to take into consideration the major shortcomings of the Nicholson-Bailey model. In Holling's model

$$N_{(t+1)} = FN_t e^{(-ATP_t/1+AhN_t)}$$

$$P_{(t+1)} = N_t(1 - e^{(-ATP_t/1+AHN_t)})$$

where

N, P, and F are as above, and
A = the "attack coefficient," the rate of search times the probability of finding a host
T = the total lifetime of a searching parasitoid
h = "handling time," the length of time spent in handling a host (stalking, ovipositing, marking, etc.)

Holling's model is more realistic than Nicholson's in that parasitization is a function of host density. Moreover, Holling suggested that handling time (h) was an important factor in determining response of a parasitoid to host density. Hassell and May (1973) have demonstrated, however, that this model is inherently unstable, leading to wild fluctuations, and, like Nicholson and Bailey's models, cannot describe more than a single host-parasitoid interaction.

Watt's model

Watt (1959) proposed the following model:

$$N_{(t+1)} = P_t q(1 - e^{(-qN_{t+1}P_t^{(1-m)})})$$

Here,

q = the "quest constant," a measurement of the efficiency of parasitoid attack
m = the "mutual interference constant," which reflects interference among searching female parasitoids.

Watt's model is more realistic than the preceding because it reflects interaction among searching parasitoids, though Diamond (1973) has shown that it leads to successively higher host equilibrium densities with additional parasitoid species; the parasitoid complex does not adjust its density to changes in host numbers.

Hassel-Varley "Quest Theory" Model

Hassel and Varley (1969) developed a model, later extended by Hassell and May (1973), that permits multiple parasitization of a single host in which there is a dynamic balance between aggregation and interference in areas of high host density:

$$AP_t^{(1-m)} = \log \frac{\sigma\, r\, N_t}{\sigma\, r\, N_t - P_{(t+1)}}$$

where

σ = host survival rate and
r = host intrinsic rate of increase

Because searching efficiency changes with host density, aggregation of parasitoids in areas of high host density adds to stability of interactions. Their model is essentially the same as the Nicholson-Bailey except for the addition of interference among searching parasitoids. Unlike the other models, the "quest theory" model does permit the existence of additional parasitoid species on a single host; interference between searching females at high host densities stabilizes the predicted oscillations (Rogers and Hassell, 1974).

Analysis of the "Quest Theory" model led Hassell and May (1973) to predict optimal parameters for a successful biological control agent:

High: a, A, T
Low: h and m ($0 < m < 1$)

Verbally, the "best" parasitoid should have a long lifespan (T), a large area of discovery (a), a high rate of encountering hosts (A), and a large number of offspring. Handling time (h) and mutual interference (m) should be minimal. Levins (1969) in a short analysis suggested the "best" biocontrol agent would also be one active earliest in the season; it would be active in the field when hosts are first available. This sentiment is echoed by Varley et al. (1973). It might be especially important in an ephemeral crop system where time is a limiting factor. Of yet more importance in an ephemeral crop system is synchrony of host and parasitoid. Synchrony of parasitoid activity with peak host numbers permits maximum utilization of this short-term resource (Dowell and Horn, 1976). All the models so far discussed assume that host and parasitoid populations are well synchronized. Many are not.

Diamond (1973) analyzed all these models and noted that all assume random search, lack of superparasitism and hyperparasitism, and that the same instar of the host is parasitized. Host defenses, such as encapsulation, likewise are not considered. Diamond (1973) noted that the "Quest Theory" was the only model in which introduction of additional parasitoid species (into the theoretical equations) would reduce the equilibrium density of host populations. This model would seem to support the empirical conclusions of the multiple-release school (of Doutt and DeBach, 1964; DeBach 1974).

Application of the quantitative models to field populations poses a number of practical problems. Measurement for any model requires accurate estimation of population densities of hosts and searching adult parasitoids, and of percentage parasitization of hosts. Although not incorporated in the models, encapsulation, hyperparasitization, and alternate hosts are often important facets in biological control, and their effects need to be accurately measured and modeled. Attack rates, handling times, interference, and aggregation are measurable in the laboratory. During the past few years, we have been looking at a field system involving an apparently successful biological control program. Our purpose has been to apply the models of adaptive strategy and quest theory to this system, to measure the relative importance of the parameters that DeBach, Force, and Hassell and May consider essential to success.

THE ALFALFA WEEVIL SYSTEM

Alfalfa is a perennial legume grown for high protein animal feed, either as a basic diet or as a feed supplement. Fields normally remain in production three to five years, and alfalfa is cut and baled or dehydrated from two to ten (or more) times yearly, depending upon local growing season. In most of Ohio, alfalfa is generally cut thrice: in late May, late July, and mid-September. Cutting is traumatic to the insect community (Casagrande and Stehr, 1973).

The alfalfa weevil, *Hypera postica* (Gyllenhal), is a serious pest, especially on the first growth of alfalfa in the United States. It was accidentally imported from Europe to the western states in 1904 and to the eastern states in 1952, from whence it has spread into all the conterminous United States and bordering Canadian provinces. Cothran (1966, 1968, 1972) has documented its spread and that of the closely-related Egyptian alfalfa weevil, *H. bruneipennis* (Boheman) in the southwestern United States.

Figure 2 depicts the life cycle of the alfalfa weevil. In Ohio, there is a single annual generation. Oviposition into standing alfalfa stems occurs mainly in

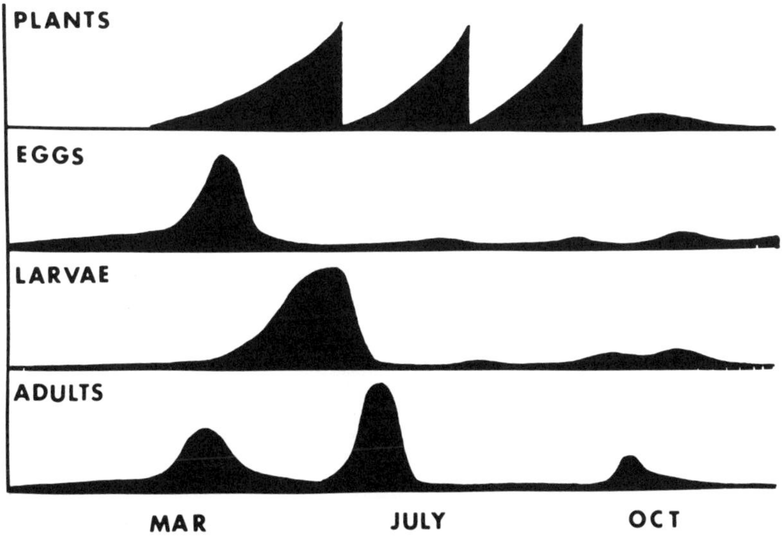

Fig. 2. Alfalfa weevil life cycle, showing seasonal density changes in plant availability, eggs, larvae, and adults, in Ohio.

March and April. (Further south, a large proportion of eggs are laid in autumn and overwinter.) Larvae hatch in one to two weeks and make their way upward to the plant tips, wherein they conceal themselves and feed on the new growth. They pass through four instars; larger larvae move out onto lateral leaves. The simultaneous feeding of numerous larvae on tips and leaves constitutes the economic injury; newly seeded stands may be killed outright, and in older plantings yields are greatly reduced. Mature larvae spin a cocoon either on the plant or in duff on the soil surface. Pupation lasts about a week. New adult weevils emerge, then enter a dispersal diapause during which they fly from the fields to aestivate and, perhaps, hibernate, in as-yet-unfound protected places. A few old adults continue oviposition through early summer in northern latitudes (Barnes, 1967; van Dreische, 1975); they are joined in autumn by some returning new adults, so that September may see a slight resurgence in larval density. this is unimportant economically but, as will be seen, provides a critical resource for at least one parasitoid species. Weevil adults overwinter both in alfalfa fields and other, unknown, places.

The alfalfa weevil was early considered a promising candidate for biological control. Chamberlin (1924, 1925, 1926) scoured much of western Europe for promising parasitoid species. Figure 3a illustrates the parasitoid complex in Europe, consisting of 27 species. Chamberlin was not clear on whether any of these were actually hyperparasitoids, nor is it at all clear in all cases which are obligate, facultative, or fortuitous. We have therefore limited our analysis

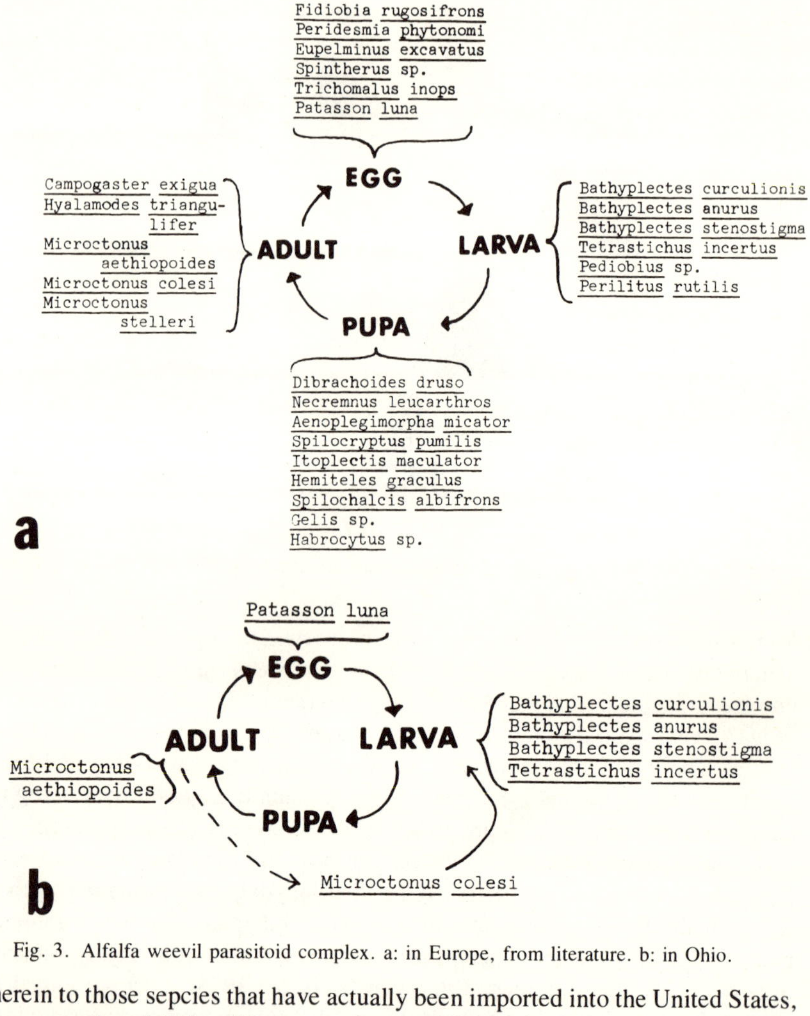

Fig. 3. Alfalfa weevil parasitoid complex. a: in Europe, from literature. b: in Ohio.

herein to those sepcies that have actually been imported into the United States, concentrating (obviously) on species established in Ohio. By ignoring the others, we admittedly may be ignoring the "best" parasitoid, but we are constrained to work on what we have. Figure 3b depicts the parasitoid complex in Ohio, and these are discussed below.

Patasson luna

This is a parasitoid of alfalfa weevil eggs and those of *Hypera punctata*; there are apparently several generations annually, and there may be other alternate hosts. *P. luna* females oviposit directly into weevil eggs in the alfalfa stem; generally 1-3 eggs/cluster are parasitized (Streams and Fuester, 1966). Niemczyk and Flessel (1970) found 0 to 12 percent of weevil eggs in Ohio parasitized during 1965–67. The habit of weevil females' ovipositing in stems apparently limits *P. luna* parasitization. Ellis (1973) found the native *Fidiobia rugosifrons* as a parasitoid of alfalfa weevil eggs in Ontario, but so far this species has not occurred in our study areas.

Bathyplectes curculionis (Thomson)

This species, introduced from Europe in 1914 and again in the early 1960s, is the most widespread and abundant parasitoid of the alfalfa weevil in the United States. It is a multivoltine, solitary endoparasitoid of alfalfa weevil larvae, and concentrates its attacks on first and seond instars. Up to 40 percent of its eggs may be encapsulated by hemocytes of alfalfa weevil larvae (Puttler 1967). It has been recorded as parasitizing over 90 percent of alfalfa weevil larvae in individual field samples (Hamlin et al., 1949; Wilson and Armbust, 1970), though lower values usually obtain when parasitization is measured through an entire season. *B. curculionis* is the only parasitoid that is almost completely sympatric with the alfalfa weevil in the United States. Michelbacher (1940) gave it credit for reducing alfalfa weevil numbers in coastal California, though this was a simple correlation analysis.

Bathyplectes anurus (Thomson)

B. anurus is very similar to *curculionis*, but is slightly larger, with a shorter ovipositor, and is univoltine. It likewise parasitizes early-instar alfalfa weevil larvae; there is no encapsulation of *B. anurus* eggs. Establishment of this species in the United States was delayed by difficulty in rearing — dormancy was difficult to terminate under laboratory conditions (Brunson and Coles, 1968). This parasitoid dispersed much more slowly after initial release, but where it occurs up to 40 percent of weevil larvae have been parasitized and alfalfa weevil population have declined somewhat (W. H. Day, personal comunications). Cocooned larvae of *B. anurus* exhibit unusual "jumping" behavior when touched or heated; this may facilitate movement to equable temperature and humidity after fields are cut, and is of some help in avoiding hyperparasitoids (Day 1969). *B. anurus* predominates in central and southern Europe, where adults are on the wing very early in spring.

Bathyplectes stenostigma

This species is common in northern Europe, and a few importations have been successful in the United States (Dysart, personal communication). *B.*

stenostigma is also a univoltine, monophagous, solitary endoparasitoid of early-instar weevil larvae. It generally emerges late in the season and its activity peaks in June when weevil larval densities are declining, both in Europe and the United States (Dysart et al., 1971). Adults are long-lived. Dysart et al. (1971) summarize its biology.

Tetrastichus incertus (Ratzeburg)

T. incertus is a multivoltine, gregarious endoparasitoid of third- and fourth-instar alfalfa weevil larvae. From 3 to 23 adults may emerge from a single parasitized host "mummy" (Streams and Fuester, 1967). There is apparently no encapsulation; the gregarious habit and rapid eclosion may simply overwhelm host defenses. *T. incertus* emerges late from overwintering, and thus concentrates its attack on summer and early fall larvae (Streams and Fuester, 1967; Horn, 1971), when up to 100 percent of larvae in field samples may be parasitized. Schroder et al. (1969) found that *T. incertus* dispersed very quickly (2000 mi^2/year) from initial releases, though recently it has become rare and has disappeared from much of its former range (Smilowitz and Hower, 1974). This may be due to decreased availability of summer host larvae.

Microctonus aethiopoides

This species parasitizes alfalfa weevil adults and also those of other *Hypera* and *Sitona* spp. (root curculios). It is a solitary endoparasitoid with two generations annually, one in early spring, one in late spring (fig. 4). Loan and

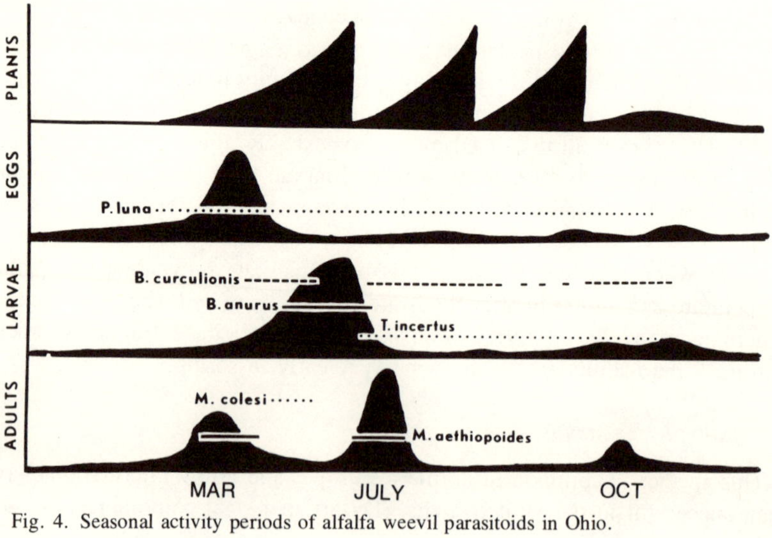

Fig. 4. Seasonal activity periods of alfalfa weevil parasitoids in Ohio.

Holdaway (1961) and van Dreische (1974) noted that *M. aethiopoides* larvae sterilize 80 percent of ovipositing alfalfa weevil adults within 72 hours of eclosion. First-instar *M. aethiopoides* larvae evidently enter diapause within summering alfalfa weevil adults. Juvenile hormone analogs, topically applied to weevils, can bring out *M. aethiopoides* adults. Where *M. aethiopoides* has been established for several years (Md., N.J., N.Y., Pa), alfalfa weevil populations have declined significantly (Dysart and Day, personal communication).

Microctonus colesi

This species was initially discovered in surveys for *M. aethiopoides*. It is thelytokous, *i.e.*, there are no males. *M. colesi* oviposit in third- and fourth-instar weevil larvae, but the larvae enter diapause with their host, to complete development and emerge from adults the following spring (Loan, 1968). There is a single annual generation. *M. stelleri*, a very similar species from Europe, has been released but not yet recovered in the United States (Drea et al., 1972).

Miscellaneous parasitoids

The preceding are the major species in our analysis but occasionally we recover additional parasitoids, most of which are fortuitous.

Gelis spp. are wingless, thelytokous solitary parasitoids of other Hymenoptera and, occasionally, alfalfa weevil pupae (Poinar and Gyrisco, 1963; Horn, 1976). They quite commonly parasitize *B. curculionis* and *B. anurus* (Day, 1969; Caldwell and Wilson, 1975), and may be important factors in their population dynamics. When given a choice, *Gelis* spp. prefer cocoons of *B. curculionis* to those of the alfalfa weevil (Horn, 1976).

Hyalamodes triangulifer (Leow), a tachinid fly, is an occasional parasitoid of alfalfa weevil adults. It has been recorded from many beetle species of several families and probably is a facultative parasitoid of any beetle of the appropriate size. As its alternate host range is so broad, we have not considered it further.

Dibrachoides druso is a bivoltine, gregarious ectoparasitoid of the alfalfa weevil pupa and pupae of related *Hypera* spp. Smith (1930) has described its biology. Large numbers have been released in both the eastern and western United States (Brunson and Coles, 1968), including Ohio, but we know of no establishment other than a presumed establishment in Washington in the 1930s. The failure to colonize *D. druso* requires further study. Smith's study shows its reproductive biology to be remarkably similar to that of *T. incertus*. Perhaps the short temporal availability of alfalfa weevil pupae prevents the second generation of *D. druso* from locating enough hosts to survive.

Field Observations

The alfalfa weevil entered Ohio in 1961 and by 1969 had spread throughout the state and has since been a serious pest. Flessel and Niemczyk (1972) released several parasitoid species in Ohio and followed up their local establishment and spread; these were augmented by spread of *B. curculionis, P. luna, M. colesi* and, recently, *M. aethiopoides*, from areas outside Ohio. They studied dynamics of alfalfa weevil eggs from 1965 through 1967 (Niemczyk and Flessel, 1970), and found *P. luna* to be a variable, low-level mortality factor (0-12%) throughout weevil oviposition.

Beginning in 1973, we (and D. R. Lewis) intensively sampled alfalfa weevil and parasitoid populations at three locations in Ohio. From March through June in each field, at intervals of approximately 44 degree-days (9°C base), we sampled the following: weevil egg density/stem, weevil larval density/9 dm^2, weevil and parasitoid adult density/standard net sweep (Armbrust et al., 1970), and parasitization of eggs, larvae, and adults (in reared sample material). We also obtained information on plant height, condition and density, and climatological data. (Lewis, 1977, records details of our sampling.)

The data were used in construction of a series of life-tables, from which Lewis and Horn (1976) have generated a series of k-values (table 2), after the

TABLE 2

K-Values for Alfalfa Weevil Paarasitoids in Three Ohio Alfalfa Fields

Location	Year	*P. luna*	*B. curculionis*	*B. anurus*	*T. incertus*	*M. aethiop.*	Total
Columbus	1973	.019	.064	0	0	0	.087
Columbus	1975	0	.018	0	0	0	.018
S. Charleston	1974	.027	.059	<.005	0	0	.091
S. Charleston	1975	.006	.048	0	0	<.005	.057
Apple Creek	1973	.054	.118	.005	<.005	0	.182
Apple Creek	1974	.037	.039	.043	0	<.005	.127
Apple Creek	1975	.016	.051	.030	0	.102	.206

method of Varley and Gradwell (1970). The k-values are the logarithm of survivorship preceding action of a mortality factor, minus the logarithm of survivorship following that factor. Total generation k-values averaged around 3.0 (Lewis, 1977). By comparing the various k-values, we conclude the following:

1. *P. luna* was a variable low-level mortality factor.
2. *B. curculionis* effect declined where *B. anurus* was established.
3. *T. incertus* declined; it was negligible after 1973.

4. Neither egg nor larval parasitization is well-correlated with total intergeneration mortality as measured by annual populations of mature larvae (the damaging stage), excepting a weak correlation with *B. anurus* parasitization (Lewis and Horn, 1976).
5. *Microctonus* spp. were present at such low levels as to be presently unimportant, but later we suggest that *M. aethiopoides* may indeed be capable of exerting substantial biological control on the alfalfa weevil.

Figures 5–8 show relationship among weevil larvae and their parasitoids throughout the season. Here we see clearly the partial replacement of *B. curculionis* by *B. anurus* (Figs. 5–7), where these co-occur, and the effective disappearance of *T. incertus*. Also of note is the decline in the second peak of *B. curculionis* parasitization where it and *B. anurus* co-occur.

ANALYSIS

We may recognize 3 guilds of alfalfa weevil parasitoids: the egg guild, comprised only of *P. luna*; the larval guild, being the 2 *Bathyplectes* spp. plus *T. incertus*; and the adult guild, both *Microctonus* spp. It might be argued that *M. colesi* overlaps larval and adult guilds because it oviposits in the larvae.

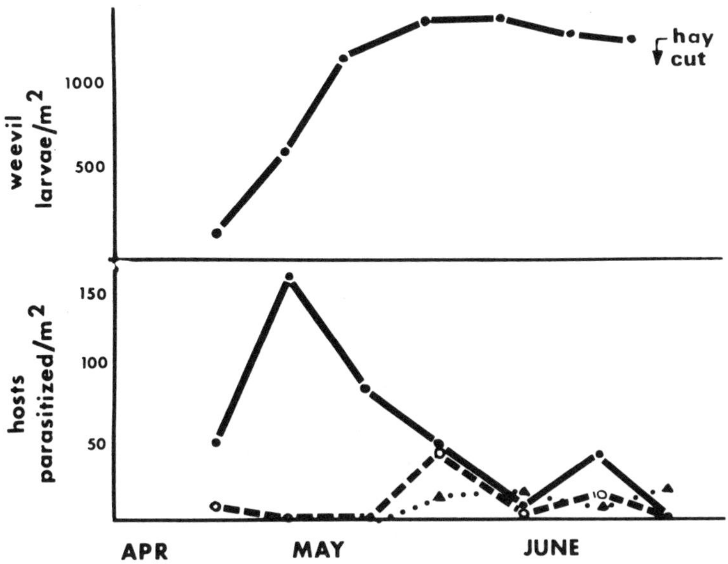

Fig. 5 Alfalfa weevil larval density and parasitization, Wooster, Ohio, 1973. Lower graph, solid line, *B. curculionis*. Dashed line, *B. anurus*. Dotted line, *T. incertus*.

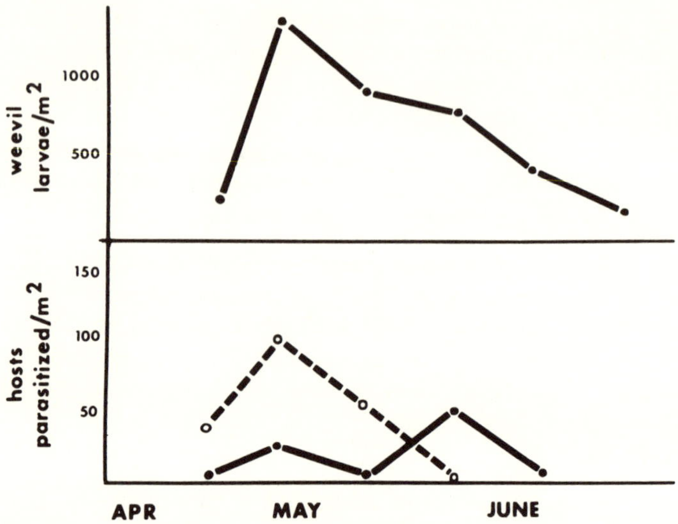

Fig. 6. Alfalfa weevil larval density and parasitization, Wooster, Ohio, 1974. Lower graph, solid line, *B. curculionis*. Dashed line, *B. anurus*.

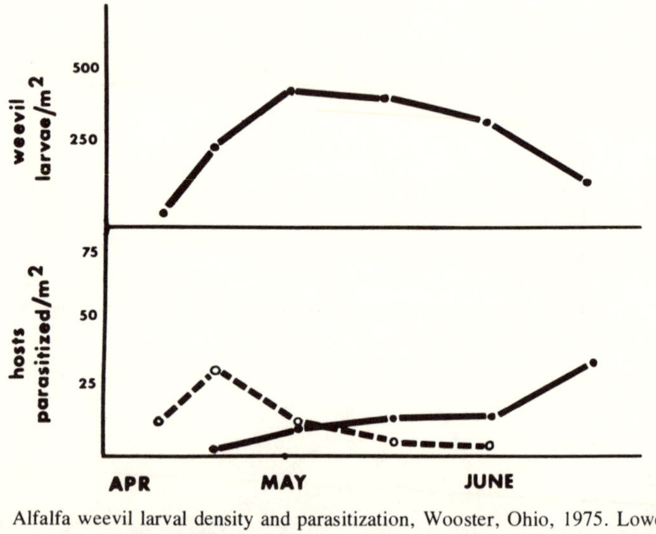

Fig. 7. Alfalfa weevil larval density and parasitization, Wooster, Ohio, 1975. Lower graph, solid line, *B. curculionis*. Dashed line, *B. anurus*. Note change of scale from figs. 5 and 6.

We suspect that it is a competitive loser to other members of the larval guild, for its larvae remain in diapause while the hosts metamorphose and aestivate. *Patasson luna*, the single member of the egg guild, is not included in our analysis.

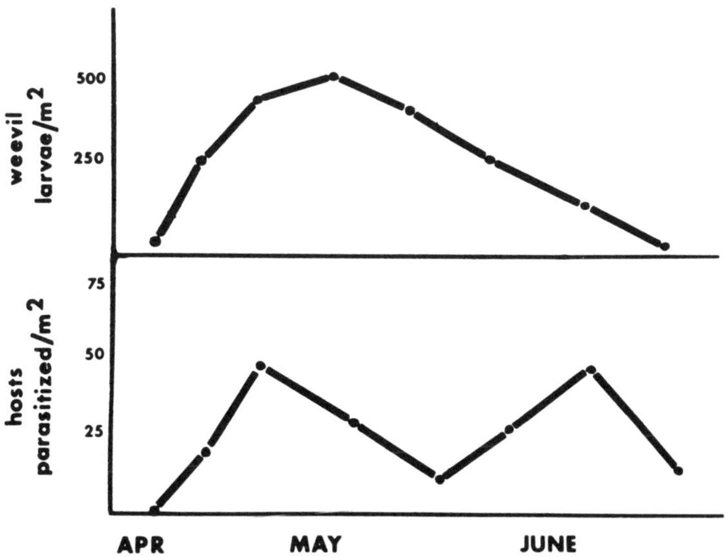

Fig. 8. Alfalfa weevil larval density and parasitization, South Charleston, Ohio, 1974. *B. curculionis* was only parasitoid present.

Larval Guild

Dowell and Horn (1976) analyzed the biological attributes of the three larval parasitoid species. We observed behavior in field and laboratory, and Dowell (1978) dissected ovaries of female parasitoids to determine oocyte maturation patterns. Table 3 shows our data together with some values from literature. We cannot adequately arrange these species along an "r-K continuum"; we found the more factors one wished to include, the cloudier became the distinction between "r" and "K". *T. incertus*, for instance, behaves like a K-strategist in its long handling time and low aggregation ratio, but has the highest potential "r" and the poorest competitive ability.

The "aggregation ratio" bears explaining. Dowell ran a series of trials in which searching parasitoids were alternately exposed to alfalfa on which weevil larvae had fed versus untouched alfalfa that had never been fed upon. The aggregation ratio is the ratio of time spent searching exposed versus untouched alfalfa. We noted that *B. curculionis* especially concentrated its searching in areas of weevil damage and frass.

Although *B. anurus* has a greater egg production, *B. curculionis* has the higher "r" by virtue of its multivoltinism. Moreover, *B. curculionis* emerges earlier in the season and is present for the first weevil larvae. (Berberet, personal communication, says that *B. anurus* precedes *B. curculionis*

TABLE 3

CHARACTERISTICS OF LARVAL PARASITOIDS OF THE ALFALFA WEEVIL

	B. curculionis	*B. anurus*	*T. incertus*
Ovarioles/eggs in oviduct at 24 hr.	23/202	32/303	12/0
Annual generations	2+	1	3+
Handling time	3.5 sec.	1.5 sec.	4-73 min.[a]
Ovipositor length	1.4 mm	1 mm	0.5 mm
Aggregation ratio	40	2.5	1
Competition	<*B.a.*>*T.i.*[b]	>*B.c.*[c]	<*B.c.*[b]
Encapsulation	yes	no	no

[a]Horn (1970).
[b]Miller (1969).
[c]Bartell (1973).

emergence in Oklahoma.) The longer ovipositor of *B. curculionis* enables it to reach well down into alfalfa tips to parasitize first-instar weevil larvae unreachable by *B. anurus*. Both species are "dumpers"; the oviducts are loaded with eggs (Dowell, 1978), and handling times are very short. This is consistent with Force's expectations of r-strategists. *T. incertus*, by contrast, is a "plodder." Capable of high egg production, it spends a long time ovipositing (Horn, 1970) and does not aggregate at high host densities. Its multivoltinism requires continual availability of host larvae throughout the growing season, and it therefore is limited to where summer larvae are available (Horn, 1971).

B. anurus is a superior competitor to *B. curculionis* when both appear in the same host at the same time. (Bartell, 1973). Competitive superiority is necessary if a univoltine species is to survive on the same host as a multivoltine species. *B. anurus* is also well synchronized with its host (Figs. 4, 7, 8). *B. curculionis* eggs and first-instar larvae are often encapsulated by hemocytes of alfalfa weevil larvae, though this varies greatly from place to place (Puttler, 1967; Gibson and Berberet, 1974). The widespread, multivoltine *B. curculionis* is probably the most genetically variable of the larval parasitoids and, as such, fulfills Force's (1972) predictions for being the most successful parasitoid of an ephemeral crop system. Though it is out-competed locally by *B. anurus,* it spreads much more rapidly to new alfalfa weevil infestations. Both *Bathyplectes* species satisfy Hassell and May's (1973) requisites for a "good" parasitoid. Rutz (1976) has shown in laboratory studies that interference between searching *B. curculionis* only becomes important at exceedingly high, artificially maintained densities.

Adult Guild

Table 4 presents a comparison between the two *Microctonus* species with respect to many of the attributes considered important to reproductive strat-

TABLE 4

CHARACTERISTICS OF ADULT PARASITOIDS OF ALFALFA WEEVIL

	M. colesi	*M. aethiopoides*
Maximum fecundity	400	45
Annual generations	1	2
Handling time	4 sec.	43 sec. — 30 min.[a]
Interference	?	Yes
Searching	Random	Directed
Alternate hosts	No	*Sitona, Hypera* spp.

[a]Includes time "stalking" hosts.

egy. Here there is no doubt but that *M. colesi* is a relative r-strategist, with a high egg production and no males. On the other hand, *M. aethiopoides*, being bivoltine, has a potential "r" close to that of *M. colesi*, if *M. aethiopoides* can locate sufficient hosts in both generations. *M. aethiopoides* "stalks" potential hosts for up to one hour before ovipositing; we include stalking in handling time.

M. colesi adults parasitize alfalfa weevil larvae, and their adaptive strategy seems geared to larval availability. They are thus similar to *Bathyplectes* spp. *M. aethiopodes*, on the other hand, produces a relatively low number of eggs but is active very early in the season. Its two peaks of activity coincide with maximum availability of alfalfa weevil adults. Loan and Holdaway (1961) and van Dreische (1975) have shown that up to 80 percent of alfalfa weevil females parasitized by *M. aethiopoides* are sterilized within 72 hours of eclosion of the parasitoid larva. This sterilizing ability of *M. aethiopoides* greatly reduces the over-all fecundity of alfalfa weevil populations where the parasitoid is established (van Dreische, 1975). From an economic standpoint, control measures will be most effective when directed at a life stage immediately *preceding* that which causes damage. Controlling eggs (or gravid adults) of the alfalfa weevil would indeed reduce larval numbers significantly, as Lewis (1977) showed for hypothetical life tables of the alfalfa weevil in which *M. aethiopoides* were presumed to have a higher population density. In fact, Day (personal communication) has noted that in areas of the eastern United States, where *M. aethiopoides* and *B. anurus* have been established for several years, equilibrium densities of alfalfa weevil larvae have declined substantially, to where the insect is rarely a problem of economic import.

Another factor in the effectiveness of *M. aethiopoides*, not considered in any of the models, is the presence of alternate hosts, *Sitona* spp., the clover root curculios. The relationship between the alternate hosts of *M. aethiopoides* has not been documented recently (since Loan and Holdaway, 1961) but it is reasonable to presume that the alternate hosts serve as a "buffer" for *M. aethiopoides* when alfalfa weevil adult populations are low. This occurs in other systems (Quednau, 1970; Horn, 1974), and may be critical to *M. aethiopoides* during periods of host scarcity. Watt (1965) recognized the potential importance of this in biological control, though he did not integrate it into his model.

In summary, there appears to be no one "ideal" parasitoid in the alfalfa weevil system yet to measure up to the empirical considerations of DeBach (1964, 1974) or the theoretical considerations of Force (1972) or Hassell and May (1973). *B. curculionis* is currently the superior disperser (though initially *T. incertus,* aided by high densities of summer larvae, and doubtless its multivoltine habit, out-dispersed all as reported by Schroder et. al., 1969). *B. anurus* is the superior competitor but a slow disperser that displaces *B. curculionis* somewhat where both are established. Neither *M. colesi* nor *T. incertus* seems well suited to biological control. *M. aethiopoides*, by virtue of its slow dispersive powers, long handling time, and (relatively) low egg production, is incapable of reducing high densities of alfalfa weevils; but once densities are low, it effectively reduces subsequent larval populations by sterilizing adult females. The presence of alternate hosts, *Sitona* spp., provides *M. aethiopoides* with alternatives during alfalfa weevil scarcity, and may well buffer, or stabilize, its populations. There is no one "best" species involved, but the sequential introduction of several parasitoid species on slightly different life stages of the host has resulted in establishment of a complex that has at least the potential for reduction of equilibrium densities of the alfalfa weevil.

THE CEREAL LEAF BEETLE SYSTEM

The cereal leaf beetle, *Oulema melanopus*, and its parasitoids comprise a system remarkably similar to that of the alfalfa weevil (fig. 9). There is again

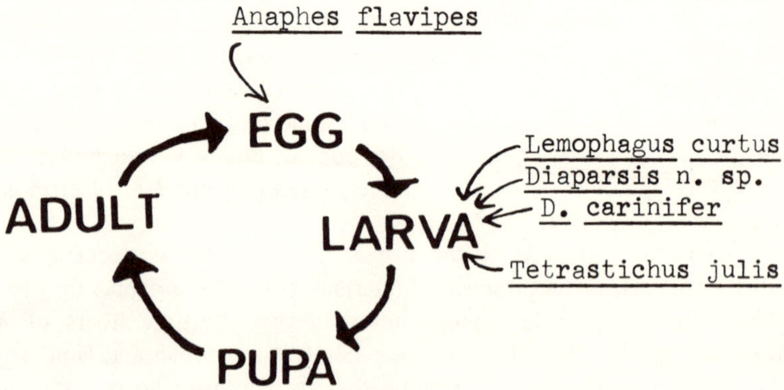

Fig. 9. Cereal leaf beetle parasitoid complex.

a single-generation larval defoliator of an herbaceous spring crop (oats). There is an egg parasitoid, and a larval guild of four species that have been introduced into the United States. The system in Europe is much simpler than that presumed for the alfalfa weevil, and essentially is similar to that now in the United States (Dysart et. al., 1973).

Table 5 shows the larval guild of cereal leaf beetle parasitoids, further extending the similarity to alfalfa weevil larval parasitoids. These data are from Dysart et. al., 1973, and our own observations with those of Leibee (1975). There are two ichneumonids, analagous to the *Bathyplectes* spp., and *Tetrastichus julis* is functionally similar to *T. incertus*. Compare table 5 with table 3. We lack field data on the interactions among these parasitoids, though Burger (personal communication to G. L. Leibee) noted that nowhere are all three sympatric; they seem to exclude one another, in Michigan. It seems, however, that one might generalize: in a seasonal, ephemeral system such as alfalfa or oats, with seasonally available larvae of a holometabolous insect, larval parasitoids have three potential combinations of attributes: (1) superior competitive ability with a closely synchronized univoltine generation, (2) superior dispersal ability with a bi- or multivoltine habit, and (3) inferior competitive ability but gregarious development to maximize utilization of each host successfully parasitized.

Anaphes flavipes, unlike *P. luna*, parasitizes a large proportion (sometimes over 90 percent) of eggs of the cereal leaf beetle. Cereal leaf beetles oviposit directly on oat leaves, so that all eggs are within reach of searching parasitoids. Burger and coworkers (cited by Graham, 1975) claim that parasitization by *A. flavipes* has been an important factor in the observed decline in densities of cereal leaf beetles in the United States. Haynes and Gage (1972) noted that *A. flavipes* was so successful that relatively few larvae were present for *T. julis* to parasitize. This might support the contention of Turnbull and Chant (1971) and Watt (1959, 1965) except that an over-all reduction in larval densities of cereal leaf beetles has indeed occurred. This is

TABLE 5

CHARACTERISTICS OF LARVAL PARASITOIDS OF THE CEREAL LEAF BEETLE

	Diaparsis spp.	*Lemophagus curtus*	*Tetrastichus julis*
Ovariolos/eggs in oviduct at 24 hr.	?	21/200+	11-12/0
Annual generations	1	2	2+
Handling time	"Rapid"	5-10 sec.	5+ min.
Ovipositor length	1.5 mm	1 mm	0.5 mm
Competition	>*T.j.*	?	<*D.c.*
Encapsulation	Yes	Yes	Yes

Data on *Diaparsis* spp. from Dysart et al. (1973).

"control" from an economic standpoint. *A. flavipes*' effect is thus reminiscent of *M. aethiopoides*'; both reduce host population density in the life stage immediately preceding that which causes damage.

PROSPECT: APPLICATION OF MODELS

Both the alfalfa weevil and cereal leaf beetle are holometabolous insects inhabiting an ephemeral crop plant; each has a complex of parasitoids despite rather short, seasonal availability of hosts. Especially in the alfalfa weevil system, the most successful parasitoids, in both distribution and effect on host populations, are those with short handling time and low interference, and high vagility and fecundity. These are characteristics of DeBach's (1964) "best" parasitoid, Force's (1972) "r-strategists," and are the parameters Hassell and May (1973) theorized to be important in selecting agents for biological control from their extension of the quest theory model. There is no obvious "best" parasitoid, though *Microctonus aethiopoides* may have the greatest impact both on equilibrium host densities and on the damaging stage. The adaptive strategy and quest theory models thus have circumstantial validity in that those parameters we have measured in the alfalfa weevil system fit the predictions of the models. Though our analysis herein is limited to Ohio alfalfa, colleagues in New York, Pennsylvania, Kentucky, Illinois, and Oklahoma suggest similar trends in field populations of weevils and parasitoids. Detailed field studies of alfalfa weevil and coevolved parasitoid populations in Europe would be valuable additions to our knowledge.

All the multiple-introduction models assume a guild on precisely the same host stage, whereas in the alfalfa weevil and cereal leaf beetle systems, and in most others, parasitoids vary slightly in instar or size preference even on the same host stage (e.g., Price 1973). They are specialists within a specialty. In effect, this increases the stability of the quest theory model by reducing mutual interference among searching parasitoids as well as reducing competition among their larvae. Parasitoids that attack later life stages may be more efficient at searching and ovipositing, increasing search area and handling time to compensate for reduced availability of hosts.

The models do not consider alternate hosts, secondary parasitism, or host defenses (though the latter could be compensated for by appropriate adjustment of parasitoid fecundity). These are important components in the alfalfa weevil system and in many others. Moreover, we know very little about the genetics of parasitic wasps. Practical and political considerations often prevent release of the largest possible genetic variance from a source population. We have very little control over the source of parasitoids received for release in Ohio.

Models of biological control are still difficult to apply adequately to field situations. We are a long way from actual field measurement of the variables required to apply the model of Hassell and May (1973) to biological control practice. We as yet have *no* meaningful estimates of the density of searching parasitoids; the sweep net method simply tells us whether or not parasitoids were present on the plants. (They might not have been searching if the weather was cloudy and cool!) It is no accident that the quest theory was developed from analysis of laboratory systems, and that Force (1972, 1974) based much of his adaptive strategy approach on intensive study of laboratory systems. Field samples have a large variance that can be stabilized only by replicating many times at huge expense. Controlled laboratory "universes" with alfalfa weevils and parasitoids at several densities might minimize some of these difficulties.

POSTSCRIPT

Our analysis ignores a number of important economic considerations that face growers of alfalfa. In the United States, most growers produce alfalfa to make a profit, so that the economic injury level depends on not only insect populations but also production costs and demand for alfalfa. Both production costs and market price for alfalfa hay have risen markedly in recent years, along with increasing problems due to potato leafhoppers (in the eastern United States). Many growers may soon find it profitable to treat alfalfa in early spring with an application of a granular, systemic carbamate insecticide, for season-long protection of foliage against weevil and leafhopper attack. *M. aethiopoides*, and in fact most parasitic Hymenoptera, are exceedingly susceptible to these insecticides. Unless economical means of nonchemical management of *all* insect, weed, and pathogen pests of alfalfa are integrated, biological control will be unable to compete with chemicals in the race to produce high profits.

ACKNOWLEDGMENTS

Donald R. Lewis, Gary L. Leibee, and John K. Flessel made major contributions of their time and talent to the work we report herein. They shared with us their ideas, their bugs, and their toil in field and laboratory. M. Gardner, M. Haimerl, A. Kolbe, and C. A. Horn helped sort, feed, and count field samples. H. D. Niemczyk provided some badly needed supplies and arranged to place facilities of the Ohio Agricultural Research and Development Center at our disposal. The work was supported fitfully by small grants from the Department of Entomology, the College of Biological Sciences, and the

Graduate School, all of the Ohio State University. R. S. Horn, L. K. Wange, H. S. Horn, D. R. Lewis, and T. K. Wood all cast their critical gaze upon this manuscript, and it is the better for that.

LITERATURE CITED

Armbrust, E. J., H. D. Neimczyk, B. C. Pass, and M. C. Wilson. 1969. Standardized procedures adopted for cooperative Ohio Valley states alfalfa weevil research. J. Econ. Entomol. 62:250–51.

Askew, R. R. 1971. Parasitic insects. American Elsevier, New York. 316 pp.

Barnes, J. R. 1967. Investigation into the source of a summer alfalfa weevil larval population in New Jersey. J. Econ. Entomol. 60:638–40.

Bartell, D. P. 1973. Biology of the immature forms of *Bathyplectes curculionis* and *B. anurus*, and their effect on growth and development of host, *Hypera postica*. Ph.D. diss., University of Kentucky.

Brunson, M. H., and L. W. Coles. 1968. The introduction, release, and recovery of parasites of the alfalfa weevil in eastern United States. USDA-ARS Production Res. Rep. 101, 12 pp.

Caldwell, D. L., and M. C. Wilson. 1975. *Gelis* sp., a hyperparasitoid attacking *Bathyplectes curculionis* cocoons in Indiana. Envir. Entomol. 4:333–36.

Casagrande, R. A., and F. W. Stehr. 1973. Evaluating the effects of harvesting alfalfa on the alfalfa weevil (Coleoptera: Curculionidae) and parasite populations in Michigan. Canad. Entomol. 105:1119–28.

Chamberlin, T. C. 1924. Introduction of parasites on the alfalfa weevil into the United States. USDA/Dept. Circular 301, 3 pp.

Chamberlin, T. C. 1925. A new parasite of the alfalfa weevil in Europe. J. Econ. Entomol. 18:597–602.

Chamberlin, T. C. 1926. The introduction and establishment of the alfalfa weevil parasite, *Bathyplectes curculionis* (Thoms.) in the United States. J. Econ. Entomol. 19:302–10.

Cothran, W. R. 1966 A bibliography of the alfalfa weevil, *Hypera postica* (Gyllenhal). Bull. Entomol. Soc. Amer. 12:151–60.

Cothran, W. R. 1966. A bibliography of the alfalfa weevil, *Hypera postica* (Gyllenhal). Bull. ment I. Bull. Entomol. Soc. Amer. 14:285–88.

Cothran, W. R. 1972. A bibliography of the alfalfa weevil, *Hypera postica* (Gyllenhal), and the Egyptian alfalfa weevil, *Hypera bruneipennis* (Boheman), supplement II. Bull. Entomol. Soc. Amer. 18:102–8.

Day, W. H. 1969. biological notes on *Dibrachys cavus*, a secondary parasite attacking parasites (*Bathyplectes* spp.) of the alfalfa weevil in the eastern United States. J. Econ. Entomol. 62:1225–26.

DeBach, P. (ed.). 1964. Biological control of insect pests and weeds. Reinhold, New York, 844 p.

DeBach, P. 1974. Biological control by natural enemies. Cambridge University Press, Cambridge, England. 323 pp.

Diamond, P. 1973. The effect of multiple parasitoid introductions upon the equilibrium value of host density. Oecologia 13:279–90.

Doutt, R. L. and P. DeBach. 1964. Some biological control concepts and questions. *In* P. DeBach (ed.). Biological control of insect pests and weeds. Pp. 118–42. Reinhold, New York.

Dowell, R. V. 1978. Ovary structure and reproductive biologies of larval parasitoids of the alfalfa weevil. Canad. Entomol. 110:507–12.

Dowell, R. V., and D. J. Horn. 1976. Adaptive strategies of larval parasitoids of the alfalfa weevil. Canad. Entomol. 109:641–48.

Drea, J. J. 1968. A new species of *Microctonus* (Hymenoptera:Braconidae) parasitizing the alfalfa weevil. Entomol. News 79:97–102.

Drea, J. J., R. J. Dysart, L. W. Coles, and C. C. Loan. 1972. *Microctonus stelleri* (Hymenoptera: Braconidae:Euphorinea), a new parasite of the alfalfa weevil introduced into the United States. Canad. Entomol. 104:1445–56.

Dysart, R. J., and L. W. Coles. 1971. *Bathyplectes stenostigma*, a parasite of the alfalfa weevil in Europe. Ann. Entomol. Soc. Amer. 64:1361–67.

Dysart, R. J., H. L. Maltby, and M. H. Brunson. 1973. Larval parasites of *Oulema melanopus* in Europe and their colonization in the Unites States. Entomophaga 18:133–67.

Ehler, L. E., and R. van den Bosch. 1974. An analysis of the natural biological control of *Trichoplusia ni* (Lepidoptera:Noctuidae) on cotton in California. Canad. Entomol. 106:1067–73.

Ellis, C. R. 1973. Parasitism of *Hypera postica* eggs at Guelph, Ontario, by *Patasson luna* and *Fidiobia rugosifrons*, J. Econ. Entomol. 66:1059–61.

Flessel, J. K., and H. D. Neimczyk. 1972. A summary of the alfalfa weevil parasite situation in Ohio—1972. Pesticide News 25(3):60–62.

Force, D. C. 1970. Competition among four hymenopterous parasites of an endemic insect host. Ann. Entomol. Soc. Amer. 63:1675–88.

Force, D. C. 1972. r- and K-strategists in endemic host-parasitoid communities. Bull. Entomol. Soc. Amer. 18:135–37.

Force, D. C. 1974. Ecology of insect host-parasitoid communities. Science 184:624–32.

Gibson, W. P., and R. C. Berberet. 1974. Histological studies on encapsulation of *Bathyplectes curculionis* eggs by larvae of the alfalfa weevil. Ann. Entomol. Soc. Amer. 67:588–90.

Graham, F. 1975. Plague the pests. Audubon 77(5):64–87.

Hamlin, J. C., F. V. Leiberman, R. W. Bunn, W. C., McDuffie, R. C. Newton, and L. J. Jones. 1949. Field studies of the alfalfa weevil and its environment. USDA Tech. Bull. 975, 84 pp

Hassell, M. P., and R. M. May. 1973. Stability in insect host-parasite models J. Anim. Ecol. 42:693–726.

Hassell, M. P., and G. C. Varley. 1969. New inductive population model for insect parasites and its bearing on biological control. Nature 223:1133–37.

Haynes, D. L., and S. H. Gage. 1972. Host parasite interaction between the cereal leaf beetle and its principal North American parasite. Paper presented at National Meeting Entomol. Soc. Amer., Montreal, Quebec, Canada.

Headley, J. C. 1972. Economics of agricultural pest control. Ann. Rev. Entomol. 17:273–86.

Headley, J. C. 1975. The economics of pest management. *In* R. L. Metcalf and W. H. Luckmann (eds.). Introduction to insect pest management. Pp. 75–99. Wiley, New York.

Holling, C. S. 1959. Some characteristics of simple types of predation and parasitism. Canad. Entomol. 91:385–98.

Horn, D. J. 1970. Oviposition behavior to *Tetrastichus incertus*, a parasite of the alfalfa weevil. J. Econ. Entomol. 63:303–4.

Horn, D. J. 1971. The relationship between a parasite, *Tetrastichus incertus* (Hymenoptera-:Eulophidae) and its host, the alfalfa weevil, *Hypera postica* (Coleoptera:Curculionidae) in New York. Canad. Entomol. 103:83–94.

Horn, D. J. 1974. Observations on primary and secondary parasitoids of California oakworm, *Phryganidia californica*, pupae. Pan-Pac. Entomol. 50:53–59.

Horn, D. J. 1976. *Gelis* species' parasitizing *Bathplectes* spp. in Ohio. Ohio J. Sci. 76:279–80.

Huffaker, C. B. (ed.) 1971. Biological control. Plenum, New York.

Leibee, G. L. 1975. The influence of tillage on survivorship of cereal leaf beetle larval parasitoids *Tetrastichus julis* and *Lemophagus curtus*. M.S. thesis, Ohio State University, Columbus, Ohio.

Levins, R. 1968. Evolution in changing environment. Princeton University Press, Princeton, N.J. 120 pp.

Levins, R. 1969. Some demographic and genetic consequences of environmental heterogeneity for biological control. Bull. Entomol. Soc. Amer. 15:237–40.

Lewis, D. R. 1977. Life table analysis of alfalfa weevil population dynamics in Ohio. Ph.D. diss. Ohio State University.

Lewontin, R. C. 1965. Selection for colonizing ability. *In* H. G. Baker and G. L. Stebbins (eds.). Genetics of colonizing species. Pp. 77–91. Academic Press, New York.

Loan, C., and F. G. Holdaway. 1961 *Microctonus aethiops* (Nees) auctt. and *Perilitus rutilus* (Nees) (Hymenoptera:Braconidae), European parasites of *Sitona* weevils (Coleoptera:Curculionidae). Canad. Entomol. 93:1057:79.

Michelbacher, A. E. 1940. Effect of *Bathyplectes curculionis* on the alfalfa weevil population in lowland middle California. Hilgardia 13:81–99.

Miller, M. C. 1969. Studies of interspecific competition between *Tetrastichus incertus* and *Bathyplectes curculionis*, larval endoparasites of the alfalfa weevil. J. Econ. Entomol. 63:719–21.

Nicholson, A. J., and V. A. Bailey. 1935. The balance of animal populations. Part I. Proc. Zool. Soc. London 551–98.

Niemczyk, H. D., and J. K. Flessel. 1970. Population dynamics of alfalfa weevil eggs in Ohio. J. Econ. Entomol. 63:242–47.

Pianka, E. R. 1970. On r- and K-selection. Amer. Nat. 100:592–97.

Pimentel, D. 1963. Introducing parasites and predators to control native pests. Canad. Entomol. 95:785–92.

Poinar, G. O. Jr. and G. G. Gyrisco. 1963. Hymenopterous parasites of the alfalfa weevil, *Hypera postica*, in New York. J. Econ. Entomol. 56:533–34.

Price, P. W. 1973. Parasitoid strategies and community organization. Env. Entomol. 2:623–26.

Price, P. W. 1975. Insect ecology. Wiley, New York. 534 pp.

Puttler, B. 1967. Interrelationship of *Hypera postica* (Coleoptera:Curculionidae) and *Bathyplectes curculionis* (Hymenoptera:Ichneumonidae) in the eastern United States with particular reference to encapsulation of the parasite eggs by the weevil larvae. Ann. Entomol. Soc. Amer. 60:1031–38.

Quednau, F. W. 1970. Competition and co-operation between *Chrysocharis laricinellae* and *Agathis pumila* on larch casebearer in Quebec, Canad. Entomol. 102:602–12.

Remington, C. L. 1968. The population genetics of insect introduction. Ann. Rev. Entomol. 13:415–26.

Rogers, D. J., and M. P. Hassell. 1974. General models for insect parasite and predator searching behavior: interference. J. Anim. Ecol. 43:239–54.

Rutz, D. A. 1976. Numerical relationships between the alfalfa weevil *Hypera postica* and the parasitoid *Bathyplectes curculionis*. M.S. thesis, Pennsylvania State University.

Schroder, R. F. W., J. L. Huggans, D. J. Horn, and G. T. York. 1969. The distribution, establishment and dispersal of *Tetrastichus incertus* in the eastern United States. Ann. Entomol. Soc. Amer. 62:812–15.

Smith, H. D. 1930. The bionomics of *Dibrachoides dynastes* (Foerster), a parasite of the alfalfa weevil. Ann. Entomol. Soc. Amer. 23:577–93.

Stern, V. M., R. F. Smith, R van den Bosch, and K. S. Hagen. 1959. The integrated control concept. Hilgardia 29:81–101.

Streams, F. A., and R. W. Fuester. 1966. *Patasson luna* in overwintering eggs of the alfalfa weevil. J. Econ. Entomol. 59:331–33.

Streams, F. A., and R. W. Fuester. 1967. Biology and distribution of *Tetrastichus incertus*, a parasite of the alfalfa weevil. J. Econ. Entomol. 60:1574–79.

Sweetman, H. L. 1936. The biology control of insects. Comstock, Ithaca, N.Y. 461 pp.

Turnbull, A. L., and D. A. Chant. 1961. The practice and theory of biological control of insect in Canada. Canad. J. Zool. 39:697–753.

vanDreische, R. G. 1975. An evaluation of *Microctonus aethiops* (Nees) (Hymenoptera-:Braconidae) as a parasitoid of the adult alfalfa weevil, *Hypera postica* (Gyll.) (Coleoptera-:Curculionidae) in central New York. Ph.D. thesis, Cornell University, Ithaca, N.Y. 321 pp.

Varley, G. C. and G. R. Gradwell. 1970. Recent advances in insect population dynamics. Ann. Rev. Entomol. 15: 1–24.

Varley, G. C., G. R. Gradwell, and M. P. Hassell. 1973. Insect population ecology: an analytical approach. University of California Press, Berkeley, California.

Watt, K. E. F. 1959. A mathematical model for the effect of densities of attacked and attacking species on the number attacked. Canad. Entomol. 91:129–44.

Watt, K. E. F. 1965. Community stability and the strategy of biological control. Canad. Entomol. 97: 887–95.

Wilson, M. C., and E. J. Armbrust. 1970. Approach in integrated control of the alfalfa weevil. J. Econ. Entomol. 63:554–57.

Index